Applied Probability
Control
Economics
Information and Communication
Modeling and Identification
Numerical Techniques
Optimization

Applications of Mathematics

4

A. A. Borovkov

Stochastic Processes in Queueing Theory

translated by Kenneth Wickwire

Springer-Verlag
New York Heidelberg Berlin
1976

A. A. Borovkov
Akademie der Wissenschaften der UDSSR
Sibirische Abteilung
Math Insistute
Novosibirsk, 90,
USSR

K. Wickwire
University of California
Systems Science Department
Los Angeles, California 90024

AMS Subject Classification
60K25

Library of Congress Cataloging in Publication Data

Borovkov, Aleksandr Alekseevich.
Stochastic processes in queueing theory.

(Applications of mathematics; 4)
Translation of Veroiatnostnye protsessy v teorii massovogo obsluzhivaniia.
Bibliography: p.
Includes indexes.
1. Queuing theory. 2. Stochastic processes.
I. Title.
T57.9.B6713 519.8'2 75-43542

The original Russian edition VEROYATNOSTNYE PROCESSY V TEORII MASSOVOGO OBSLUŽIVANIYA was published in 1972 by NAUKA, Moscow.

Softcover reprint of the hardcover 1st edition 1976

ISBN-13:978-1-4612-9868-7 e-ISBN-13:978-1-4612-9866-3

DOI: 10.1007/978-1-4612-9866-3

Preface

The object of queueing theory (or the theory of mass service) is the investigation of stochastic processes of a special form which are called queueing (or service) processes in this book. Two approaches to the definition of these processes are possible depending on the direction of investigation. In accordance with this fact, the exposition of the subject can be broken up into two self-contained parts. The first of these forms the content of this monograph.

The definition of the queueing processes (systems) to be used here is close to the traditional one and is connected with the introduction of so-called *governing* random sequences. We will introduce algorithms which describe the governing of a system with the aid of such sequences. Such a definition inevitably becomes rather qualitative since under these conditions a completely formal construction of a stochastic process uniquely describing the evolution of the system would require introduction of a complicated phase space not to mention the difficulties of giving the distribution of such a process on this phase space.

If the governing sequence and algorithms are not too complicated, an investigation of the *explicit* distributions of functionals of the process is justified (for example, the waiting time or queue length). The investigations of this sort carried out in this book pertain to the systems described in § 1. This is a sufficiently wide class of systems. Progress in problems with more complicated governing algorithms is, as a rule, only possible for certain very special and simple forms of governing sequences. This is true, for example, for exponentially distributed sequences, where one can use differential equation methods. Various special problems of this type are treated in a voluminous literature, a review of which can be found in the books by Syski [68], Saaty [58] and Gnedenko and Kovalenko [30].

The contents of the present book comprise 8 chapters, 4 appendices and bibliographical notes. Chapter 1 is devoted to the study of so-called single-channel queueing systems (see § 1 for a description of the types of systems) under quite general assumptions on the nature of the governing sequences. Along with ergodic theorems under conditions close to being necessary Chapter 1 presents methods of finding limiting and prelimiting distributions of the processes under consideration, the behavior of these distributions under conditions of light and heavy traffic, estimates of convergence rates and a number of other results.

If the terms of the governing sequence are independent and stationary, single-channel systems can be investigated rather completely. The problems of Chapter 1 lead in this case to boundary problems for random walks described by processes or sequences with independent increments. Chapters 2–4 are devoted to the study of these. It should be noted that these problems go beyond the framework of queueing theory proper and are in our view also of general interest for probability theory. In Chapter 2 we consider processes continuous from below with independent increments and in Chapters 3 and 4 sequences with independent increments. Chapter 4 is devoted mainly to new theorems, but Chapter 3 has more of a review character: with the aid of a single analytic method we obtain there almost all of the presently known general results on boundary problems for sums of independent random variables. Although Chapters 2–4 are essentially a continuation of Chapter 1, they will be referred to often in the subsequent exposition, which treats other types of service systems.

In Chapters 5 and 6 we investigate multi-channel queueing systems; in Chapter 6 the number of service channels is assumed infinite. Chapter 7 treats systems with refusals and Chapter 8 systems with autonomous service. The problems of these chapters have a natural character and are on the whole the same as those in Chapter 1. These comprise above all ergodic theorems, methods of finding distributions of various system characteristics, estimates of convergence rates, etc.

The appendices contain information and results which are used in the main text and they complement it. However, their character is such that if placed in the main presentation, they would lead too far-afield. There are four of them. The first is concerned with renewal theory. In Appendix 2 we study various questions connected with factorizations of elements of the ring of the Fourier–Stieltjes transformations of functions of bounded variation. In Appendix 3 we present theorems on the asymptotic behavior of the coefficients of functions which can be represented as absolutely convergent series. The results of Appendices 2 and 3 are in our opinion of general interest from the point of view of analysis. Appendix 4 is related to the theory of large deviations for sums of independent random variables.

The book is concluded with bibliographical remarks which refer to other work connected with the results presented. We note that in view of their brevity, these remarks cannot pretend to reflect completely the accomplishments of various mathematicians in the development of queueing theory and the related aspects of probability theory.

The contents of this monograph are based on the approach to the understanding and definition of queueing processes referred to at the beginning. However, another approach to the definition of queueing processes is also possible. We intend to present this second approach and the results connected with it in another book, in which queueing processes will be defined as arbitrary vector stochastic processes with nondecreasing step components (the input process, the number of requests refused, the number of served requests). (See [14].) Following Kolmogorov, such processes can be called *processes with discrete interference of chance*. Their components are chosen so that the basic system characteristics which are usually studied turn out to be sufficiently simple functionals of the process. The main goal of such an approach is the establishment of general asymptotic laws when, say, the number of busy channels or the queue length is large. Conditions imposed on the components of the process are of a quite general nature and turn out to be sufficiently easy to verify in concrete problems. The limiting laws obtained under these conditions can be used effectively to study very complex service systems, including those with a control. As an example, for so-called systems in heavy traffic (compare with §§ 8, 25, 40) the distribution of the normalized queue length is, as the examples show, well approximated under quite general conditions by the corresponding distribution of a diffusion process.

On the whole, this book and its intended continuation are an attempt to set forth the contemporary state (as it appears to us) of the mathematical theory of queues from the most unified and general standpoint possible. The author has not striven to cover the greatest possible range of questions, hoping that the discussion of the general methods given here will allow the reader to orientate himself on his own in various concrete problems.

This book grew out of a two year special course of lectures at the Novosibirsk State University. A large part of the results has been obtained in the last few years and is published either for the first time or for the first time in book form.

In the choice of references placed in the main text the author has preferred monographic sources to original papers whenever possible, assuming that the reader finds the former easier to use than the latter. In order to avoid frequent reference to the bibliography, references to monographs are usually accompanied by those to their authors as well.

In a first reading the more difficult sections relating to convergence of processes (such as Subsections 2 and 3 of § 3 and parts of Subsections 2 and 3 of § 6) can be omitted. The results of a more general character will leave a more unified impression if one also omits Chapters 2–4 and the latter sections of Chapters 5–7.

Among the other books connected with queueing theory we want to mention above all the monographs of Hinčin [32], Takács [70], Beneš [3] and Gnedenko and Kovalenko [30]. An acquaintance with these will aid the reader considerably. The most suitable among the general texts are obviously those of Feller [24] and [25]. A knowledge of the fundamentals of probability theory contained in these two volumes is quite sufficient for a preliminary reading of this monograph.

The author was greatly assisted by N. P. Leont'eva and I. Ahmarov in getting the book ready for press. §§ 21 and 26–28 were written with their participation. To both of them I want to express my sincere gratitude. I am also indebted to my colleagues Yu. V. Prohorov, B. A. Rogozin and S. V. Nagaev for useful discussions on various aspects of the book which doubtless contributed to its improvement.

Translator's Preface

In recent years, the literature of queueing theory has grown to such an extent that the number and variety of problems and approaches can be positively discouraging. This is the first volume (a second volume is scheduled) of a work which aims to provide a unifying view of the field by discussing the structures common to apparently diverse systems.

For this translation Professor Borovkov has provided nearly one hundred pages of additional material, including several detailed numerical examples of the theory. I wish to thank him here for a careful reading of the manuscript and for suggesting a number of improvements in exposition and terminology.

In a few cases it has still proved difficult to select popular translations of certain technical terms. In Russian, queueing systems are usually called "systems of mass service" and the term "queue" is reserved for systems in which demands can actually accumulate to form a waiting line. In Anglo-American usage it is customary to refer to most service systems as queues regardless of whether or not a waiting line can actually form. I have generally followed this usage. "Lot" or "group" has been used for what is also called a "batch", "channel" for "server" and "request" or "customer" for "demand"; what I have translated as a system with "refusals" is sometimes called a system with "blocking" or "balking". In view of these differences in usage and of the large amount of new material, I have seen fit to compile a new index. Finally, in a few instances bibliographical references to texts in Russian have been retained where it was impracticable to find corresponding references to translations.

Kenneth Wickwire

Table of Contents

Introduction 1

§ 1. Classifications. Some Notation 1

Chapter 1. *Systems with Queues and Service of Type One* 4

§ 2. Cases in Which the Systems $\langle G \rangle$ Can be Described by Means of Recursion Equations. Equivalence to the System $\langle G, G, G, 1 \rangle$ 4

§ 3. The Basic Equation. Properties of the Solution as a Process. Ergodic Theorems 7

§ 4. Interrupted Governing Sequences 19

§ 5. On Systems Governed by Sequences of Independent Random Variables 20

§ 6. The Virtual Waiting Time. A Continuous Analogue of the System Equation. Properties of the Solution 22

§ 7. Further Properties of the Process $w(t)$. Beneš' Equation . . 29

§ 8. The Stationary Solution of Beneš' Equation. Approximation Formulae for Heavy and Light Traffic 33

§ 9. The Processes $X(t)$ and $Y(t)$ with Stationary Increments Corresponding to Governing Sequences with Independent Terms. The Connection between the Distributions of $w^c(t)$ and w^k . 44

§ 10. Estimates of the Rate of Convergence of the Distributions of w_n and $w(t)$ to Stationarity. Connection with the Queue Length 48

§ 11. Theorems on the Stability of the Stationary Waiting Time under a Change of the Governing Sequences. 52

Chapter 2. *Some Boundary Problems for Processes Continuous from below with Independent Increments. Their Connection with the Distribution of* $w(t)$ 64

§ 12. Boundary Problems for Processes Continuous from below with Independent Increments 64

§ 13. Properties of the Distribution of $w(t)$. The Busy Period . . 75
§ 14. Discrete Time 80

Chapter 3. *Boundary Problems for Sequences with Independent Increments and Factorization Identities* 85

§ 15. Preliminary Remarks. 85
§ 16. The First Factorization Identity and Its Consequences . . 88
§ 17. The Second Factorization Identity and Its Consequences . 95

Chapter 4. *Properties of the Supremum of Sums of Independent Random Variables and Related Problems of Queueing Theory* . . 103

§ 18. Uniqueness Theorems 103
§ 19. Methods of Finding the Distribution of $\bar{Y}$ 105
§ 20. Explicit Formulae for the Distribution of $\bar{Y}$ under the Conditions of Queueing Theory 110
§ 21. Stability Theorems. The Rate of Convergence . . . 113
§ 22. Asymptotic Properties of the Distributions of $\bar{Y}$ and θ . . 123
§ 23. Inequalities for the Distributions of $\bar{Y}_n$ and $\bar{Y}$. The Rate of Approach of the Distributions of w_n and w^1 139
§ 24. Comparison Theorems 141
§ 25. Conditions for Heavy Traffic. Transitional Phenomena . . 147
§ 26. The Relation between the Waiting Time and Queue Length Distributions 155

Chapter 5. *Multi-Channel Queueing Systems* 161

§ 27. Classes of Systems Which Can Be Described by Recursion Equations. Existence Theorems for a Stationary Solution in the Systems $\langle G, G, G/m, 1\rangle$. The Relation between the Waiting Time and the Queue Length 161
§ 28. The Systems $\langle G_I, G_I, G_I/m, 1\rangle$. Stability Theorems. Connection between the Waiting Time and Queue Length. Estimates of Rates of Convergence 169
§ 29. The Systems $\langle G_I, 1, E/m, 1\rangle$ and $\langle E, G_I, G_I/m, 1\rangle$ . . . 178

Chapter 6. *The Systems $\langle G, G, G/\infty, 1\rangle$ with an Infinite Number of Service Channels* 185

§ 30. Theorems on Convergence to Stationary Processes. . . 185
§ 31. Stability Theorems 188
§ 32. The Systems $\langle G_I, G_I, G_I/\infty, 1\rangle$ 193
§ 33. The Systems $\langle E, 1, G_I/\infty, 1\rangle$ 195
§ 34. The Systems $\langle G_I, 1, E/\infty, 1\rangle$ 199

Chapter 7. *Systems with Refusals* 202

§ 35. The Systems $\langle G, G, G/m, 1\rangle_R$. General Theorems . . . 202
§ 36. Stability Theorems 209
§ 37. The Systems $\langle G_I, 1, G_I/m, 1\rangle_R$ 211
§ 38. The Systems $\langle G_I, G_I, E/m, G_I\rangle_R$ 217
§ 39. The Systems $\langle G\rangle_R$ 219
§ 40. Asymptotic Analysis of Multi-Channel Systems . . . 225

Chapter 8. *Systems with Autonomous Service* 235

§ 41. General Properties 235
§ 42. Methods of Calculating the Stationary Distributions . . 240

Appendices 243

Appendix 1. Some Theorems from Renewal Theory . . 243

Appendix 2. Factorization in the Ring $\mathfrak{B}$ and Some Theorems Associated with It 249

Appendix 3. The Wiener-Lévy Theorems and the Asymptotic Behavior of the Coefficients of Absolutely Convergent Series 258

Appendix 4. Estimates for the Distributions of Sums of Independent Random Variables 261

List of Basic Notation. 267

Bibliographical Notes 269

Bibliography 273

Author Index 276

Subject Index 277

Introduction

§ 1. Classifications. Some Notation

Let $(\Omega, \mathfrak{B}, \mathsf{P})$ be the basic probability space on which are given four sequences of nonnegative random variables

$$\{\tau_i^e, \nu_i^e, \tau_i^s, \nu_i^s; \quad 0 \leqslant i < \infty\}, \tag{1}$$

where ν_i^e and ν_i^s are integer-valued. We will call this the *governing sequence*. The first two components define the so-called *input stream* e: at times $\tau_0^e, \tau_0^e + \tau_1^e, \tau_0^e + \tau_1^e + \tau_2^e, \ldots$ groups of "requests" (or customers) of sizes $\nu_0^e, \nu_1^e, \nu_2^e \ldots$, resp. enter some system. The second pair of components $\{\tau_i^s\}$ and $\{\nu_i^s\}$ describes the "service" process s of these requests: requests are served in groups of sizes $\nu_0^s, \nu_1^s, \ldots$ (or groups of smaller size if there is an insufficient number of requests in the system); the service of the k-th group requires time τ_k^s. Already served requests are no longer taken into consideration. By the state of the system we usually understand various numerical characteristics connected with requests located in the system at a given moment of time (i.e., arrived in the system, but not yet served). The study of the stochastic processes describing the behavior in time of these characteristics is the main object of queueing or service theory.

It is not difficult to see that giving the governing sequences does not uniquely define the behavior of the system. One must also indicate, for example, a rule which would determine the time service of subsequent groups of requests is to begin. We will consider the following types of systems:

Queueing Systems. Requests arriving in the system and not immediately accepted for service accumulate forming a "queue" waiting for service. In the sequel requests are served in the order of their arrival (we can assume that the members of each arriving group have been ordered in some way). We will say that a system is *busy* or *occupied* at time t if at this time there is a queue or the next group of requests is being served. Otherwise, the system is said to be *free* at time t. In classifying queueing systems we will differentiate between two types of service.

Type I. If the system is free, then it starts operating (to serve a group of requests) as soon as at least one request arrives. If the system is busy, then service of the following group begins as soon as service of the previous group is completed.

Type II. *Systems with autonomous service.* Here service can begin only at times $0, \tau_1^s, \tau_1^s+\tau_2^s, \dots$.

Systems with refusals.[1] Requests entering the system and not immediately accepted for service "are refused" and removed from consideration. In systems with refusals one usually considers only the first type of service.

In each of the three types of systems (two with queues and one with refusals), prescription of the governing sequence completely determines the system's evolution. In other words, each elementary event $\omega \in \Omega$ (for simplicity we can assume that Ω is the set of values of 4-dimensional sequences with σ-algebra generated by Borel cylinder sets) uniquely determines the state of the system at an arbitrary moment of time t.

Notation. What has been said makes it natural to denote a service system (if its form is fixed) by a set of four symbols corresponding to the one-dimensional governing sequences. If, for example, the distributions of these sequences are arbitrary, then the corresponding system will be denoted by $\langle G, G, G, G\rangle$. If $v_i^e=1$ w.p.1, and the nature of the remaining sequences is arbitrary, then we have $\langle G, 1, G, G\rangle$. The same change is carried out with the last symbol if $\mathsf{P}(v_i^s=1)=1$.

If it is known that the sequence $\{\tau_k^e\}$ is generated by i.i.d. random variables independent of the remaining sequences, then the system will be denoted by

$$\langle G_I, \bullet, \bullet, \bullet\rangle .$$

In the important special case where in addition to independence,

$$\mathsf{P}(\tau_i^e>x)=e^{-\alpha x}, \quad \alpha>0,$$

we will write $\langle E, \bullet, \bullet, \bullet\rangle$. This notation will be retained when τ_i^e is discrete and

$$\mathsf{P}(\tau_i^e=k)=p^k(1-p), \quad k=0, 1,\dots; \quad p>0. \tag{2}$$

The symbols G_I or E in the remaining three places will denote the same things w.r.t. the sequences $\{v_j^e\}$, $\{\tau_j^s\}$ and $\{v_j^s\}$. The letter E in the second and fourth places will always correspond to the distribution (2).

The type of system (governing algorithm) will be denoted by a subscript on the cornered brackets: systems with queues and autonomous service (type II) will be provided with the index A:

$$\langle \bullet, \bullet, \bullet, \bullet\rangle_A;$$

those with refusals with the index R:

$$\langle \bullet, \bullet, \bullet, \bullet\rangle_R .$$

The absence of an index will indicate queueing systems of Type I.

In our treatment of the various types of systems we must also distinguish an important sub-class formed by the so-called *multi-channel systems.* These are systems in which service can be carried out simultaneously in $m>1$ channels, so

[1] Also called queueing systems in the Anglo-American literature.—Transl.

that service of the next group of requests can begin before that of the preceding one is completed.

The governing algorithms for multi-channel systems look similar to the previous ones for all three types of service (each channel functions as an independent servicing mechanism). It is merely necessary to supplement them by indicating into which channel requests should be directed if several channels are simultaneously free. Here, as before, the service of the i-th group (of size $\leqslant \nu_i^s$) requires time τ_i^s, and in each channel, service of the following group can begin only after that of the previous one is completed.[2]

If the sequences governing service are arbitrary, we denote such multi-channel systems by

$$\langle \bullet, \bullet, G/m, G \rangle$$

with the same possible values for the symbol G as before.

The three types of service we mentioned above will also be indicated in multi-channel systems by the indices A and R or by the absence of an index.

For brevity we will denote the systems $\langle G, G, G, G \rangle_{\bullet}$ and $\langle G, G, G/m, G \rangle_{\bullet}$, resp. by

$$\langle G \rangle_{\bullet} \quad \text{and} \quad \langle G/m \rangle_{\bullet}.$$

A separate treatment for systems in which the input stream is multi-channel obviously makes little sense in the case of indistinguishable customers.

Naturally, more complicated governing algorithms than those above are also possible. They give rise to an unlimited variety of service systems which to a certain extent hampers the development of general methods.

In the sequel, in order to make the nature of the elements of the governing sequence more homogeneous, it will often be expedient to separate from it initial conditions—the value of one characteristic or another at time $t=0$ (for example, the queue length, which we could take as ν_0^e for $\tau_0^e=0$) and the service times of customers in the system at time $t=0$. In such cases we will assume that in addition to the governing sequence $\{\tau_j^e, \nu_j^e, \tau_j^s, \nu_j^s; 1 \leqslant j < \infty\}$, a suitably chosen vector of initial conditions is given in an appropriate way on the basic probability space.

The numbering of the formulae and theorems is self-contained in each chapter. References to formulae of other chapters will be made by means of the double designation ($\bullet$) § $\bullet$ without additional clarification. For example, (15) § 5 means Formula (15) in § 5. Numbering of the sections (§) proceeds uniformly through the book. The symbol □ denotes the end of a proof.

[2] To simplify the nature of the governing sequences it is sometimes convenient to give on the space $(\Omega, \mathfrak{B}, \mathsf{P})$ not two, but $m+1$ two-dimensional governing sequences (τ_i^e, ν_i^e), $(\tau_i^{s1}, \nu_i^{s1})$, …, $(\tau_i^{sm}, \nu_i^{sm})$, so that the k-th service channel is governed by the sequence $(\tau_i^{sk}, \nu_i^{sk})$, $k=1, \ldots, m$. For example, τ_i^{sk} is the service time of the group of requests which is i-th in the k-th channel.

Chapter 1

Systems with Queues and Service of Type One

§ 2. Cases in Which the Systems $\langle G \rangle$ Can Be Described by Means of Recursion Equations. Equivalence to the System $\langle G, G, G, 1 \rangle$

The most natural and interesting characteristics of the status of queueing systems are of the following two types:

1) *The waiting times* until service of requests the numbers of which generate some (possibly random) increasing sequence.

2) *The queue lengths* at instants of time which generate some unbounded increasing sequence.

First we pick out classes of systems for which there exists a characteristic of one of these two types satisfying some sufficiently simple equation which will depend, of course, on the elements of the governing sequences.

Roughly speaking, the following exposition is connected with the fact that the system

$$\langle \bullet, \bullet, \bullet, \bullet \rangle \tag{1}$$

corresponds to the classes referred to if at least one of the positions in (1) is occupied by the symbol E. In order to avoid excessively cumbersome explanations, we will limit ourselves to consideration of the systems

$$\begin{array}{ll} \langle G, G, G, 1 \rangle, & \langle E, G, G, G \rangle, \\ \langle G, 1, G, G \rangle, & \langle G, G, E, G \rangle, \end{array} \tag{2}$$

where in the second and fourth places E is replaced by unity.

1. *The Systems* $\langle G, G, G, 1 \rangle$. For these, as well as all other systems in this chapter, we will assume that the arrival time of the first group of customers coincides with $t=0$. We denote by w_n, $n \geqslant 2$, the time from arrival until the start of service of the first customer in the n-th group, i.e., at time $\tau_1^e + \cdots + \tau_{n-1}^e$ (knowing the sequence w_n, we can easily determine the waiting time for the other customers). Let w_1 be the waiting time until the system has served the first succession of requests. The initial condition w_1 is given along with the governing sequence on the underlying probability space. Then

$$w_2 = \max(0, w_1 + \tau_1^s + \cdots + \tau_{\nu_1^e}^s - \tau_1^e),$$

since the time required to free the system with account taken of requests arriving at time 0 is now $w_1+\tau_1^s+\cdots+\tau_{\nu_1^e}^s$. The waiting time for requests from the second group is reckoned from the moment τ_1^e. In exactly the same way we show that

$$w_3=\max(0, w_2+\tau_{\nu_1^e+1}^s+\cdots+\tau_{\nu_1^e+\nu_2^e}^s-\tau_2^e),$$

and in general,

$$w_n=\max(0, w_{n-1}+S_{n-1}-\tau_{n-1}^e), \tag{3}$$

where S_n is the time spent servicing the n-th successive group of requests. This is the well-known equation for the characteristic w_n. It coincides with the equation for a random walk on the half-line $[0, \infty]$ with a delaying barrier at the point 0, jump sizes

$$\xi_1=S_1-\tau_1^e,$$
$$\xi_2=S_2-\tau_2^e, \ldots$$

and initial value $w_1 \geqslant 0$.

2. *The Systems* $\langle G, 1, G, G\rangle$. In this subsection w_n will have a different meaning. Let w_n be the waiting time of the first request to be served in the n-th group (its number will be denoted by a_n). What is the connection between the variables w_n and w_{n-1}? The service of request a_n begins τ_{n-1}^s time units later than the request a_{n-1}. But it also arrives later. If all ν_{n-1}^s customers of the lot to which a_{n-1} belongs are served, then the delay in arrival is equal to

$$E_{n-1}=\tau_{a_{n-1}}^e+\cdots+\tau_{a_{n-1}+\nu_{n-1}^s-1}^e.$$

In this case

$$w_n=\max(0, w_{n-1}-E_{n-1}+\tau_{n-1}^s). \tag{4}$$

However, in distinction to the previous case (see (3)), this equation will be violated if ν_{n-1}^s is not fully exploited. Then in place of $w_{n-1}-E_{n-1}$ in (4), we must substitute the first negative value in the sequence

$$w_{n-1}-\tau_{a_{n-1}}^e-\cdots-\tau_{a_{n-1}+k}^e,\ k=0, 1, 2, \ldots, \nu_{n-1}^s-1.$$

Denoting this value by $[w_{n-1}-E_{n-1}]^*$ we obtain finally

$$w_n=\max(0, [w_{n-1}-E_{n-1}]^*+\tau_{n-1}^s).$$

Hence, if we study the evolution of the state of the system until the first time the queue dissolves (i.e., until $w_n<\tau_{n-1}^s$), then this evolution will be described by Eq. (4), an equation of the same form as (3).

3. *The Systems* $\langle E, G, G, G\rangle$. Let q_n be the length of the queue at time t_n, the beginning of service of the n-th group of requests, and b_n the number of the first group of requests to arrive after the time t_n. Then, as is easy to see, q_n and q_{n-1} are related by the following equation:

$$q_n=\max(0, q_{n-1}+\nu_{b_{n-1}}^e+\cdots+\nu_{b_{n-1}+N_{n-1}-1}^e-\nu_n^s), \tag{5}$$

where N_n is the number of groups of requests arriving in the system during time

τ_n^s. If α is the parameter of the exponential distribution of τ_k^e, then, on the basis of the properties of the latter, the conditional distribution of N_n is Poisson:

$$\mathsf{P}(N_n=k \mid \tau_n^s=u)=\frac{(u\alpha)^k}{k!}e^{-u\alpha}, \quad k=0, 1, \ldots .$$

Hence, the conditional and unconditional characteristic functions are equal, resp. to

$$\mathsf{M}_{\tau_n^s} e^{i\lambda N_n}=\exp\{\tau_n^s\alpha(e^{i\lambda}-1)\} \quad \text{and} \quad \mathsf{M}\, e^{i\lambda N_n}=\varphi_n\left(\alpha\frac{e^{i\lambda}-1}{i}\right), \tag{6}$$

where $\varphi_n(\lambda)$ is the characteristic function of τ_n^s. An analogous result holds if τ_k^e and τ_k^s are integral and τ_k^e has the distribution (2), § 1.

We thus see that the basic equation (5) for the system $\langle E, G, G, G\rangle$ is the integer analogue of Equations (3) and (4). It also corresponds to a random walk on the half-line with a delaying barrier at the point 0 and the known jump sizes. As initial condition one must give here q_1.

4. *The Systems* $\langle G, G, E, G\rangle$ do not differ from the systems $\langle E, G, G, G\rangle$ from the point of view of interest to us. Indeed, let q_n now denote the queue length previous to the moment $t_n=\tau_1^e+\cdots+\tau_{n-1}^e$ of the arrival of the n-th lot of service requests and b_n the number of the first group of requests to begin being serviced after time t_n. Then, as before, we have

$$q_n=\max(0, q_{n-1}+v_{n-1}^e-v_{b_{n-1}}^s-\cdots-v_{b_{n-1}+N_{n-1}-1}^s), \tag{7}$$

where N_n is the number of groups of requests which can be served during the time τ_n^e by continuous operation of the system. As in (6) we find that

$$\mathsf{M}_{\tau_n^e} e^{i\lambda N_n}=\exp\{\tau_n^e\alpha(e^{i\lambda}-1)\} \quad \text{and} \quad \mathsf{M}\, e^{i\lambda N_n}=\varphi_n\left(\alpha\frac{e^{i\lambda}-1}{i}\right),$$

where α is the parameter of the exponential distribution of τ_n^s and φ_n is the characteristic function of τ_n^e.

We have thus shown, at least for three of the systems at (2), that there is a characteristic of the system's state described by the essentially identical equations (3), (5) and (7).

The study of these equations, which we will write in the form

$$w_n=\max(0, w_{n-1}+\xi_{n-1}), \quad n\geqslant 2, \tag{8}$$

will be the subject of several of the following sections. The possibility of such an investigation will obviously be determined by the properties of the sequence $\{\xi_n; 1\leqslant n<\infty\}$. For the sake of simplicity and definiteness in the sequel, we will interpret the various results on the sequence w_n in terms of their relation to the systems $\langle G, G, G, 1\rangle$. Evidently, these systems find the widest application among those listed at (2). We will also carry out a more complete investigation of the systems $\langle G, G, G, 1\rangle$ going beyond the framework of Eq. (8). However, in this connection it is always necessary to keep in mind that analogous results will also hold for the other systems at (2).

§ 3. The Basic Equation. Properties of the Solution as a Process. Ergodic Theorems

1. In the previous section we established that the study of the systems (2) in terms of the waiting time for some systems and in terms of queue length for others, leads to the investigation of Eq. (8) with some initial value w_1. The sequence $\{\xi_n\}$ is random on the basic probability space $(\Omega, \mathfrak{B}, \mathsf{P})$ since, as we have seen, for all four types of system under consideration each value of ξ_n is defined in a unique and measurable way by a finite number of terms of the governing sequences. In each case this was the difference of two nonnegative random variables. For example, in the system $\langle G, G, G, 1\rangle$, $\xi_n = S_n - \tau_n^e$, where S_n is the time spent serving the n-th arriving group of requests.

Let us write

$$X_n = \xi_1 + \cdots + \xi_n, \quad X_0 = 0.$$

We then have

Theorem 1.

$$w_n = X_{n-1} - \min(-w_1, X_1, \ldots, X_{n-1}). \tag{9}$$

Proof. We will use induction. The equality holds for $n=1$. Hence the assertion will be proved if we can show that (9) implies

$$w_{n+1} = X_n - \min(-w_1, X_1, \ldots, X_n).$$

To this end we consider first the ω-set on which

$$X_n = \min(-w_1, X_1, \ldots, X_n). \tag{10}$$

Then

$$X_{n-1} + \xi_n \leqslant \min(-w_1, X_1, \ldots, X_{n-1})$$

and by (9) $w_n + \xi_n \leqslant 0$. Using (8) we get

$$w_{n+1} = \max(0, w_n + \xi_n) = 0 = X_n - \min(-w_1, X_1, \ldots, X_n).$$

If (10) does not hold, then

$$X_{n-1} + \xi_n > \min(-w_1, X_1, \ldots, X_{n-1}), \qquad w_n + \xi_n > 0,$$

so that by (8) and (9)

$$\begin{aligned} w_{n+1} &= w_n + \xi_n = X_n - \min(-w_1, X_1, \ldots, X_{n-1}) \\ &= X_n - \min(-w_1, X_1, \ldots, X_n). \quad \square \end{aligned}$$

Formula (9) provides us with an explicit expression for w_n in terms of the elements of the governing sequence. On the other hand, if we are interested only in the *distribution* of w_n and not in the form of the dependence of w_n on $\{\xi_n\}$ for each ω, we can also obtain a more convenient representation. However, to do this it is necessary to assume that the sequence ξ_n is homogeneous.

2. In this paragraph we will assume that the sequence $\{\xi_n; n \geqslant 1\}$ is *strictly stationary*, i.e., the distributions of finite-dimensional random variables $(\xi_{k+n_1}, \xi_{k+n_2}, \ldots, \xi_{k+n_j})$ do not depend on k for arbitrary j and $n_1, \ldots, n_j$. This property will be possessed, for example, by the systems $\langle G, G, G, 1\rangle$, whose governing sequences $\{\tau_j^e\}$, $\{v_j^e\}$ and $\{\tau_j^s\}$ are stationary and mutually independent. (The "strictly" in "strictly stationary" will often be omitted since sequences stationary in any other sense are not considered here.)

It will be convenient in the sequel to supplement the sequence $\{\xi_n; n \geqslant 1\}$ so that it is stationary on the whole time-axis. In other words, we will assume that the governing by the sequence w_n takes place by giving on $(\Omega, \mathfrak{B}, \mathsf{P})$ an initial condition w_1 and a stationary sequence

$$\{\xi_n; -\infty < n < \infty\}.^1 \tag{11}$$

Write

$$Y_n = \xi_{-1} + \cdots + \xi_{-n},\ n \geqslant 1; \qquad Y_0 = 0,$$
$$\overline{Y}_n = \max(Y_0, \ldots, Y_n), \qquad Y = \overline{Y}_\infty = \sup_{k \geqslant 0} Y_k.$$

Theorem 2. *If the sequence* (11) *is stationary and for an arbitrary fixed interval* Δ

$$\lim_{n \to \infty} \mathsf{P}(X_n \in \Delta) = 0, \tag{12}$$

then for arbitrary $x \geqslant 0$ *and arbitrary proper (i.e., finite w.p.1) initial condition* $w_1 \geqslant 0$

$$\lim_{n \to \infty} \mathsf{P}(w_n > x) = \mathsf{P}(Y > x)$$

exists and

$$\mathsf{P}(w_{n+1} > x) - \mathsf{P}(Y > x) = \mathsf{P}(\max(X_n - X_0, \ldots, X_n - X_n) \leqslant x,\ X_n + w_1 > x) - \mathsf{P}(\overline{Y}_n \leqslant x,\ Y > x).$$

If $w_1 \equiv 0$, *then the condition* (12) *is superfluous.*

Proof. Set

$$p_{x,n} = \mathsf{P}(\overline{Y}_{n-1} \leqslant x,\ Y_n > x).$$

Then

$$\mathsf{P}(Y > x) = \textstyle\sum_{n=1}^{\infty} p_{x,n}$$

[1] We can proceed in this way without loss of generality since by Kolmogorov's theorem on the extension of compatible distributions we can construct the required distribution of the sequence (11) on the corresponding measurable space using the compatible finite-dimensional distributions of the original sequence.

and the series on the right converges. Moreover, since

$$w_{n+1} = \max(X_n + w_1, X_n - X_1, \ldots, X_n - X_n), \tag{13}$$

we can write

$$\begin{aligned} \mathsf{P}(w_{n+1} > x) &= \mathsf{P}(\max(X_n - X_0, \ldots, X_n - X_n) > x) \\ &\quad + \mathsf{P}(\max(X_n - X_0, \ldots, X_n - X_n) \leqslant x, X_n + w_1 > x) \\ &= \mathsf{P}(\overline{Y}_n > x) + \mathsf{P}(\max(X_n - X_0, \ldots, X_n - X_n) \leqslant x, X_n + w_1 > x). \end{aligned}$$

Since

$$\mathsf{P}(\overline{Y}_n > x) = \mathsf{P}(Y > x) - \textstyle\sum_{k>n} p_{x,k}$$

and

$$\mathsf{P}(\max(X_n - X_0, \ldots, X_n - X_n) \leqslant x, X_n + w_1 > x) \leqslant \mathsf{P}(x - w_1 < X_n \leqslant x),$$

the theorem is proved. □

It is easy to see that if (12) does not hold, then $\lim_{n\to\infty} \mathsf{P}(w_n > x)$ may not exist. For example, if w_1 is constant (nonrandom) and the distribution of ξ_n is symmetric, $\xi_{n+1} = -\xi_n$, $n = \ldots, -1, 0, 1 \ldots$, then

$$w_n \underset{d}{=} \begin{cases} \max(0, \xi_{-1}, w_1), & \text{if } n > 1 \text{ is even}; \\ \max(0, \xi_{-1} + w_1), & \text{if } n > 1 \text{ is odd}. \end{cases}$$

Here, the notation $\xi \underset{d}{=} \eta$ means that the distributions of the random variables ξ and η coincide.

3. We put

$$w^k = \sup(0, \xi_k, \xi_k + \xi_{k-1}, \xi_k + \xi_{k-1} + \xi_{k-2}, \ldots); \qquad -\infty < k < \infty \tag{14}$$

and consider the sequence $\{w^k; -\infty < k < \infty\}$, which along with $\{\xi_k\}$ will be strictly stationary. The assertion of Theorem 2 says that the limiting distribution of w_n as $n \to \infty$ coincides with the distribution of w^k for arbitrary k. (It is not difficult to convince oneself that w^k *satisfies an equation of the form* (8).)

If it were proved that not only the one-dimensional, but also *arbitrary finite-dimensional* distributions of the sequences $\{w_{n,k}; k \geqslant 0\} = \{w_{k+n}; k \geqslant 0\}$, depending on n, converge for $n \to \infty$ to the corresponding distributions of the sequence $\{w^k; k \geqslant 0\}$ then we would obtain a representation for the *stationary* waiting time (or queue length) *process* in the form (14). We would have the *convergence of* $\{w_{k+n}; k \geqslant 0\}$ *as a process* to $\{w^k; k \geqslant 0\}$.

The existence of such a representation and the occurrence of such a convergence are important in the search for the distribution of various functionals of the stationary waiting-time process or of the process $\{w_{n+k}; k \geqslant 0\}$ for large n. We will be interested, for example, in the probability that the waiting time in a system operating in a stationary regime will not exceed a given level during some time interval.

Results on these questions are contained in Theorems 3–5.

Theorem 3. *Under the hypotheses of Theorem 2 the finite-dimensional distributions of the sequences* $\{w_{n,k}; k \geqslant 0\} = \{w_{k+n}; k \geqslant 0\}$ *converge for* $n \to \infty$ *to the corresponding distributions of* $\{w^k; k \geqslant 0\}$.

In other words, for arbitrary j and $n_1, \ldots, n_j$ the joint distribution of the random variables $(w_{n+n_1}, \ldots, w_{n+n_j})$ converges to the distribution of $(w^{n_1}, \ldots, w^{n_j})$.

Proof. In order to simplify the computations, we will limit ourselves to consideration of the distributions of the variables (w_n, w_{n+1}). (Using the same line of argument, we can then carry out the proof for arbitrary distributions.)

Set

$$A_n = \{\overline{Y}_{n-1} > x), \qquad B_n = \{\max(0, \overline{Y}_{n-1} + \xi_0) > y\},$$
$$A_n^x = \{\max(X_{n-1} - X_0, \ldots, X_{n-1} - X_{n-1}) \leqslant x, X_{n-1} + w_1 > x\}$$

and

$$B_n^y = \{\max(X_n - X_0, \ldots, X_n - X_n) \leqslant y, X_n + w_1 > y\}.$$

Then, by (13)

$$\mathsf{P}(w_n > x, w_{n+1} > y) = \mathsf{P}(A_n B_n) + R_1; \qquad R_1 < \mathsf{P}(A_n^x) + \mathsf{P}(B_n^y).$$

As we have already seen in the proof of Theorem 2, $\mathsf{P}(A_n^x) \to 0$ and $\mathsf{P}(B_n^y) \to 0$ for $n \to \infty$. On the other hand,

$$\mathsf{P}(Y > x, \max(0, Y + \xi_0) > y) = \mathsf{P}(A_n B_n) + R_2,$$

where

$$0 \leqslant R_2 \leqslant \mathsf{P}(\overline{Y}_{n-1} \leqslant x, Y > x) + \mathsf{P}(\max(0, \overline{Y}_n + \xi_0) \leqslant y, \max(0, Y + \xi_0) > y).$$

Turning again to the proof of Theorem 2, we see that R_2 also tends to zero for $n \to \infty$. We thus convince ourselves that in the notation of (14), as $n \to \infty$

$$\mathsf{P}(w_n > x, w_{n+1} > y) = \mathsf{P}(w^{-1} > x, w^0 > y) + o(1). \quad \square$$

Assuming that

$$X_n \xrightarrow[\text{a.s.}]{} -\infty \quad \text{and} \quad Y_n \xrightarrow[\text{a.s.}]{} -\infty \tag{15}$$

for $n \to \infty$, we can strengthen the statement of this theorem.

Let $\mathbf{w}_n$ and $\mathbf{w}$ denote realizations of the sequences $\{w_{n,k}\}$ and $\{w^k\}$ (in other words, $\mathbf{w}_n$ and $\mathbf{w}$ are infinite-dimensional vectors: $\mathbf{w}_n = (w_{n,0}, w_{n,1}, \ldots)$, $\mathbf{w} = (w^0, w^1, \ldots)$) and let B be a set of sequences for which $\{\mathbf{w}_n \in B\}$ and $\{\mathbf{w} \in B\}$ are events.[2]

Theorem 4. *For arbitrary B satisfying the formulated conditions,*

$$|\mathsf{P}(\mathbf{w}_{n+1} \in B) - \mathsf{P}(\mathbf{w} \in B)| < \mathsf{P}(\min_{0 < j \leqslant n} X_j > -\max(w_1, Y + \xi_0)). \tag{16}$$

[2] That is, measurable sets.—Transl.

This follows from the fact that the right side of (16) can be estimated by

$$\mathsf{P}(\cup_{k\geqslant 0}\{w_{n+k+1}\neq w^{n+k}\}), \tag{17}$$

where the sequence $\{w^{n+k}; k\geqslant 0\}$ obviously has the same distribution as $\mathbf{w}$.

Proof. The event appearing in (17) after the P-symbol is equal to

$$\cup_{k\geqslant 0}\{\max(X_{n+k}+w_1, X_{n+k}-X_1, \ldots, X_{n+k}-X_{n+k-1}, 0) \neq \sup(0, \xi_{n+k}, \xi_{n+k}+\xi_{n+k-1}, \ldots)\}.$$

Hence, we obtain as an estimate for (17)

$$\mathsf{P}(\cup_{k\geqslant 0}\{X_{n+k}+w_1>\max_{0<j\leqslant n+k}(X_{n+k}-X_j)\}\cup\{\sup_{j\geqslant 0}(X_{n+k}+Y_j^0) > \max_{0<j\leqslant n+k}(X_{n+k}-X_j)\}),$$

where we have written $Y_j^0=Y_j+\xi_0$. The event following the P-symbol is equal to

$$\cup_{k\geqslant 0}\{w_1>-\min_{0<j\leqslant n+k}X_j\}\cup\{Y+\xi_0>-\min_{0<j\leqslant n+k}X_j\};$$

since

$$\min_{k\geqslant 0}[-\min_{0<j\leqslant n+k}X_j]=-\min_{0<j\leqslant n}X_j,$$

the assertion is proved. □

It is not hard to see that when (15) is satisfied, the Y appearing in the derived formulae is a proper random variable (w.p.1 finite).

As an application of the theorems proved we can find, for example, the joint limiting distribution of w_n and the "time" (in terms of the number of jumps of τ_j^e) the system has spent in the busy state up to the moment of arrival of the n-th group of requests (we have the system $\langle G, G, G, 1\rangle$ in view). We can call this the *busy "period" in the backward direction* or *the "backward busy period"*.

Indeed, we set

$$v_n=\min\{k\geqslant 0: w_{n-k}=0\}$$

and assume that $Y=\sup_{k\geqslant 0}Y_k$, as well as the first time

$$\theta=\min\{k\geqslant 0: Y_k=Y\}$$

this supremum is attained, is a proper random variable. The following assertion then holds:

Theorem 5. *Under the assumptions made above and assuming the validity of the conditions of Theorem 2, the limit*

$$\lim_{n\to\infty}\mathsf{P}(w_n>x, v_n=k)=\mathsf{P}(Y>x, \theta=k)$$

exists.

Proof. By Theorem 3

$$\lim_{n\to\infty} \mathsf{P}(w_n>x, v_n=k) = \lim_{n\to\infty} \mathsf{P}(w_n>x, w_{n-1}>0, \ldots, w_{n-k+1}>0, w_{n-k}=0)$$

$$= \mathsf{P}(w^0>x, w^{-1}>0, \ldots, w^{-k+1}>0, w^{-k}=0) = \mathsf{P}(Y>x, \theta=k). \quad \square$$

The proof of Theorem 5 can also be obtained somewhat differently using the fact that the joint distribution of (w_n, v_n) coincides with the joint distribution of $(\max(X_{n-1}+w_1, X_{n-1}-X_1, \ldots, X_{n-1}-X_{n-1}); \theta_n)$, where

$$\theta_n = \min\{k \geqslant 0 : X_{n-1} - X_{n-1-k} = w_n\}.$$

On the ω-sets where v_n and θ_n are not defined, we can set them equal, for example, to ∞.

4. *We now formulate more general conditions for the convergence of the distributions of the w_n to a stationary distribution.* As one can see, to this end it is not at all necessary to require the stationarity of $\{\xi_k\}$; it suffices that this sequence merely *converge* to one that is stationary. It is clear that such a convergence condition for w_n is in a certain sense minimal. It will be satisfied for a considerably wider class of governing sequences which contains almost all cases of practical interest.

We will thus assume that the sequence $\{\xi_{N,k} = \xi_{N+k}; k>0\}$ converges as a *process* as $N\to\infty$ to a strictly stationary sequence $\{\xi^k; k>0\}$.

Suppose, for example, that for an arbitrary measurable set B

$$|\mathsf{P}(\boldsymbol{\xi}_N \in B) - \mathsf{P}(\boldsymbol{\xi} \in B)| < \varepsilon(N), \tag{18}$$

where $\boldsymbol{\xi}_N$ and $\boldsymbol{\xi}$ are realizations of $\{\xi_{N,k}\}$ and $\{\xi^k\}$ and $\varepsilon(N)\to 0$ for $N\to\infty$. Supplementing the sequence $\{\xi^k; k>0\}$ to obtain $\{\xi^k; -\infty<k<\infty\}$, stationary on the entire time-axis, we will also require that

$$X_n \xrightarrow[\mathrm{p}]{} -\infty \quad \text{and} \quad Y^n \xrightarrow[\text{a.s.}]{} -\infty \tag{19}$$

for $n\to\infty$, where $Y^n = \xi^{-1} + \cdots + \xi^{-n}$; $Y_0 = 0$.

Theorem 6. *Under the formulated conditions*

$$\lim_{n\to\infty} \mathsf{P}(w_n<x) = \mathsf{P}(\sup_{n\geqslant 0} Y^n < x) \tag{20}$$

exists and is proper.

Proof. The assertion is implied by the following relations. For arbitrary fixed $N>0$

$$\lim_{n\to\infty} \mathsf{P}(w_{n+1}<x) = \lim_{n\to\infty} \mathsf{P}(\max(X_n+w_1, X_n-X_1, \ldots, X_n-X_n)<x)$$

$$= \lim_{n\to\infty} \mathsf{P}(\max(X_n-X_N, \ldots, X_n-X_n)<x);$$

$$\lim_{n\to\infty} \mathsf{P}(\max(X^n-X^N, \ldots, X^n-X^n)<x) = \mathsf{P}(\sup_{n\geqslant 0} Y^n<x),$$

where $X^n = \xi^1 + \cdots + \xi^n$. It remains to note that by (18) $\mathsf{P}(\max(X_n-X_N, \ldots,$

$X_n - X_n) < x)$ differs negligibly from $\mathsf{P}(\max(X^n - X^N, \ldots, X^n - X^n) < x)$ for sufficiently large N. □

Conditions for the convergence of the sequences $\xi_{N,k}$ can also be given in a weaker form than (18). We can, for example, simply assume that

the finite-dimensional distributions of $\{\xi_{N,k}\}$ *converge.* (*)

However, in addition to (19) the sequence X_n must then be such that

$$\lim_{N\to\infty} \lim_{n\to\infty} \mathsf{P}\big(\max_{n>k\geq N} (X_n + w_1, X_n - X_{n-k}) > 0\big) = 0\,. \tag{21}$$

The proof of this fact is left to the reader.

When the conditions (18) and (19) or (*), (19) and (21) hold, then just as in Theorem 3, along with the convergence of (20) we also have the convergence of the finite-dimensional distributions of the process $\{w_{n,k}\}$.

What has been said above allows us to deduce, in particular, the following fact: *In the study of the properties of the limiting distribution of* $\{w_{n,k}\}$ *for* $n\to\infty$, *we can assume from the outset that the system is governed by a stationary sequence* $\{\xi^k\}$. *As we have seen, this distribution will be the same in the case of a stationary governing sequence* $\{\xi^k\}$ *as in the case in which the distributions of the sequences* $\{\xi_{N,k} = \xi_{N+k}\}$ *merely converge for* $N\to\infty$ *to stationary ones.*

5. We turn now to conditions under which the sequence $\{w^k\}$ will be *proper*, i.e., $\mathsf{P}(w^k < \infty) = 1$. It is intuitively clear that if $\mathsf{M}\xi_n > 0$ exists (the service time of the n-th lot is larger on average than the interval of time τ_n^e before the appearance of the following lot), then $w_n \to \infty$ w.p.1, so that for arbitrary x

$$\lim_{n\to\infty} \mathsf{P}(w_n < x) = \mathsf{P}(w^0 < x) = \mathsf{P}(Y < x) = 0\,.$$

It turns out that this law can be extended under wide assumptions to the case $\mathsf{M}\xi_n = 0$ (in the sequel we will exclude the trivial case $\xi_n = 0$).

If $\mathsf{M}\xi_n < 0$, then w^k is finite w.p.1.

In order to formulate the corresponding assertion we will need the notion of the *metric transitivity* or *ergodicity* of a process, which we will now review. Denote by $\mathfrak{M} \subset \mathfrak{B}$ the σ-algebra generated by the sequence $\{\xi_n, -\infty < n < \infty\}$. As is well-known ([23]), with each such stationary sequence we can associate a one-to-one measure-preserving shift transformation T of sets $A \in \mathfrak{M}$ such that

$$T\{\omega : \xi_n \in B_0, \ldots, \xi_{n+k} \in B_k\} = \{\omega : \xi_{n+1} \in B_0, \ldots, \xi_{n+k+1} \in B_k\}$$

for an arbitrary collection of Borel sets $B_0, \ldots, B_k$. The corresponding transformation on $\mathfrak{M}$-measurable random variables will be denoted by the letter U so that

$$\xi_{n+1}(\omega) = U\xi_n(\omega)\,.$$

The transformations T^{-1} and U^{-1} are defined analogously.

If $\zeta(\omega)$ is an $\mathfrak{M}$-measurable random variable, then the sequence $\{U^n\zeta;\ -\infty < n < \infty\}$ will obviously be strictly stationary.

In order to extend the class of random variables measurable w.r.t. $\mathfrak{M}$ it will be convenient to understand $\mathfrak{M}$ as a possibly larger σ-algebra w.r.t. which T, as before, can be considered as a measure-preserving shift transformation of sets from $\mathfrak{M}$. For example, we can always assume that an arbitrary stationary sequence of independent random variables $\{\zeta_n\}$, not depending on the $\{\xi_n\}$, is $\mathfrak{M}$-measurable since we can view $\mathfrak{M}$ in this case as the σ-algebra generated by $\{\zeta_n\}$ and $\{\xi_n\}$ and define the shift T by means of the relations

$$T\{\omega: \xi_n \in B_0, \ldots, \xi_{n+k} \in B_k; \zeta_m \in A_0, \ldots, \zeta_{m+l} \in A_l\}$$
$$= \{\omega: \xi_{n+1} \in B_0, \ldots, \xi_{n+k+1} \in B_k; \zeta_{m+1} \in A_0, \ldots, \zeta_{m+l+1} \in A_l\}.$$

Hence, if, for example, w_1 and $\{\xi_n\}$ are independent, then the assumption of $\mathfrak{M}$-measurability of w_1 entails no loss of generality.

The set $A \in \mathfrak{M}$ is called *invariant* w.r.t. T if A and TA coincide up to a set of measure 0.

A sequence $\{\xi_n\}$ is said to be *metrically transitive* if the shift T has no other invariant sets than those of probability 0 or 1.

A sequence $\{\xi_n\}$ is *ergodic* iff for an arbitrary $\mathfrak{M}$-measurable random variable ζ having finite expectation we have

$$\frac{1}{n} \sum_{k=1}^{n} U^k \zeta \to \mathsf{M}\zeta$$

w.p.1.

A sequence $\{\xi_n\}$ is ergodic iff it is metrically transitive (see Doob [23], Chapter 10).

For example, sequences of independent or weakly dependent random variables are always ergodic. However, a sequence of random variables defined on the states of a reducible Markov chain will, generally speaking, not be ergodic.

We now formulate a theorem which gives conditions for the existence of a proper limiting sequence $\{w^k\}$.

Theorem 7. *Assume that the sequence $\{\xi_n; -\infty < n < \infty\}$ is stationary and ergodic. Then either $\mathsf{P}(Y=\infty)=1$ or $\mathsf{P}(Y<\infty)=1$.*

1. *If $\mathsf{M}\xi_n \neq 0$ exists then*

$$\lim_{n\to\infty} \mathsf{P}(w_n > x) = \mathsf{P}(Y > x)$$

exists and

$$\mathsf{P}(Y=\infty)=1 \quad \text{if} \quad \mathsf{M}\xi_n > 0,$$
$$\mathsf{P}(Y<\infty)=1 \quad \text{if} \quad \mathsf{M}\xi_n < 0.$$

2. *Let $\mathsf{M}\xi_n = 0$. Then $\mathsf{P}(Y=\infty)=1$ iff*

$$\lim_{n\to\infty} \mathsf{P}(w_n < x) = 0$$

for arbitrary $x > 0$.

In order for the second possibility $\mathsf{P}(Y<\infty)=1$ *to hold it is necessary and sufficient that there exist a stationary sequence of nonnegative* $\mathfrak{M}$*-measurable random variables* $\{\eta_n=U^n\eta_0;\ -\infty<n<\infty\}$ *such that* w.p.1.

$$\xi_n=\eta_{n+1}-\eta_n;$$

here we necessarily have $\eta_0=Y$ w.p.1.

If $\mathsf{P}(Y<\infty)=1$ *and the random variables* $w_1-\eta_1$ *and* η_n *are asymptotically independent* ($\lim_{n\to\infty}\mathsf{P}(w_1-\eta_1\in A, \eta_n\in B)=\mathsf{P}(w_1-\eta_1\in A)\mathsf{P}(\eta_1\in B)$), *then there exists*

$$\lim_{n\to\infty}\mathsf{P}(w_n<x)=\mathsf{P}(\eta+\max(w_1-\eta_1, 0)<x),$$

where η *is distributed like* η_1 *and is independent of* $w_1-\eta_1$.

Remark 1. If the asymptotic independence of $w_1-\eta_1$ and η_n in Part 2 of the theorem is not required, then the limiting distribution of w_n may not exist. This is shown by the following example: As Ω we take the two-point set (ω_1, ω_2) and put $\mathsf{P}(\omega_1)=\mathsf{P}(\omega_2)=\frac{1}{2}$, $T\omega_1=\omega_2$, $T\omega_2=\omega_1$, $\xi_0(\omega_1)=-1$ and $\xi_0(\omega_2)=1$. Then the sequence $\{\xi_n=U^n\xi_0=\xi_0(T^n\omega);\ -\infty<n<\infty\}$ will be stationary and metrically transitive. As is easy to see, its elements can be written as

$$\xi_n=\eta_{n+1}-\eta_n,$$

where $\eta_0(\omega_1)=1$, $\eta_0(\omega_2)=0$ and $\eta_n=U^n\eta_0=\eta_0(T^n\omega)$. The sequence $\{\eta_n\}$ is clearly also stationary and metrically transitive. In this example

$$Y_n=\eta_0-\eta_{-n}, \qquad X_n=\eta_{n+1}-\eta_1$$

and

$$Y=\sup_{n\geqslant 0}Y_n=\eta_0-\inf_{n\geqslant 0}\eta_{-n}=\eta_0.$$

If $w_1\geqslant 1$, then by Theorem 1

$$\begin{aligned}
w_n&=X_{n-1}-\min(-w_1, X_1, \dots, X_{n-1})\\
&=\eta_n-\min(-w_1+\eta_1, \eta_2, \dots, \eta_n)=\eta_n-\min(-w_1+\eta_1, 0)\\
&=\eta_n+\max(w_1-\eta_1, 0)=\eta_n+w_1-\eta_1.
\end{aligned}$$

This means that $w_n=w_1$ if n is odd and $w_n=w_1+1-2\eta_1$ if n is even.

Remark 2. It should also be noted that the assumption of weak dependence among the ξ_k (we exclude complete independence of the ξ_k) does not preclude the possibility that $\mathsf{P}(Y<\infty)=1$ when $\mathsf{M}\xi_k=0$.

It is sufficient to treat the example where the η_k are independent and $\xi_k=\eta_{k+1}-\eta_k$. Then the collections of random variables $\{\cdots, \xi_{n-1}, \xi_n\}$ and $\{\xi_{n+2}, \xi_{n+3}, \dots\}$ will clearly also be independent although the variable Y, equal to η_0 w.p.1., is finite.

The relation $\mathsf{P}(Y=\infty)=1$ can be guaranteed by various conditions on the "decay" of the distributions of X_n or Y_n as $n\to\infty$. For example, the condition

$$\sup_y \limsup_{n\to\infty}\mathsf{P}(|X_n|>y)>0 \tag{22}$$

is incompatible with the representation $X_n=\eta_{n+1}-\eta_n$ and consequently, with the possibility that $\mathsf{P}(Y<\infty)=1$.

Proof of Theorem 7. By the finiteness of ξ_{-1}, we have with accuracy up to a ω-set of measure 0

$$\{Y=\infty\}=\{\sup_{n\geqslant 1}(Y_n-\xi_{-1})=\infty\}=T^{-1}\{Y=\infty\}.$$

Hence, the event $\{Y=\infty\}$ is invariant under T. By metric transitivity, this implies that $\mathsf{P}(Y=\infty)$ is either 0 or 1.

Furthermore, w.p.1,

$$\lim_{n\to\infty}\frac{Y_n}{n}=\mathsf{M}\xi_1. \tag{23}$$

Thus, for $\mathsf{M}\xi_1\neq 0$ the conditions of Theorem 2 hold. In this connection, if $\mathsf{M}\xi_1>0$, then

$$\mathsf{P}(Y<x)\leqslant\mathsf{P}\left(\frac{Y_n}{n}<\frac{x}{n}\right)\to 0 \quad \text{as} \quad n\to\infty.$$

If $\mathsf{M}\xi_1<0$, then (23) says that the sequence $Y_1, Y_2, \ldots$ contains w.p.1 only a finite number of positive terms so that $Y=\sup_{k\geqslant 0}Y_k$ is finite w.p.1.

It remains to consider the case $\mathsf{M}\xi_1=0$.

Assume that $\mathsf{P}(Y<\infty)=1$ and set

$$Y_{n,k}=\sum_{j=1}^{k}\xi_{n-j}, \qquad Y^n=\sup_{k\geqslant 0}Y_{n,k}$$

so that $Y_{0,k}=Y_k$ and $Y^0=Y$. Clearly, we can also write

$$Y_{n,k}=U^nY_k \quad \text{and} \quad Y^n=U^nY.$$

We set

$$\xi_n^*=Y^{n+1}-Y^n.$$

Then, as is easy to verify directly,

$$\xi_n^*=\max(\xi_n, -Y^n)\leqslant\max(\xi_n, 0).$$

This means that $\mathsf{M}\xi_n^*$, finite or infinite, exists. Moreover, ξ_n^* is measurable w.r.t. $\mathfrak{M}$ and $\xi_{n+1}^*=U\xi_n^*$. From this it follows that the sequence $\{\xi_n^*\}$ is stationary and ergodic. Since Y^0 and Y^n are almost everywhere finite, we have

$$\frac{1}{n}\sum_{k=0}^{n-1}\xi_k^*=\frac{1}{n}(Y^n-Y^0)\xrightarrow[\mathsf{P}]{}0.$$

This implies in turn that $\mathsf{M}\xi_n^*=0$. Combined with the relations $\mathsf{M}\xi_n=0$ and $\xi_n\leqslant\xi_n^*$, this leads to the conclusion that w.p.1

$$\xi_n=\xi_n^*=Y^{n+1}-Y^n.$$

The remaining assertions of the theorem are consequences of the following

Lemma. *Assume that the sequence $\{\xi_n; -\infty<n<\infty\}$ is stationary and metrically transitive with $\mathsf{M}\xi_k=0$. Let $\eta=\eta(\omega)$ be an arbitrary $\mathfrak{M}$-measurable random variable. Then for the stationary sequence $\eta_n=U^n\eta$ we have*

$$\sup_{n\geqslant 0} U^n\eta(\omega)=\sup_{n\geqslant 0} U^{-n}\eta(\omega)=\operatorname{vrai\,sup}_{\omega}\eta(\omega)\ ^3.$$

Proof. Put $\operatorname{vrai\,sup}_\omega \eta(\omega)=b$. Then it is clear that $\sup_{n\geqslant 0}\eta_n(\omega)\leqslant b$ w.p.1. Now suppose that the assertion relative to $\sup_{n\geqslant 0}\eta_n$ is not true. This would imply that there exists a $b_1<b$ such that the event

$$B=\{\sup_{n\geqslant 0}\eta_n\leqslant b_1\}$$

has positive probability. By the definition of the transformation T the set

$$B_1=TB=\{\omega\colon \sup_{n\geqslant 0} U\eta_n\leqslant b_1\}=\{\sup_{n\geqslant 1}\eta_n\leqslant b_1\}\supset B\,.$$

Together with the relation $\mathsf{P}(TB)=\mathsf{P}(B)$ this implies that $TB=B$ except possibly for a set of measure 0, which says that B is invariant. Hence, by the metric transitivity, $\mathsf{P}(B)=1$. But then $\mathsf{P}(\eta\leqslant b_1)=1$ and $b=\operatorname{vrai\,sup}_\omega\eta(\omega)\leqslant b_1$ which contradicts $b_1<b$. The second assertion of the lemma is established analogously. □

We turn now to the proof of the theorem. If it is assumed that $\xi_n=\eta_{n+1}-\eta_n$, then we obtain $Y_n=\eta_0-\eta_{-n}$, so that

$$Y=\sup_{n\geqslant 0}(\eta_0-\eta_{-n})=\eta_0-\inf_{n\geqslant 0} U^{-n}\eta_0=\eta_0-\operatorname{vrai\,inf}_{\omega}\eta_0$$

is a proper random variable.

In the same way[4] it is not hard to prove that $\operatorname{vrai\,inf}_\omega Y(\omega)=0$.

If $\mathsf{P}(Y<\infty)=1$ and the variables $w_1-\eta_1$ and η_n are asymptotically independent, then by Theorem 1

$$w_n=X_{n-1}-\min(-w_1, X_1, \ldots, X_{n-1})=\eta_n-\min(-w_1+\eta_1, \eta_2, \ldots, \eta_n);$$

$$\lim_{n\to\infty}\mathsf{P}(w_n<\infty)=\lim_{n\to\infty}\mathsf{P}(\eta_n+\max(w_1-\eta_1, 0)<x)$$

$$=\mathsf{P}(\eta+\max(w_1-\eta_1, 0)<x)\,,$$

where η is distributed like η_1 and is independent of $w_1-\eta_1$. □

Remark 3. The lemma we used in the preceding proof also allows us to establish that for the stationary sequence $\{\xi_n\}$ in the case $\mathsf{M}\xi_k\leqslant 0$, the random variables $Y=\sup_{n\geqslant 0}Y_n$ and $X=\sup_{n\geqslant 0}X_n$ have, generally speaking, different distributions. This is revealed most simply in the case $\xi_n=\eta_{n+1}-\eta_n$, where $\{\eta_k\}$ is stationary.

[3] The number $b=\operatorname{vrai\,sup}_\omega\eta(\omega)$ is the *essential upper bound* of the values of $\eta(\omega)$ and is equal to $\inf_{U,\mathsf{P}(U)=1}\sup_{\omega\in U}\eta(\omega)$, so that $\mathsf{P}(\eta>b)=0$ and $\mathsf{P}(\eta>b-\varepsilon)>0$ for arbitrary $\varepsilon>0$.

[4] Of interest are the following relations for $Z=\inf_{n\geqslant 0}Y_n$, which follow from the lemma:

$$Z=\inf_{n\geqslant 0}Y_n=\inf_{n\geqslant 0}(Y-Y^{-n})=Y-\sup_{n\geqslant 0}Y^{-n}=Y-\operatorname{vrai\,sup}_\omega Y(\omega)$$

These imply in particular, that Y is an unbounded random variable iff $\mathsf{P}(Z=-\infty)=1$. However, if Z is finite, then $\operatorname{vrai\,sup}_\omega Z(\omega)=0$ and $\operatorname{vrai\,sup}_\omega Y(\omega)=-\operatorname{vrai\,inf}_\omega Z(\omega)$.

Since here $Y_n = \eta_0 - \eta_{-n}$ and $X_n = \eta_{n+1} - \eta_1$, we have

$$Y = \eta_0 - \inf_{n \geq 0} \eta_{-n} = \eta_0 - \operatorname{vrai\,inf}_{\omega} \eta_0(\omega)$$

and

$$X = \sup_{n \geq 0} \eta_{n+1} - \eta_1 = \operatorname{vrai\,sup}_{\omega} \eta_1(\omega) - \eta_1 .$$

If, for example, $\eta_0 \geq 0$ and is unbounded from above, then Y will be a proper random variable and $X = \infty$ w.p.1.

The assertion of Theorem 7 can be extended in the following way:

Theorem 8. *If the sequence X_n satisfies the strong law of large numbers (for example, $\{\xi_n\}$ is metrically transitive), then as $n \to \infty$*

$$\frac{w_n}{n} \to \max(0, \mathsf{M}\xi_1), \quad \text{w.p.1.}$$

Proof. If $\mathsf{M}\xi_1 > 0$, then by the strong law of large numbers the sequence

$$w_{n+1} = X_n - \min(-w_1, X_1, \ldots, X_n) \geq w_1 + X_n$$

has only a finite number of zero terms. That is, there exists a proper random variable $\nu(\omega)$ such that on the set $\Omega_N = \{\nu(\omega) < N\}$

$$w_n > 0, \quad \text{and} \quad w_{n+1} = w_n + \xi_n$$

for all $n \geq N$. Consequently, for each N and a.e. on Ω_N for $n \to \infty$

$$\frac{w_n}{n} = \frac{w_N + \xi_{N+1} + \cdots + \xi_n}{n} \to \mathsf{M}\xi_1 .$$

Since $\mathsf{P}(\cup_{N=1}^{\infty} \Omega_N) = 1$, the required convergence is proved.

Now assume that $\mathsf{M}\xi_1 = 0$. In addition to $\{\xi_n\}$ we consider the sequence $\{\xi_n^* = \xi_n + \delta_n\}$, where the δ_n are mutually independent and also independent of the sequence $\{\xi_n\}$, $\mathsf{P}(\delta_n = 1) = p$, $\mathsf{P}(\delta_n = 0) = 1 - p$ and we denote by $\{w_n^*\}$ the waiting time process constructed according to the sequence $\{\xi_n^*\}$, $(w_1^* = w_1)$. Then we obviously have $w_n \leq w_n^*$, $\mathsf{M}\xi_n^* = p > 0$ and the sequence $\{\xi_n^*\}$ satisfies the strong law of large numbers. Hence, from the part of the theorem already proved, $w_n/n \leq w_n^*/n \to p$ w.p.1. Since p can be chosen arbitrarily small,

$$w_n/n \to 0 \quad \text{w.p.1} .$$

The case $\mathsf{M}\xi_1 < 0$ is treated analogously. □

Remark 4. It is necessary to remark that the assertion of the theorem in the case $\mathsf{M}\xi_1 < 0$ is not obvious. For example, for a stationary sequence of independent random variables v_n, the relation $v_n/n \to 0$ holds a.e. iff $\mathsf{M}|v_1| < \infty$. At the same time, for the sequence w_n with a stationary initial distribution of w_1 and independent terms ξ_k in the governing sequence, $w_n/n \to 0$ holds a.e. according to Theorem 8, although, as we will see in Chapter 4, $\mathsf{M}w_1 = \infty$ when $\mathsf{M}\xi_k^2 = \infty$ and $\mathsf{M}\xi_k < 0$.

§ 4. Interrupted Governing Sequences

Stationary sequences do not exhaust the class of sequences for which the limiting distribution of w_n exists and can be investigated. We dwell here on the important case in which the service times, the intervals between customer arrivals, etc. generate sequences "interrupted" whenever the system becomes free. Such a situation arises in a number of applications. In order to describe it, we assume that on $(\Omega, \mathfrak{B}, \mathsf{P})$ there is given a family

$$\{\xi_n^{(1)}; -\infty < n < \infty\}, \{\xi_n^{(2)}; -\infty < n < \infty\}, \dots \tag{24}$$

of sequences of arbitrarily related random variables. It is assumed that these sequences are independent and have the same distributions. We can also assume that they are generated by the corresponding family of initial four-dimensional governing sequences.

The sequence $\{\xi_n\}$, which governs through the solution of Eq. (8), § 2, is defined as follows:

$$\xi_1 = \xi_1^{(1)}, \dots, \xi_{\eta_1} = \xi_{\eta_1}^{(1)},$$

$$\xi_{\eta_1+1} = \xi^{(2)}, \dots, \xi_{\eta_1+\eta_2} = \xi_{\eta_2}^{(2)}, \quad \text{etc.},$$

where $\eta_1 = \min\{k > 0 : w_{k+1} = 0\}$, $\eta_1 + \eta_2 = \min\{k > \eta_1 : w_{k+1} = 0\}$, etc.

Such a system is governed by the first sequence at (24) until it becomes free for the first time. Its subsequent evolution is determined by the second sequence of the family until it again becomes free. The underlying sequence is then the third one, etc. In this way, the development of the system is decomposed into independent, uniquely determined cycles, "generated" by visits of the sequence w_n to 0.

We assume for the sake of simplicity that $w_1 = 0$. Then what has been said implies, in particular, that the random variables $\eta_1, \eta_2, \dots$ are i.i.d. Naturally, in order for this scheme to be interesting, the sequences (24) must satisfy conditions which guarantee the finiteness of the η_j.

The random variables η_j, $j \geqslant 1$, can also be defined by means of the equations

$$\eta_j = \min\{n \geqslant 1 : \textstyle\sum_{k=1}^n \xi_k^{(j)} \equiv X_n^{(j)} \leqslant 0\}$$

as the first indices for which the sums $X_n^{(j)}$ become nonpositive.

Theorem 9. *If* $\mathsf{M}\eta_1 = \infty$, *then*

$$\lim_{n\to\infty} \mathsf{P}(w_n = 0) = 0 .$$

If $\mathsf{M}\eta_1 < \infty$ *and the greatest common divisor (g.c.d.) of the possible values of* η_1 *equals* 1, *then*

$$\lim_{n\to\infty} \mathsf{P}(w_n \leqslant x) = \frac{1}{\mathsf{M}\eta_1}\left[1 + \textstyle\sum_{k=1}^\infty \mathsf{P}\left(\min\limits_{1\leqslant i\leqslant k} X_i^{(1)} > 0, X_k^{(1)} \leqslant x\right)\right].$$

Proof. Denote by γ_n the difference between n and the time of the last change in the sequences (24) previous to n:

$$\gamma_n = n - \sum_{k=1}^{\eta(n)} \eta_k, \; \eta(n) = \max\{k : \sum_{j=1}^{k} \eta_j \leqslant n\}.$$

This is the so-called "defect" up to the level n in the random walk $\eta_1, \eta_2, \cdots$.

By the formula of total probability

$$\mathsf{P}(w_n \leqslant x) = \sum_{k=0}^{n} \mathsf{P}(\gamma_n = k)\mathsf{P}(w_n \leqslant x \mid \gamma_n = k) = \sum_{k=0}^{n} \mathsf{P}(\gamma_n = k)\mathsf{P}(w_{k+1} \leqslant x \mid \eta_1 > k).$$

Again using the total probability formula for $\mathsf{P}(\gamma_n = k)$, we obtain for $k < n$

$$\mathsf{P}(\gamma_n = k) = \mathsf{P}(\eta_1 = n-k)\mathsf{P}(\eta_2 > k) + \mathsf{P}(\eta_1 + \eta_2 = n-k)\mathsf{P}(\eta_3 > k)$$
$$+ \ldots = \mathsf{P}(\eta_1 > k) \sum_{l=1}^{\infty} \mathsf{P}\left(\sum_{j=1}^{l} \eta_j = n-k\right) = \mathsf{P}(\eta_1 > k) h(n-k).$$

This equality obviously remains valid for $k = n$ if we set $h(0) = 1$. As is well-known, $h(n) \to 1/\mathsf{M}\eta_1$ as $n \to \infty$.

Since $\mathsf{P}(w_n = 0) = \mathsf{P}(\gamma_n = 0)$, it follows that assertion of the theorem holds in the case $\mathsf{M}\eta_1 = \infty$. Moreover, by the fundamental renewal theorem (Appendix 1),

$$\lim_{n \to \infty} \mathsf{P}(w_n \leqslant x) = \frac{1}{\mathsf{M}\eta_1} \sum_{k=0}^{\infty} \mathsf{P}(w_{k+1} \leqslant x, \eta_1 > k)$$
$$= \frac{1}{\mathsf{M}\eta_1} \left[1 + \sum_{k=1}^{\infty} \mathsf{P}\left(\min_{1 \leqslant j \leqslant k} X_j^{(1)} > 0, X_k^{(1)} \leqslant x\right)\right] \quad \text{exists.} \quad \square$$

There remains the case in which the initial value w_1 is >0; we will leave it to the reader to fill in this gap.

Under the conditions of Theorem 9 it is obviously not difficult to find the joint limiting distribution of w_n and the "time" γ_n the system spends in the busy state:

$$\lim_{n \to \infty} \mathsf{P}(w_n \leqslant x, \gamma_n = k) = \frac{1}{\mathsf{M}\eta_1} \mathsf{P}\left(\min_{1 \leqslant j \leqslant k} X_j^{(1)} > 0, X_k^{(1)} \leqslant x\right), \quad k \geqslant 1. \tag{25}$$

§5. On Systems Governed by Sequences of Independent Random Variables

We assume here that the governing sequence is such that the sequence

$$\{\xi_n; -\infty < n < \infty\}$$

is stationary and the ξ_k are independent. This is a very important and doubtless the most completely investigated case. It will occur, in particular, if all governing sequences of the system are mutually independent and are formed from i.i.d. variables.

We have seen that distributions connected with w_n reduce to such functionals of the sequence as $\max_{0 \leqslant k \leqslant n} Y_k$, $\sup_{k \geqslant 0} Y_k$ and the time θ that this supremum is attained.

These characteristics are closely connected with a certain number of other so-called boundary functionals, so that the investigation of the distribution of w_n turns out to be related to many general questions. Hence, the exposition of our studies of the systems considered in this section will be placed in separate chapters devoted to boundary problems for random walks.

The goal of the present section is mainly to find out what the conditions of Theorems 1–9 will be for independent ξ_k. To avoid trivial difficulties, we will assume that ξ_k assumes positive and negative values with positive probability.

Using the fact that the distributions of the sequences $(Y_0, Y_1, \ldots)$ and $(X_0, X_1, \ldots)$ coincide in our case (we recall that $Y_n=\sum_{k=1}^{n} \xi_{-k}$, $X_n=\sum_{k=1}^{n} \xi_k$), we can formulate the assertions of Theorems 2–9 in terms of the sequence $\{X_n; n\geqslant 0\}$ and the random variables $X=\sup_{k\geqslant 0} X_k$, $\theta=\min\{k\geqslant 0: X_k=X\}$.

Condition (12) is obviously always satisfied since for sums of i.i.d. terms $\mathsf{P}(|X_n|<N)\to 0$ when $n\to\infty$ for arbitrary $N>0$.

Hence,

$$\lim_{n\to\infty} \mathsf{P}(w_n<x)=\mathsf{P}(X<x)$$

always exists and the distribution of the sequence $\{w_{k+n}; k\geqslant 0\}$ *converges for* $n\to\infty$ *to the distribution of* $\{w^k; k\geqslant 0\}$, *where*

$$w^k=\sup(0, \xi_k, \xi_k+\xi_{k+1}, \ldots)=U^k w^0 .$$

The joint distribution of (w_n, v_n) *converges to that of* (X, θ).

Subsection 4 of § 3 allows us to consider questions of the existence and determination of the limiting distribution of $\{w_n\}$ when the ξ_n are not identically distributed. Conditions on the convergence of the distributions of ξ_n (one-dimensional) guaranteeing the validity of the conditions of this subsection are a separate question.

Here we indicate simple conditions sufficient for the validity of the requirements (∗) and (21) of Subsection 4 § 3. Indeed, *Conditions* (∗) *and* (21) *will hold if* $\mathsf{P}(\xi_n<x)\Rightarrow\mathsf{P}(\xi<x)$ (the symbol $\Rightarrow$ denotes weak convergence of distributions) *and if starting from some n*

$$\mathsf{M}\xi_n\leqslant a<0 \quad\text{and}\quad \mathsf{D}\xi_n\leqslant\sigma^2<\infty .$$

In fact, the convergence of the finite-dimensional distributions of $\{\xi_{N,k}=\xi_{N+k}; k\geqslant 0\}$ for $N\to\infty$ is obvious. Condition (21) follows from the inequality

$$\mathsf{P}(\max_{n>k\geqslant N}(X_n+w_1, X_n-X_{n-k})>0)\leqslant r_n+r_{n,N}+\mathsf{P}\left(\max_{n>k\geqslant N}(X_{n-N}-X_{n-k})>-\frac{aN}{2}\right),$$

where $r_n\leqslant\mathsf{P}(X_n+w_1>0)\to 0$ for $n\to\infty$ and $r_{n,N}\leqslant\mathsf{P}(X_n-X_{n-N}>aN/2)$ converges to 0 for $N\to\infty$ uniformly in n. Finally, the probability

$$\mathsf{P}\left(\max_{n>k\geqslant N}(X_{n-N}-X_{n-k})>-\frac{aN}{2}\right)\leqslant\sum_{j=1}\mathsf{P}\left(\max_{2^{j-1}N\leqslant k<2^jN}(X_{n-N}-X_{n-k})>-\frac{aN}{2}\right)$$

can be estimated with the aid of Kolmogorov's inequalities by which it does not exceed

$$\sum_{j=1} \frac{(2^j-1)N\sigma^2}{\left(\frac{aN}{2}+(2^{j-1}-1)aN\right)^2} = \frac{4\sigma^2}{a^2N}\sum_{j=1}\frac{1}{2^j-1} < \frac{8\sigma^2}{a^2N}. \quad \square$$

With the help of more precise calculations we can convince ourselves that the condition imposed on the variance can be weakened to the uniform boundedness of $\mathsf{M}|\xi_n|^{1+\varepsilon}$ for some $\varepsilon>0$.

A stationary sequence of independent random variables is always metrically transitive so that by Theorem 7, $\mathsf{P}(X=\infty)=1$ if $\mathsf{M}\xi_k>0$, and $\mathsf{P}(X<\infty)=1$ if $\mathsf{M}\xi_k<0$. If $\mathsf{M}\xi_k=0$, then, as we will see in Chapter 3, $\mathsf{P}(X=\infty)=1$ always holds. This also follows from the considerations of § 3, since in our case $\mathsf{P}(|X_n|\leqslant y)\to 0$ as $n\to\infty$ for arbitrary fixed y; consequently, (23) holds and by it the finiteness of X and Y is impossible. By Theorem 8, $w_n/n\to\max(0, \mathsf{M}\xi_1)$ w.p.1.

Finally, systems governed by a sequence of independent ξ_k can be considered as systems with interrupted sequences ($\sum_{k=1}^m \eta_k$ is a random variable independent of the future since the event $\{\sum_{k=1}^m \eta_k=j\}$ does not depend on the σ-algebra generated by the random variables $\xi_{j+1}, \xi_{j+2}, \ldots$). The hypothesis of the theorem imposed on the greatest common divisor of the possible values of η_1 is always fulfilled since $\mathsf{P}(\eta_1=1)=\mathsf{P}(\xi_1\leqslant 0)>0$. Comparing Theorems 9 and 2, we obtain

$$\mathsf{P}(X=0)=\frac{1}{\mathsf{M}\eta_1}.$$

Consequently, $\mathsf{M}\eta_1$ is finite when $\mathsf{P}(X=0)>0$. For the latter inequality it is in turn sufficient that $\mathsf{P}(X<\infty)=1$. Indeed, in this case there exists an integer x_0 such that $\mathsf{P}(X<x_0)>0$. Assuming without loss of generality that $\mathsf{P}(\xi<-1)>0$, we get

$$\mathsf{P}(X=0)>\mathsf{P}(\xi_1<-1)\ldots\mathsf{P}(\xi_{x_0}<-1)\mathsf{P}(X<x_0)>0\,.$$

Thus, the assertion of Theorem 9 holds if $\mathsf{P}(X<\infty)=1$. Comparing (25) with the assertion of Theorem 5, we find that

$$\lim_{n\to\infty}\mathsf{P}(w_n\geqslant x, v_n=k)=\frac{1}{\mathsf{M}\eta_1}\mathsf{P}(\min_{1\leqslant j\leqslant k} X_j>0, X_k\geqslant x)=\mathsf{P}(X\geqslant x, \theta=k)\,.$$

§ 6. The Virtual Waiting Time. A Continuous Analogue of the System Equation. Properties of the Solution

1. Let us consider the system $\langle G, G, G, 1\rangle$. Another possible characteristic of the performance of the system which is closely connected with $\{w_n\}$ is the so-called *virtual waiting time* $w(t)$ defined as the time required to free the system of requests arriving up to time t. Less precisely, $w(t)$ is the time a request arriving at time t would have to wait until the start of his service. If we denote by n_t the number of the last group of requests arriving in the system before time t, then in the notation of § 2 (see (3))

$$w(t)=w_{n_t}+S_{n_t}-(t-\tau_1^e-\cdots-\tau_{n_t-1}^e).$$

Denote by $S(t)$ the time resulting from summation of the service times of requests arriving in the system up to time t. It will be convenient to assume that the function $S(t)$ is left-continuous: $S(t)=S(t-0)$, so that $S(\tau_1^e)=S(+0)=w_1+S_1$ and $S(\tau_1^e+0)=S(+0)+S_2$. The process

$$\{X(t)=S(t)-t;\ t\geqslant 0\}$$

together with the sequence (1), §1, will be called a *governing sequence*.

The process $X(t)$ can be considered as a left-continuous stochastic process on the space $D(0,\infty)$ of functions on $[0,\infty]$ having no discontinuities of the second kind with σ-algebra $\mathcal{N}_D$ generated by the Borel cylinder sets. The distribution of this process is uniquely determined by the given distribution of the governing sequence and is a distribution on $(D,\mathcal{N}_D)$ generated by a measurable mapping of $(\Omega,\mathfrak{B},\mathsf{P})$ onto $(D,\mathcal{N}_D)$ (this mapping is given essentially by the definition of $S(t)$.)

The realizations of $X(t)$ have the form shown in Fig. 1. For the sequel we note that if we choose the governing sequence so that $S_1=0$, we can make the process $X(t)$ continuous at the point $t=0$.

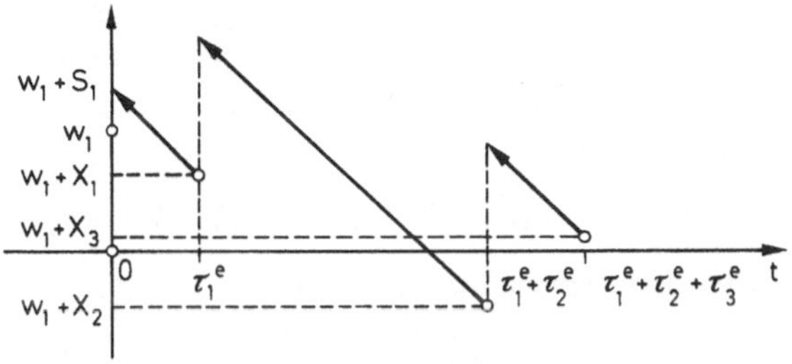

Fig. 1. A Realization of $X(t)$

The analogue of the fundamental Eq. (8) is here

$$dw(t)=dS(t)-dt(1-\delta(w(t)))=dX(t)+dt\,\delta(w(t)), \tag{26}$$

which obviously holds for any $w\in\Omega$. $\delta(x)$ is defined by

$$\delta(x)=\begin{cases}1 & \text{if } x=0;\\ 0 & \text{if } x>0.\end{cases}$$

Integrating (26) from 0 to t and using the fact that $w(0)=X(0)$, we get

$$w(t)=X(t)+\textstyle\int_0^t \delta(w(u))\,du, \tag{27}$$

which also has the following very simple interpretation: the waiting time $w(t)$ is $S(t)-t$ plus the sum of the idle times during $[0,t]$. Obviously, Eqs. (26) and (27) uniquely determine the function $w(t)$ for each realization $X(t)$ and initial value $w(0)$.

Theorem 10.

$$w(t)=X(t)-\inf_{0\leqslant u\leqslant t}(0, X(u)). \tag{28}$$

Proof. It suffices to show that

$$-\inf_{0\leqslant u\leqslant t}(0, X(u))=\textstyle\int_0^t \delta(w(t))\,dt, \tag{29}$$

where $w(t)$ is defined by (28). Both sides of (29) are continuous in t and coincide for $t=0$. Hence, it is sufficient to prove that at points of continuity of $X(t)$

$$d\inf_{0\leqslant u\leqslant t}(0, X(u))=-\delta(w(t))\,dt.$$

If $X(t)>\inf_{[0,t]}(0, X(u))$, then $d\inf_{[0,t]}(0, X(u))=0$ and by (28) $\delta(w(t))=0$. If $X(t)=\inf_{[0,t]}(0, X(u))$, then $d\inf_{[0,t]}(0, X(u))=-dt$ and $\delta(w(t))=1$. □

2. Following the line of argument which we used to investigate Eq. (8), we will consider in the sequel *homogeneous* governing processes $X(t)$. We will in fact assume given on $(\Omega, \mathfrak{B}, \mathsf{P})$ (or on $(D(-\infty, \infty), \mathcal{N}_D)$ if it is assumed that Ω coincides with $D(-\infty, \infty)$) a process

$$\{X(t);\ -\infty<t<\infty\}$$

whose increments are strictly stationary, with $\{X(t), t\geqslant 0\}$ as a "subprocess", and that this subprocess is governing. A process with stationary increments (the word "strictly" will often be omitted, as before) is a process for which the joint distribution of the increments

$$X(t+t_1)-X(t+u_1), \ldots, X(t+t_k)-X(t+u_k)$$

is independent of t for arbitrary $t_1, u_1, \ldots, t_k, u_k$.

Set

$$Y(t)=X(0)-X(-t), \qquad \bar{Y}(t)=\sup_{0\leqslant u\leqslant t} Y(u) \quad \text{and} \quad \bar{Y}=\bar{Y}(\infty).$$

Theorem 11. *If, in addition to the assumptions already made,*

$$\lim_{t\to\infty}\mathsf{P}(X(t)-X(0)\in\Delta)=0, \tag{30}$$

for an arbitrary fixed interval Δ, then for arbitrary $x\geqslant 0$ and an arbitrary proper random initial condition $w(0)$,

$$\lim_{t\to\infty}\mathsf{P}(w(t)>x)=\mathsf{P}(\bar{Y}>x)$$

exists and

$$\mathsf{P}(w(t)>x)-\mathsf{P}(\bar{Y}>x)=\mathsf{P}\Big(\sup_{0\leqslant u\leqslant t}(X(t)-X(u))\leqslant x,$$
$$X(t)-X(0)+w(0)>x\Big)-\mathsf{P}(\bar{Y}(t)\leqslant x,\ \bar{Y}>x).$$

If $w(0)=0$, then Condition (30) is superfluous.

Theorem 12. *Under the conditions of Theorem* 11, *the finite-dimensional distributions of the sequence of processes* $\{w_t(u)=w(u+t);\, u\geqslant 0\}$ *converge for* $t\to\infty$ *to the corresponding distributions of the process* $\{w^c(u);\, u\geqslant 0\}$, where

$$w^c(u)=\sup_{v\geqslant -u}(Y(v)-Y(-u))=\sup_{v\geqslant u}(X(u)-X(v)).$$

The proofs of these theorems will be omitted since they are completely analogous to the discrete case (see Theorems 2, 3). The reader can obtain them without difficulty using the equality

$$w(t)=\sup_{0\leqslant u\leqslant t}(X(t)-X(0)+w(0),\, X(t)-X(u)).$$

However, the situation is different when we are dealing with the convergence of distributions of functionals $f(w_t)$ of the trajectory $w_t(u)$, on, say, some interval $[t_1, t_2]$. For example, the functionals

$$\int_{t_1}^{t_2} w_t(u)\,du \quad \text{or} \quad \sup_{t_1\leqslant u\leqslant t_2} w_t(u). \tag{31}$$

As is well known, for the convergence of the distributions of these functionals it is generally not sufficient that merely the finite-dimensional distributions converge; one needs special additional properties of the processes. On the other hand, in the investigation of these convergence questions, it is reasonable to select a class of functionals which is not too large.

One of the most natural notions here is that of weak convergence of distributions in the metric space $D(t_1, t_2)$. The space D is metrized by means of the Skorohod–Prohorov distance

$$\rho_D(w, v)=\inf_{\lambda}\left[\sup_t |w(t)-v(\lambda(t))| + \sup_t |t-\lambda(t)|\right],$$

where the infimum is taken over all continuous monotone functions $\lambda(t)$ for which $\lambda(t_1)=t_1$ and $\lambda(t_2)=t_2$. The class of functionals f to be considered is continuous in the metric $\rho_D(f(v_n)\to f(v)$ if $\rho_D(v_n, v)\to 0$ for $n\to\infty$). It's easy to verify that the functionals at (31) are ρ_D-continuous. Since the σ-algebra $\mathcal{N}_D$ coincides with the σ-algebra of Borel sets (generated by ρ_D-open sets), $f(w_t)$ will be a random variable. We say that *the sequence* P_t *of distributions of the processes* w_t *converges weakly to the distribution* P *of the process* w^c *if for an arbitrary* ρ_D*-continuous functional* f

$$\mathsf{P}_t(f(w_t)<x)\Rightarrow \mathsf{P}(f(w^c)<x).$$

Conditions sufficient for weak convergence in D are known (I. I. Gihman and A. V. Skorohod [29]). Examining these conditions, it is not difficult to establish that *if the process* $X(t)$ *has* w.p.1 *a finite number of jumps on any finite interval, then for arbitrary* t_1 *and* t_2 *the distribution of the process* $\{w_t(u);\, t_1\leqslant u\leqslant t_2\}$ *converges weakly in* $D(t_1, t_2)$ *to the distribution of* $\{w^c(u);\, t_1\leqslant u\leqslant t_2\}$.

Such an approach to the study of distributions of functionals is related to general theorems on the convergence of distributions in metric spaces. However, it is appropriate here to deal with processes w_t of a very special form. For them we can obtain stronger assertions under much wider conditions.

Theorem 13. *If the process* $\{X(t);\, -\infty<t<\infty\}$ *has the following property:* w.p.1

$$\lim_{t\to\infty}(X(t)-X(0))=\lim_{t\to\infty} Y(t)=-\infty, \tag{32}$$

then

$$\mathsf{P}(\sup_{u\geqslant 0} | w_t(u) - w_t^c(u) | \neq 0) \to 0 \tag{33}$$

for $t\to\infty$, *where* $\{w_t^c(u); u\geqslant 0\}$ *is the process*

$$w_t^c(u) = w^c(t+u) = \sup_{v\leqslant t+u} (X(u+t) - X(v)),$$

whose distribution is independent of t and coincides with that of the proper stationary process $\{w^c(u), u\geqslant 0\}$.

If f is an *arbitrary* $(D, \mathcal{N}_D)$-measurable functional, then (33) implies that for each x

$$\mathsf{P}(f(w_t) < x) \to \mathsf{P}(f(w^c) > x) \quad \text{as} \quad t\to\infty. \tag{34}$$

The assertion of this theorem can be formulated without difficulty in terms of $(D, \mathcal{N}_D)$-measurable sets (events). Indeed, if B is any such set, then for $t\to\infty$

$$\mathsf{P}(w_t \in B) \to \mathsf{P}(w^c \in B).$$

Hence, the behavior of the waiting time $w_t(u) = w(t+u)$, observed after the time has been increased by t, differs very little from the behavior of the stationary process $w^c(u)$.

Proof. By Theorem 10 it is sufficient to show that the probability of the event

$$\bigcup_{u\geqslant 0} \{ \sup_{v\leqslant u+t} (X(u+t) - X(v)) \neq \sup_{0\leqslant v\leqslant u+t} (X(u+t), X(u+t) - X(v))\}$$

converges to zero for $t\to\infty$. This event implies either

$$\bigcup_{u\geqslant 0} \{X(u+t) > 0\}$$

or

$$\bigcup_{u\geqslant 0} \{\sup_{v<0} (X(u+t) - X(v)) > 0\}.$$

Thus, the probability to be estimated does not exceed

$$\begin{aligned}\mathsf{P}(\{\sup_{u\geqslant 0} X(u+t) > 0\} \cup \{ \sup_{u\geqslant 0, v<0} (X(u+t) - X(v)) > 0\})\\ \leqslant \mathsf{P}(\sup_{u\geqslant 0} X(u+t) > 0) + \mathsf{P}\left(\sup_{u\geqslant 0}\left(X\left(\frac{t}{2}+u\right) - X(0)\right) > 0\right)\\ + \mathsf{P}\left(\sup_{v\geqslant 0} Y\left(\frac{t}{2}+v\right) > 0\right).\end{aligned} \tag{35}$$

When (32) holds, this sum converges to 0. It is clear that (32) also implies the finiteness of

$$w^c(u) = \sup_{v\geqslant -u} (Y(v) - Y(-u)). \quad \square$$

As in the discrete case, we can obviously obtain other estimates for the probability (33). For example,

$$\mathsf{P}(\min_{0\leqslant v\leqslant u+t} X(v) > \min(0, \inf_{v<0} X(v)). \tag{36}$$

We remark in this connection that we simultaneously obtain *estimates for the rate of convergence* of the distributions of arbitrary functionals of $w(t)$. Indeed, the difference between the distributions at (34) does not exceed (35) and (36).

We also note that when (32) is satisfied, $\theta = \inf\{u\colon Y(u) = \overline{Y}\}$ will also be a proper random variable since

$$\mathsf{P}(\theta \geqslant t) \leqslant \mathsf{P}(\sup_{v \geqslant t} Y(v) > 0).$$

As in the discrete case, from the arguments above we can also show the convergence of the joint distribution of the variables $(w(t), v(t))$ to the distribution of $(\overline{X}, \theta)$, where $v(t) = \inf\{u \geqslant 0\colon w(t-u) = 0\}$ is the *actual busy period* of the system *looking backward* from time t.

Theorem 14. *When* (32) *holds, the distribution of* $(w(t), v(t))$ *converges for* $t \to \infty$ *to the distribution of* $(\overline{Y}, \theta)$.

Proof. This follows immediately from Theorem 13. Indeed, for fixed $x \geqslant 0$, $y \geqslant 0$ and $t \to \infty$

$$\mathsf{P}(w(t) \geqslant x, v(t) \geqslant y) = \mathsf{P}(w(t) \geqslant x, \inf_{0 \leqslant u \leqslant y} w(t-y) > 0)$$
$$\to \mathsf{P}(w^c(0) \geqslant x, \inf_{0 \leqslant u \leqslant y} w^c(-u) > 0) = \mathsf{P}(\overline{Y} \geqslant x, \theta \geqslant y). \quad \square$$

It is clear that we can find the limiting distribution of the *process* $(w_t(u), v_t(u)) = (w(t+u), v(t+u))$ for $t \to \infty$ in exactly the same way. For finite t and $w(0) = 0$ the distribution of $(w(t), v(t))$ coincides with that of $(\overline{Y}(t), \theta(t))$,

$$\theta(t) = \inf\{u \geqslant 0\colon \overline{Y}(u) = \overline{Y}(t)\}.$$

3. The analogy with the discrete case is also completely preserved when we consider *more general conditions guaranteeing the convergence of the distribution of* $w(t)$ *for* $t \to \infty$. Here also, stationarity of the increments of the process $X(t)$ need not be required and one is restricted merely by conditions for convergence of the processes $\{X_t(u) = X(t+u) - X(t);\, u \geqslant 0\}$ to the process $X^c(u)$ with stationary increments.

We now indicate as in § 3 two groups of conditions which guarantee the required convergence of the distributions of $w(t)$.

1. Assume that there exists a process $\{X^c(u);\, -\infty < u < \infty\}$ with strictly stationary increments such that for arbitrary $B \in \mathcal{N}_D$

$$|\mathsf{P}(X_t \in B) - \mathsf{P}(X^c \in B)| < \varepsilon(t),$$

where $\varepsilon(t) \to 0$ for $t \to \infty$; X_t and X^c are realizations of the processes $\{X_t(u);\, u \geqslant 0\}$ and $\{X^c(u),\, u \geqslant 0\}$. Moreover, assume that for $t \to \infty$

$$X(t) \xrightarrow[\mathrm{p}]{} -\infty \quad \text{and} \quad X^c(-t) \xrightarrow[\text{a.s.}]{} \infty. \tag{37}$$

2. Conditions of the second type assume the convergence of the finite-dimensional distributions of $\{X_t(u)\}$ to the distribution of $X^c(u)$ or the convergence of the distributions of ρ_D-continuous functionals of these processes on an arbitrary finite interval. In addition, along with (37) we should also have

$$\lim_{V\to\infty}\lim_{t\to\infty}\mathsf{P}(\sup_{t\geqslant v\geqslant V}(X(t)-X(t-v))>0)=0\,.$$

When Conditions 1 *or* 2 *hold, then there exists*

$$\lim_{t\to\infty}\mathsf{P}(w(t)<x)=\mathsf{P}(\sup_{u\geqslant 0}(X^c(0)-X^c(-u))<x)\,.$$

The proof of this assertion is analogous to the proof of Theorem 6 and we leave it to the reader.

Thus here, as in the discrete case, we can replace the governing process $X(t)$ in our investigation of the properties of the *limiting* distribution of $w_t(u)$ for $t\to\infty$ from the outset by the *stationary* process to which $X(t)$ converges (for example, in the sense of Conditions 1 and 2).

We turn now to conditions which guarantee the validity of (32).

Consider the sequence

$$\{\xi_n=X(nh)-X((n-1)h);\ -\infty<n<\infty\}$$

and denote by $\mathfrak{M}$ the σ-algebra generated by it. With this stationary sequence we can associate, as before, a shift transformation T_h of sets A from $\mathfrak{M}$ (or a shift transformation U_h of random variables measurable w.r.t. $\mathfrak{M}$). Then, if

$$\mathsf{M}\,Y(1)=\mathsf{M}(X(1)-X(0))=a$$

exists and the sequence $\{\xi_n\}$ is ergodic (T_h has no proper invariant sets), one has w.p.1

$$\lim_{t\to\pm\infty}\frac{Y(t)}{t}=\lim_{t\to\pm\infty}\frac{X(t)-X(0)}{t}=a\,. \tag{38}$$

These relations obviously follow from the inequalities

$$\textstyle\sum_{k=1}^{[t]+1}\xi_k+h\geqslant X(th)-X(0)\geqslant\sum_{k=1}^{[t]}\xi_k-h$$

and the ergodicity of the sequence $\{\xi_n\}$.

If $\{\xi_n\}$ is metrically transitive for arbitrary $h>0$, then the process $\{X(t)\}$ with stationary increments is said to be *metrically transitive*. For example, a process with independent increments is metrically transitive. However, a random process $X(t)$ whose trajectories for each ω are periodic with period t_0 is obviously not metrically transitive, since T_{t_0} is the identity.

Since it follows immediately from (38) for $a<0$ that $\lim_{u\to\infty}(X(u)-X(0))=-\infty$ and $\lim_{u\to\infty}Y(u)=-\infty$, we may assert that *in order for Condition* (32) *of Theorem* 13 *to hold it is sufficient that* $\mathsf{M}(Y(1)-Y(0))=a<0$ *exist and that the transformation* T_h *be metrically transitive for at least one* $h>0$.

Conditions under which $w(t)\to\infty$ for $t\to\infty$ w.p.1 will obviously coincide with the corresponding ones for the discrete case.

However, the continuous case possesses a peculiarity. It consists of the existence of equations describing the distribution of $w(t)$ in terms of certain conditional distributions of the process $X(t)$.

We will treat these equations in the next section.

4. For the systems $\langle G, G, G_I, 1\rangle$, where the τ_j^s are mutually independent, also don't depend on $\{\tau_j^e, v_j^e\}$, and have a distribution whose characteristic function is

$$\prod_{j=1}^{m}\left(1+\frac{i\lambda}{\alpha_j}\right)^{-1}, \quad \alpha_j>0$$

(the sum of exponentially distributed variables), one can obtain explicit formulae as well as other results analogous to the preceding (see Theorems 10–13) for the *queue length* $q(t)$ at time t. These results will be presented in § 41.

§ 7. Further Properties of the Process $w(t)$. Beneš' Equation

1. The specific character of the process $w(t)$ and especially, its continuity from below ($w(t+\Delta)-w(t)\geqslant -\Delta$), allow the derivation of equations connecting the distribution of $w(t)$ with the functions

$$\mathsf{P}(X(t)<x) \quad \text{and} \quad \mathsf{P}(X(t)-X(u)<x \mid w(u)=0); \quad 0\leqslant u\leqslant t.$$

It would possibly be more natural to describe the condition $\{w(u)=0\}$ here and in the sequel in terms of the process X as $\{X(u)=\inf_{0\leqslant v\leqslant u}(0, X(v))\}$. However, this notation would require more space.

The following theorem holds:

Theorem 15. *The distribution of* $w(t)$ *satisfies*

$$\mathsf{P}(w(t)<x)=\mathsf{P}(X(t)<x)-\frac{\partial}{\partial x}\int_0^t \mathsf{P}(X(t) - X(u)<x \mid w(u)=0)\mathsf{P}(w(u)=0)\,du. \tag{39}$$

For $-t\leqslant x\leqslant 0$ *this equation goes over into an equation for* $\mathsf{P}(w(u)=0)$:

$$0=\mathsf{P}(X(t)<x)-\frac{\partial}{\partial x}\int_0^{t+x}\mathsf{P}(X(t)-X(u)<x \mid w(u)=0)\mathsf{P}(w(u)=0)\,du.$$

(On the right $\int_0^t$ is replaced by $\int_0^{t+x}$ since $\mathsf{P}(X(t)-X(u)<x)=0$ for $t+x\leqslant u\leqslant t$, $x\leqslant 0$.)

This theorem can be proved in various ways. We present first a proof which is in our opinion more direct and better sets off the useful properties of the process $w(t)$.

Another method, due to Beneš, consists in the use of a certain identity which is of independent interest. This variant of the proof will be presented in Theorem 16.

Proof of Theorem 15. Since $\{w(t)<x\} \subset \{X(t)<x\}$, we get

$$\mathsf{P}(X(t)<x)=\mathsf{P}(w(t)<x)+\mathsf{P}(w(t)\geqslant x,\, X(t)<x)$$

and by (27)

$$\mathsf{P}(w(t)<x)=\mathsf{P}(X(t)<x)-\mathsf{P}(0>X(t)-x\geqslant -\textstyle\int_0^t \delta(w(u))\,du)\,.$$

Consequently, we must prove that

$$\mathsf{P}(0>X(t)-x\geqslant -\textstyle\int_0^t \delta(w(u))\,du)=\dfrac{\partial}{\partial x}\displaystyle\int_0^t \mathsf{P}(X(t)-X(u)<x,\, w(u)=0)\,du\,. \quad (40)$$

Set $(dx)=(x-dx,\, x)$, $\underline{X}(u)=\inf_{0\leqslant v\leqslant u}(0,\, X(v))$ and let $I(A)$ be the indicator of the set A:

$$I(A)=\begin{cases}1\,, & \text{if } \omega\in A\,,\\ 0\,, & \text{if } \omega\notin A\,.\end{cases}$$

Then the following identity holds:

$$\begin{aligned}L(\omega)&=\textstyle\int_0^t I(X(t)-X(u)\in(dx),\, w(u)=0)\,du\\ &=I(0>X(t)-x\geqslant -\textstyle\int_0^t \delta(w(u))\,du)\,dx\,.\end{aligned} \quad (41)$$

Indeed,

$$\begin{aligned}L(\omega)&=\textstyle\int_0^t I(X(t)-X(u)\in(dx),\, X(u)=\underline{X}(u))\,du\\ &=\textstyle\int_E I(\underline{X}(u)\in(X(t)-x,\, X(t)-x+dx))\,du\,,\end{aligned} \quad (42)$$

where $E=\{u\colon X(u)=\underline{X}(u)\}\cap[0,\, t]$. On this set, $X(u)$ runs monotonically through all numbers from 0 to $-\operatorname{mes} E=-\int_0^t \delta(w(u))\,du$. Since $d\underline{X}(u)/du=-1$ on E, either there is a $u\leqslant t$ such that $X(u)=\underline{X}(u)=X(t)-x$ and then the integrand in (42) is different from 0 on a subset of E of measure dx; or $X(t)-x\notin[\underline{X}(t),\, 0]$ and then the integrand in (42) is zero everywhere on E. This implies the validity of the identity (41).

In order to obtain (40) it is necessary to take the expectations of both sides of (41) and divide them by dx. □

We note now that the assertion of Theorem 15 is equivalent to

$$\mathsf{M}\, e^{-\lambda w(t)}=\mathsf{M}\, e^{-\lambda X(t)}-\lambda\textstyle\int_0^t \mathsf{M}\,[e^{-\lambda(X(t)-X(u))}\,\delta(w(u))]\,du\,; \quad \operatorname{Re}\lambda\geqslant 0, \quad (43)$$

which can be obtained by multiplying both sides of (39) by $e^{-\lambda x}$ and integrating from $-t$ to ∞.

Integrating (39) from $-t$ to x, we get the *integral form of Beneš' equation*:

$$\begin{aligned}\textstyle\int_0^x \mathsf{P}(w(t)<y)\,dy&=\textstyle\int_{-t}^x \mathsf{P}(X(t)<y)\,dy\\ &\quad-\textstyle\int_0^{t+\min(0,x)} \mathsf{P}(X(t)-X(u)<x,\, w(u)=0)\,du\,.\end{aligned} \quad (44)$$

We spoke above about another way to prove Theorem 15. It consists in the proof of the following identity (see Beneš [3]):

Theorem 16. *If the process $X(t)$ has* w.p.1 *a finite number of jumps on* $[0, t]$, *then* w.p.1

$$\exp\{-\lambda \int_0^t \delta(w(u))\,du\} = 1 - \lambda \int_0^t e^{\lambda X(u)}\,\delta(w(u))\,du\,. \tag{45}$$

This relationship combined with (27) reduces immediately to (43).

Proof. Let (t_i, u_i) be the i-th interval on which $w(t)=0$. For $t \leqslant t_1$ Eq. (45) is obvious. Assume that it holds for $t \leqslant t_n$. We will show that it then also holds for $t \leqslant t_{n+1}$. On the interval (t_n, u_n), $X(t)$ has no jumps and $w(t_n)=0$. Thus, for $t_n < t \leqslant u_n$

$$X(t) = X(t_n) - (t - t_n)\,, \quad X(t_n) + \int_0^{t_n} \delta(w(u))\,du = 0$$

and for the prescribed values of t

$$\begin{aligned}\exp\{\lambda X(t)\} &= \exp\{\lambda(X(t_n) - t + t_n)\}\\ &= \exp\{-\lambda \int_0^{t_n} \delta(w(u))\,du - \lambda(t - t_n)\} = \exp\{-\lambda \int_0^t \delta(w(u))\,du\}\,.\end{aligned}$$

On the interval (u_n, t_{n+1}), $\delta(w(t)) = 0$. Hence, for $t_n < t < t_{n+1}$

$$-\lambda\,\delta(w(t)) \exp\{\lambda X(t)\} = -\lambda\,\delta(w(t)) \exp\{-\lambda \int_0^t \delta(w(u))\,du\}\,.$$

This says that the derivatives w.r.t. t of both sides of (45) coincide on the interval (t_n, t_{n+1}) so that (45) holds for $t \leqslant t_{n+1}$. □

2. In order to illustrate the use of the results obtained, we take a governing process $X(t)$ with independent increments. $X(t)$ will be such a process if τ_j^e and S_j in (3) are independent and $\mathsf{P}(\tau_j^e \geqslant x) = e^{-\alpha x}$ (this will hold, for example, in the system $\langle E, G_I, G_I, 1\rangle$). If $\mathsf{M}\, e^{i\lambda S_j} = f(\lambda)$, then by well known properties of generalized Poisson processes

$$\mathsf{M} \exp\{i\lambda(X(t) - X(0))\} = \exp\{t\psi(\lambda)\},\ \psi(t) = -i\lambda + \alpha(f(\lambda) - 1)\,. \tag{46}$$

Proceeding to characteristic functions in (39) and putting $p(u) = \mathsf{P}(w(u) = 0)$ for the sake of brevity, we immediately obtain the well known formula

$$\mathsf{M} \exp\{i\lambda w(t)\} = \mathsf{M} \exp\{i\lambda w(0) + t\psi(\lambda)\} + i\lambda \int_0^t p(u)\, e^{(t-u)\psi(\lambda)}\,du\,. \tag{47}$$

We need only find the function $p(u)$ here. Using (39) again with $x \leqslant 0$, for the function

$$P(\mu) = \int_0^\infty e^{-\mu t} p(t)\,dt\,,$$

we get

$$\int_0^\infty \mathsf{P}(X(t) < x)\, e^{-\mu t}\,dt = P(\mu) \frac{\partial}{\partial x} \int_0^\infty \mathsf{P}(X(t) - X(0) < x)\, e^{-\mu t}\,dt\,. \tag{48}$$

As we will see in Chapter 2 there exist simpler and more convenient expressions for $P(u)$. We will also find the explicit form of $p(u)$, which allows the direct use of (39) in the search for $\mathsf{P}(w(t) < x)$.

Matters are somewhat more complicated in the case in which $X(t)$ is an *inhomogeneous* process with independent increments (the intensity $1/\alpha$ of the occurrence of requests, as well as the jump variables $S_1, S_2, \ldots$, depend on time). Under these conditions

$$\mathsf{P}(w(t)<x)=\mathsf{P}(X(t)<x)-\frac{\partial}{\partial x}\int_0^t \mathsf{P}(w(u)=0)\mathsf{P}(X(t)-X(u)<x)\,du\,,$$

where $\mathsf{P}(w(t)=0)$ is the solution of the Volterra equation of the first kind

$$\int_{-t}^{x} \mathsf{P}(X(t)<u)\,du=\int_0^{t+x} \mathsf{P}(w(u)=0)\mathsf{P}(X(t)-X(u)<x)\,du, \quad -t\leqslant x\leqslant 0\,.$$

3. Relations analogous to those derived above can also be obtained in the case of *discrete time* (the random variables τ_j^e and S_j are integers). In this case Eq. (39) will have the form (n, x integers)

$$\mathsf{P}(w(n)<x)=\mathsf{P}(X(n)<x)-\sum_{k=0}^{n-1} \mathsf{P}(X(n)-X(k)=x-1\,;\,w(k)=0)\,.$$

Let us show that this is in fact true. Since the processes $X(t)$ and $w(t)$ are now defined by their values at integer points, we will consider in their place the sequences $\{X(n), w(n); n=0, 1, \ldots,\}$, assuming here that $X(n)=X(n+0)$ and $w(n)=w(n+0)$.

Under such an agreement $w(0)=w(+0)$ and $w(0)$ loses its previous meaning. It is now the initial value of the waiting time $w(-0)$ plus S_1. Hence, here we always have $w(0)>0$.

The previous arguments lead us under the new conditions to the equation

$$w(n)=X(n)+\sum_{k=0}^{n-1}\delta(w(k))\,,$$

whose solution is

$$w(n)=X(n)-\min_{k\leqslant n-1}(0, X(k)-1)\,. \tag{49}$$

Thus,

$$\{w(n)=0\}=\{X(n)<\min_{k\leqslant n-1}(1, X(k))\}\,.$$

Using Eq. (49), the reader can obtain analogues of Theorems 11–14 on the convergence of the distributions of $w(t)$. If we assume in this connection that the sequence $Y(n)$ is defined as before by the equation $Y(n)=X(0)-X(-n)$, then

$$\lim \mathsf{P}(w(n)=x)=\mathsf{P}(\sup_{k\geqslant 0} Y^*(k)=x)\,,$$

where $Y^*(k)=Y(k+1)+1$.

Without difficulty one can verify the identity

$$I(0>X(n)-x\geqslant -\sum_{k=0}^{n-1}\delta(w(k)))=\sum_{k=0}^{n-1} I(X(n)-x+1=X(k)<\min_{v\leqslant k-1}(1, X(v)))\,,$$

which is the analogue of (41) and leads to

Theorem 15A. *The distribution of $w(n)$ satisfies the system of equations*

$$\mathsf{P}(w(n)<x)=\mathsf{P}(X(n)<x)-\sum_{k=0}^{n-1} \mathsf{P}(X(n)-X(k)=x-1,\,w(k)=0)$$
$$(n=0, 1, 2, \ldots)\,.$$

Moreover, it is not difficult to see that $\{X(n)\}$ will be a sequence with *independent increments* if τ_j^e and S_j are independent and $\mathsf{P}(\tau_j^e=k)=qp^{k-1}$; $k=1, 2, \ldots$; $p+q=1$ for some $p\geqslant 0$ (the initial process $X(t)$ will of course not possess independent increments under these assumptions). If again $\mathsf{M}\exp\{i\lambda S_j\}=f(\lambda)$, $|e^{i\lambda}|\leqslant 1$, then

$$\mathsf{M}\exp\{i\lambda(X(n)-X(0))\}=e^{-i\lambda n}(qf(\lambda)+p)^n,$$
$$\mathsf{M}\, e^{i\lambda w(n)}=\mathsf{M}\, e^{i\lambda w(0)}(qf(\lambda)+p)^n-(1-e^{i\lambda})\sum_{k=0}^{n-1}p(k)[e^{-i\lambda}(qf(\lambda)+p)]^{n-k}. \quad (50)$$

The generating function of the sequence $p(k)$ is defined by an equation analogous to (48). A simpler formula for $p(k)$ will be given in Chapter 2.

If $X(n)$ is an *inhomogeneous* sequence with independent increments (the increment at time n equals Z_n with probability $\pi(n)$ and 0 with probability $1-\pi(n)$, where $Z_1, Z_2, \ldots$; $\pi(1), \pi(2), \ldots$ are sequences of independent random variables and nonnegative numbers, resp.), then

$$\mathsf{P}(w(n)<x)=\mathsf{P}(X(n)<x)-\sum_{k=0}^{n-1}\mathsf{P}(w(k)=0)\mathsf{P}(X(n)-X(k)=x-1),$$

where $\mathsf{P}(w(k)=0)$, $k=0, 1, \ldots, n-1$, satisfies the system of equations

$$\mathsf{P}(X(n)<x)=\sum_{k=0}^{n+x-1}\mathsf{P}(w(k)=0)\mathsf{P}(X(n)-X(k)=x-1), \quad x=0, -1, \ldots, -n+1.$$

This system is triangular, which guarantees the existence and uniqueness of its solution and also simplifies the determination of the latter.

§ 8. The Stationary Solution of Beneš' Equation. Approximation Formulae for Heavy and Light Traffic

1. We turn now to the *stationary solution* of the equation of Beneš. We will assume in this section that $\{X(t);\ -\infty<t<+\infty\}$ is a process with strictly stationary increments and will suppose, whenever necessary, that $a=\mathsf{M}(X(1)-X(0))$ exists and that the process $Y(t)$ satisfies the strong law of large numbers: w.p.1

$$\lim_{t\to\pm\infty}\frac{Y(t)}{t}=\lim_{t\to\pm\infty}\frac{X(t)-X(0)}{t}=a. \quad (51)$$

As already mentioned, it is sufficient for this that the sequence $\{X(n)-X(n-1);\ -\infty<n<\infty\}$ be ergodic.

Theorem 17. *Under the formulated conditions*

$$\lim_{t\to\infty}\mathsf{P}(w(t)=0)=\mathsf{P}(\overline{Y}=0)=\max(0, -a). \quad (52)$$

Proof. First let $a<0$. Using Eq. (44) in integral form for $x=0$, we find that

$$0=\int_{-t}^{0}\mathsf{P}(X(t)<y)\,dy-\int_0^t\mathsf{P}(w(u)=0)\,du+\int_0^t\mathsf{P}(X(t)-X(u)\geqslant 0,\quad w(u)=0)\,du.$$

On the basis of the law of large numbers, we have for $t \to \infty$

$$\int_{-t}^{0} \mathsf{P}(X(t)<y)\,dy \sim -at\,,$$
$$\int_0^t \mathsf{P}(X(t)-X(u)\geqslant 0)\,du = o(t)\,,$$
$$\int_0^t \mathsf{P}(w(u)=0)\,du \sim pt\,,$$

where

$$p = \lim_{t\to\infty} \mathsf{P}(w(t)=0) \tag{53}$$

(this limit exists by § 6). We have obtained the equation $0=-at-pt+o(t)$ from which $p=-a$ follows.

Now let $a=0$. Assume first that $w(0)=0$. Then by Theorem 11, the limit (53) exists. Putting $x=\varepsilon t$, $\varepsilon>0$, in (44), we get

$$\pi(\varepsilon)t = \int_{-t}^{\varepsilon t} \mathsf{P}(X(t)<y)\,dy - \int_0^t \mathsf{P}(w(u)=0)\,du$$
$$+\int_0^t \mathsf{P}(X(t)-X(u)\geqslant \varepsilon t,\, w(u)=0)\,du\,,$$

where $\pi(\varepsilon)\leqslant\varepsilon$ and

$$\mathsf{P}(X(t)-X(u)\geqslant \varepsilon t)=o(1)\,,\qquad \int_{-t}^{\varepsilon t} \mathsf{P}(X(t)<y)\,dy\sim \varepsilon t$$

for $t\to\infty$. This means that for arbitrary $\varepsilon>0$

$$\pi(\varepsilon)t = tp+\varepsilon t+o(t)\,.$$

Since p does not depend on ε, one has necessarily that $p=0$.

It remains to remark that if $w_1(t)$ and $w_0(t)$ are two processes corresponding to a single $\{X(t)-X(0)\}$ and merely having different initial conditions

$$w_1(0)\geqslant w_0(0)=0\,,$$

then by Theorem 10

$$\mathsf{P}(w_1(t)=0)\leqslant \mathsf{P}(w_0(t)=0) \quad\text{and}\quad \lim_{t\to\infty}\mathsf{P}(w_1(t)=0)=0\,.$$

The case $a>0$ requires no special consideration since

$$w(t)\geqslant X(t) \quad\text{and}\quad X(t)\xrightarrow[\text{a.s.}]{}\infty \quad\text{for}\quad t\to\infty\,. \;\square$$

We remark that the assertion of this theorem can also be carried over by Subsection 3 § 6 to "prestationary" governing processes $X(t)$. In this connection, Eq. (52) is a very natural result. For example, if the governing sequence is such that $\{\tau_j^e\}$ and $\{S_j\}$ (see (3)) are two independent sequences of independent random variables distributed like some τ^e and S, then as we will see in § 9, for the stationary process $X^c(t)$ to which $X_t(u)=X(t+u)$ converges,

$$a=\mathsf{M}(X^c(1)-X^c(0))=\frac{\mathsf{M}S}{\mathsf{M}\tau^e}-1\,,$$

so that by Theorem 17

$$\lim_{t\to\infty}\mathsf{P}(w(t)=0)=-a=1-\frac{\mathsf{M}S}{\mathsf{M}\tau^e}\,,$$

when $a \leqslant 0$. On the other hand, during a long time t there will have arrived approximately $t/\mathsf{M}\tau^e$ groups of requests whose service requires approximately $t\mathsf{M}S/\mathsf{M}\tau^e$ time units. Consequently, for a time $t(1-\mathsf{M}S/\mathsf{M}\tau^e)$ the system will be idle, so that the mean fraction of the time interval t spent idle is

$$1-\frac{\mathsf{M}S}{\mathsf{M}\tau^e}=-a\,.$$

2. We have thus found the stationary solution of (44) for $\mathsf{P}(w^c(t)=0)$. We will now determine the form of the solution $\mathsf{P}(w^c(t)<x)$ for arbitrary x in the most interesting case $a<0$. For such a the proper process $w^c(t)$ can be represented as

$$w^c(t)=\sup_{v\leqslant t}(X(t)-X(v))\,.$$

A simple calculation shows that this process satisfies (27) in the form

$$w^c(t)=w^c(0)+X_0(t)+\textstyle\int_0^t \delta(w^c(u))\,du\,,$$

where $X_0(t)=X(t)-X(0)$. This allows us to obtain the *stationary Beneš equation* for $w^c(t)$: for all x

$$\begin{aligned}\mathsf{P}(w^c(t)<x)&=\mathsf{P}(X_0(t)+w^c(0)<x)\\&\quad+a\frac{\partial}{\partial x}\int_0^t \mathsf{P}(X(t)-X(u)<x\mid X(u)=\inf_{v\leqslant u}X(v))\,du\,.\end{aligned}\tag{54}$$

This is derived in complete analogy to (39). For example, with the help of the identity

$$1-\lambda\textstyle\int_0^t \exp\{\lambda(X_0(u)+w^c(0))\}\,\delta(w^c(u))\,du=\exp\{-\lambda\int_0^t \delta(w^c(u))\,du\}\,,$$

whose proof does not differ from that of Theorem 16. Using the stationary equation it is not difficult to obtain

Theorem 18. *If the process $X(t)$ satisfies the conditions at the beginning of the section and $a<0$, then for arbitrary $x>0$ the conditional renewal function of $X(t)$ exists:*

$$H_0(x)=\textstyle\int_0^\infty \mathsf{P}(0\leqslant X(u)-X(0)<x\mid X(0)=\inf_{v\leqslant 0}X(v))\,du<\infty$$

and

$$\mathsf{P}(w^c(t)\geqslant x)=-a\frac{dH_0(x)}{dx},\qquad \mathsf{P}(w^c(t)=0)=-a\,.\tag{55}$$

Proof. Integrating (54) from 0 to x, we get

$$\begin{aligned}\textstyle\int_0^x \mathsf{P}(w^c(t)<y)\,dy&=\textstyle\int_0^x \mathsf{P}(X_0(t)+w^c(0)<y)\,dy\\&\quad+a\textstyle\int_0^t \mathsf{P}(0\leqslant X(t)-X(u)<x\mid X(u)=\inf_{v\leqslant u}X(v))\,du\,.\end{aligned}$$

Using the finiteness of $w^c(0)$ and the stationarity of the increments of $X(t)$, we can assert that

$$\lim_{t\to\infty} a \int_0^t \mathsf{P}(0 \leqslant X(u) - X(0) < x \mid X(0) = \inf_{v \leqslant 0} X(v))\, du$$
$$= \int_0^x \mathsf{P}(w^c(0) < y)\, dy - x \quad \text{exists.}$$

This equation obviously concludes the proof. □

Corollary 1. *As implied by* (55) (*the left side is right-continuous in* x), *at "break" points of the monotone function* $H_0(x)$ *we must understand* $dH_0(x)/dx$ *as the right derivative.* $H_0(x)$ *itself turns out to be convex on* $[0, \infty]$. *It also follows from* (55) *that the mean time* $H_0(x)$ *spent by the process* $\{X_0(u); u \geqslant 0\}$ *on the half-open interval* $[0, x)$, *under the condition* $\inf_{v \leqslant 0} X_0(v) = 0$, *satisfies for small* x *the relation* $H_0(x) \sim -x/a$.

Corollary 2. *Integrating* (55) *from* 0 *to* ∞, *we obtain*

$$\mathsf{M}w^c(t) = -aH_0(\infty) = -a \int_0^\infty \mathsf{P}(X(u) \geqslant X(0) \mid X(0) = \inf_{v \leqslant 0} X(v))\, du\,.$$

Using this equation, we can also indicate what conditions must be satisfied by a process $X(t)$ with weakly dependent increments in order that $\mathsf{M}w^c(t)$ be finite. Under quite general weak mixing hypotheses, the required condition could be the finiteness of $\mathsf{M}|X(1) - X(0)|^{2+\gamma}$ for some $\gamma > 0$ (it is necessary to use the estimate in [34] (Ibragimov and Linnik) for $\mathsf{M}|X(u) - X(0) - ua|^{2+\gamma}$). The proof of the sufficiency of the condition $\mathsf{D}(X(1) - X(0)) < \infty$ requires a more precise approach.

Corollary 3. *If*

$$\mathsf{P}(0 \leqslant X(u) - X(0) < x \mid X(0) = \inf_{v \leqslant 0} X(v)) = \mathsf{P}(0 \leqslant X(u) - X(0) < x)\,,$$

which holds, for example, for processes with independent increments, then the distribution of $w^c(t)$ *coincides with the unconditional renewal density*:

$$\mathsf{P}(w^c(t) \geqslant x) = -a\frac{dH(x)}{dx}, \qquad H(x) = \int_0^\infty \mathsf{P}(0 \leqslant X(u) - X(0) < x)\, du\,.$$

In connection with this relation we remark that for the estimation of the asymptotic behavior of the distribution of $w^c(0)$ for $x \to \infty$ we can use an extension to processes with independent increments of the results in [65] on the asymptotic renewal density of a sequence of sums of independent variables with negative expectations.

Moreover, if $X(t)$ is a homogeneous process with independent increments, then it follows from (47) ($a = \alpha\mathsf{M}S - 1$) that

$$\mathsf{M}\, e^{i\lambda w^c(t)} = \frac{i\lambda}{1 - \exp t\psi(\lambda)} \int_0^t (-a)\, e^{v\psi(\lambda)}\, dv = \frac{i\lambda a}{\psi(\lambda)} = \frac{1 - \alpha\mathsf{M}S}{1 - \alpha\mathsf{M}S\dfrac{f(\lambda) - 1}{i\lambda\mathsf{M}S}}\,.$$

This is the well-known formula of Hinčin. Since $(f(\lambda)-1)/i\lambda \mathsf{M}S$ is the characteristic function of a distribution with density $\mathsf{P}(S \geqslant x)/\mathsf{M}S$, and $\alpha \mathsf{M}S<1$, on the basis of this formula the density of $w^c(t)$ for $x>0$ can be represented as an exponentially converging series of densities of sums of random variables with density $\mathsf{P}(S \geqslant x)/\mathsf{M}S$.

As we have already mentioned, Chapter 2 is devoted to a more detailed investigation of $w(u)$ and $w^c(u)$ when $X(t)$ is a process with independent increments.

3. If the *time is discrete*, then the stationarity assumption must naturally be relative to the increments of the sequence $\{X(n);\ -\infty<n<\infty\}$ (see **3** § 7). Assumption (51) will be retained. Under these conditions we have

Theorem 17 A.

$$\lim_{n\to\infty} \mathsf{P}(w(n)=x)=\mathsf{P}(w^c(k)=x)$$

exists, and

$$\mathsf{P}(w^c(k)=0)=\mathsf{P}(X(0)<\inf_{k<0} X(k))=\min(0,\ -a)\,.$$

If $a<0$, *then*

$$\mathsf{P}(w^c(k)\geqslant x)=-a\sum_{n=1}^{\infty} \mathsf{P}(X(n)-X(0)=x-1 \mid X(0)<\inf_{k<0} X(k))\,.$$

On the basis of Theorems 17 and 18 we obtained a series of corollaries. Similar corollaries also hold here. In particular, if $X(n)$ is a sequence with independent increments, then $\mathsf{P}(w^c(k)\geqslant x)=-ah(x)$, where $h(x)$ is the renewal density of the random variable $X(1)\dot{-}X(0)$. By (50) $(a=-1+q\mathsf{M}S_j)$

$$\mathsf{M}\,e^{i\lambda w^c(k)}=a(1-e^{i\lambda})\frac{e^{-i\lambda}(qf(\lambda)+p)}{1-e^{-i\lambda}(qf(\lambda)+p)}=-\frac{a}{p}\frac{qf(\lambda)+p}{1-\dfrac{q}{p}\dfrac{(1-f(\lambda)\,e^{-i\lambda})\,e^{i\lambda}}{1-e^{i\lambda}}}$$

$$=-\frac{a}{p}(qf(\lambda)+p)\sum_{k=0}^{\infty}\left(\frac{q(\mathsf{M}S-1)}{p}\right)^k f_0^k(\lambda)\,,$$

where $q(\mathsf{M}S-1)/p<1$ and $f_0(\lambda)$ is the characteristic function of a distribution related in a known way to that of S.

4. *Approximate Formulae for Heavy and Light Traffic* We will say that a system has *light traffic* if a is close to -1 and *heavy traffic* if a is close to 0. In these cases, there exist still simpler explicit approximation formulae for the stationary distribution of $w(t)$.

Systems with light traffic. An assumption on the smallness of $a+1$, means, roughly speaking, that either the jumps of the process $X(t)$ occur rarely—the times between their appearances are large, or that the jump sizes themselves are small. In this case $\mathsf{P}(w^c(t)=0)=-a$ is near 1. How does $\mathsf{P}(w^c(t)\geqslant x)$ look when $x>0$? Let χ be the time of the first jump of $X(t)$ on $[0,\infty]$ and S the size of this jump. Also let $\tau_{(dx)}$ be the time spent by the process $X_0(t)=X(t)-X(0)$ in the interval $(dx)=(x, x+dx)$, $x>0$. Then

$$\mathsf{P}(w^c(t)\geqslant x)=-a\frac{\mathsf{M}(\tau_{(dx)}\mid A)}{dx}, \qquad \mathsf{M}(\tau_{(dx)}\mid A)\geqslant \mathsf{P}(S>\chi+x\mid A)\,dx\,,$$

where $A=\{X(0)=\inf_{u\leqslant 0}X(u)\}$. Hence, for $x>0$

$$\mathsf{P}(w^c(t)\geqslant x)\geqslant -a\mathsf{P}(S>\chi+x\mid A)\,.$$

It is easy to see that under wide conditions, both sides of this inequality will be asymptotically close to one another when $a\to -1$. The right side will in turn be close to $\mathsf{P}(S>\chi+x)$ since $\mathsf{P}(A)=-a$ is near 1.

However, to make this assertion more precise, it is necessary to make a much more precise statement of the problem. For this we need to introduce a *family* (*double sequence*) of governing processes $X^a(t)$ depending on some parameter which, without loss of generality, can be taken as a, so that $\mathsf{M}(X^a(1)-X^a(0))=a\to -1$. (Our actual system with light traffic must now be considered as an element of such a family.) A possible variant is, for example, the investigation of the sequence $X^a(t)$, obtained from the fixed process $X(t)$, $\mathsf{M}(X(1)-X(0))=0$ by means of

$$X^a(t)=X((1+a)t)+at, \quad a\to -1\,.$$

Considering the corresponding family of service systems (these will obviously be systems with light traffic) and of processes $w_a^c(t)$, we can obtain as an example the following result:

If[5] $\mathsf{P}_{\mathfrak{M}_0}(\chi<\varepsilon)<\delta(\varepsilon)$ a.s. *and* $\delta(\varepsilon)\to 0$ *for* $\varepsilon\to 0$, *then as* $a\to -1$

$$\mathsf{P}_{\mathfrak{M}_0}(w_a^c(t)\geqslant x)\sim \mathsf{P}\left(S>\frac{\chi}{a+1}+x\right). \tag{56}$$

Here $\mathsf{P}_{\mathfrak{M}_0}$ denotes the conditional distribution w.r.t. the σ-algebra $\mathfrak{M}_0$ generated by the trajectories of the process $X(t)$ on $(-\infty, 0)$. The random variables S and χ correspond to the initial process $X(t)$. The proof of (56) under the formulated conditions is left to the reader.

We remark that if the sequence $\{S_j\}$ does not depend on $\{\tau_j^e\}$, then the distribution of S coincides with that of S_j and the problem of calculating the right side of (56) reduces to the search for the distribution of the overshoot variable χ.

5. *Systems with heavy traffic.* We will assume here that the process $X(t)$ (more precisely, its increments) satisfies a certain weak dependence condition, for example, in the form of a strong mixing condition or in some other form, which is such that the existence of constants $c<\infty$, $\gamma>0$ and $\delta>0$ is guaranteed, with[6]

$$\mathsf{M}(|X_0(t)-at|^{2+\gamma}|A)<ct^{1+\gamma/2} \tag{57}$$

and

$$\mathsf{P}\left(\frac{X_0(t)-at}{\sigma\sqrt{t}}<x\mid A\right)=\Phi(x)(1+o(1)) \tag{58}$$

[5] Essentially, it is necessary to require the smallness of the conditional probability that χ is small given that $X(t)$ has had a jump for small values of $t<0$.

[6] Condition (57) for $\gamma>0$ is superfluous in a number of cases.

for $|x| \leqslant N$ and arbitrary fixed N. Here $\Phi(x)$ is the normal distribution function. Conditions sufficient for (57) and (58) are quite broad and can be found in the monograph of Ibragimov and Linnik [34].

It turns out that when (57) *and* (58) *hold and* $a \to 0$, *then*

$$\mathsf{P}\left(w^c(t) > \frac{x}{|a|}\right) \sim e^{-2x/\sigma^2},$$

so that for heavy traffic, the values of $w^c(t) = w_a^c(t)$ are comparable with $1/|a|$.

This formulation requires some refinement. As in the previous section, to do this it is necessary, introducing a double sequence, to consider a family of governing processes $X^a(t)$ depending on a parameter $a \to 0$. We could, for example, assume for the sake of simplicity that our governing process is a member of a sequence of processes obtained from the fixed process $X(t)$, $\mathsf{M}X(t) = 0$, by means of the equation[7]

$$X^a(t) = \frac{1}{1-a}(X(t) + at), \quad a < 0 .$$

Let the sequence of processes $X^a(t)$ for which Conditions (57) and (58) are uniformly fulfilled be given in this way. In essence, we must show that the conditional renewal density for $X^a(t)$ behaves asymptotically like the renewal density of a Gaussian process with parameters $(at, \sigma\sqrt{t})$, equaling for $x > 0$

$$-\frac{1}{a} e^{2ax/\sigma^2} = \int_0^\infty \frac{1}{\sigma\sqrt{2\pi t}} e^{-(x-at)^2/2\sigma^2 t} \, dt .$$

First of all, we can assume without loss of generality that there exists a sufficiently slowly increasing function $N(t) \uparrow \infty$ for $t \uparrow \infty$ such that (58) holds for $X^a(t)$ when $|x| < N(t)$ and the remainder is uniformly estimated by $o(1/N(t))$. Assuming $x = y/|a|$, where $y > 0$ is fixed, we write

$$\textstyle\int_0^\infty \mathsf{P}(X^a(t) - X^a(0) \geqslant x \mid A) \, dt = \int_0^{M_1} + \int_{M_1}^{M_2} + \int_{M_2}^\infty , \tag{59}$$

where (M_1, M_2) is an interval of t-values for which

$$\frac{1}{\sigma\sqrt{t}}\left(\frac{y}{|a|} - at\right) < N(t) .$$

For a suitable choice of $N(t)$ we have, writing $N(1/a^2) = N$ for the sake of brevity, $M_1 < 1/a^2 N$, $M_2 > N^2\sigma^2/a^2$,

$$\int_0^{M_1} = O\left(\frac{1}{a^2 N}\right) \quad \text{and} \quad \int_{M_2}^\infty \leqslant \int_{M_2}^\infty \frac{ct^{1+\gamma/2}}{|at|^{2+\gamma}} \, dt = \frac{2c}{\gamma a^2 N^\gamma \sigma^\gamma} = O\left(\frac{1}{a^2 N^\gamma}\right).$$

[7] From the applied point of view it does not make a great deal of difference how we chose the sequence of processes X^ε; $\varepsilon \to 0$ one of whose members is to be considered as the actual process X^a with small a. It is merely necessary that (57) be valid for not very large values of c and that (58) hold with sufficiently high accuracy.

Estimates of the nearness of $\mathsf{P}(w^c(t) > x/|a|)$ and e^{-2x/σ^2}, which depend only on $|a|$ and σ, cannot be given here.

Analogously, we find

$$\int_0^{M_1}\left[1-\Phi\left(\frac{x-at}{\sigma\sqrt{t}}\right)\right]dt=O\left(\frac{1}{a^2N}\right) \quad\text{and}\quad \int_{M_2}^{\infty}\left[1-\Phi\left(\frac{x-at}{\sigma\sqrt{t}}\right)\right]dt=O\left(\frac{1}{a^2N}\right).$$

All of these estimates are obviously uniform in x for $x<M/|a|$ and arbitrary fixed M. Hence, (59) equals, with accuracy up to $O(1/a^2N^{\gamma})$ (we will assume that $\gamma\leqslant\frac{1}{2}$),

$$\int_0^{\infty}\int_x^{\infty}\frac{1}{\sigma\sqrt{2\pi t}}e^{-(v-at)^2/2\sigma^2 t}\,dv\left(1+o\left(\frac{1}{N(t)}\right)I(M_1, M_2)\right)dt,$$

where $I(M_1, M_2)$ is the indicator of the interval (M_1, M_2). Interchanging the order of integration, we obtain for this integral the value

$$\frac{\sigma^2}{2a^2}e^{2ax/\sigma^2}\left(1+o\left(\frac{1}{\sqrt{N}}\right)\right).$$

Using now Theorem 18, we can write

$$\mathsf{P}(w_a^c(t)\geqslant x)=e^{2ax/\sigma^2}+\frac{\partial F(x, a)}{\partial x}, \qquad F(x, a)=O\left(\frac{1}{aN^{\gamma}}\right), \tag{60}$$

where the estimate is uniform in $x<M/|a|$.

From this it follows that

$$f(x, a)=\frac{\partial F(x, a)}{\partial x}=o(1)$$

uniformly on $[\delta/|a|, M/|a|]$ for arbitrary $\delta>0$ and $M>0$. Indeed, assuming the existence of

$$x=x(a)>\delta/|a| \quad\text{and}\quad \varepsilon>0$$

such that for all a, $f(x, a)>\varepsilon$, we obtain by the first equation at (60) and the monotonicity of its left side that for sufficiently small a, $f(v, a)>\varepsilon/2$ for all $v\in[x-\theta/|a|, x]$ and $\theta=o(1)$. Consequently,

$$F(x, a)>F\left(x-\frac{\theta}{|a|}, a\right)+\frac{\varepsilon}{2}\frac{\theta}{|a|}.$$

This means that either $|F(x, a)|$ or $|F(x-\theta/|a|, a)|$ exceeds $\varepsilon\theta/4|a|$. For $\theta=N^{-\gamma/2}$ this contradicts the second relation at (60). In an analogous way we arrive at a contradiction when we assume that $f(x, a)<-\varepsilon$.

The required relation

$$\lim_{a\to 0}\mathsf{P}\left(w_a^c(t)\geqslant\frac{y}{|a|}\right)=e^{-2y/\sigma^2}$$

is proved. □

A similar analysis can also be carried out for a *prestationary* solution using Equation (39) for large t. From these considerations, or from the indicated calculations, it is not difficult to conclude that *for small a the entry of the system into a stationary regime is concluded during the interval of time* $(1/a^2, M/a^2)$ *for large M.* More exact estimates for the number M and for the distribution of $w(t)=w_a(t)$ for $a\to 0$ and $t\to\infty$ will be obtained in § 25 for governing sequences with independent terms.

6. In conclusion we give a numerical example illustrating the precision of the approximate relations obtained above under conditions of heavy traffic. We treat the systems $\langle G_I, 1, G_I, 1\rangle$, where the τ_j^e are distributed like the sum of the independent random variables $\zeta+\nu$ with $\mathsf{P}(\zeta<x)=x$, $x\in[0,1]$, $\mathsf{P}(\nu=0)=\mathsf{P}(\nu=2)=\frac{1}{2}$ and the τ_j^s are distributed like $\frac{1}{2}+\tau^*$ with $\mathsf{P}(\tau^*>x)=e^{-\alpha x}$ and we consider the two values $\alpha_1=\frac{20}{17}$ and $\alpha_2=\frac{20}{19}$ for α. Then $\mathsf{M}\tau_j^e=1.5$ and in the first case

$$\mathsf{M}\tau_j^s=\frac{1}{2}+\frac{1}{\alpha}=\frac{1}{2}+\frac{17}{20}=1.35$$

and

$$a_1=\mathsf{M}(X(1)-X(0))=\frac{\mathsf{M}\tau^s}{\mathsf{M}\tau^e}-1=-0.1\,.$$

In the second case,

$$\mathsf{M}\tau_j^s=\frac{1}{2}+\frac{19}{20}=1.45 \quad\text{and}\quad a_2=-\frac{0.1}{3}\cdot$$

We will view our systems as those in heavy traffic. Conditions (57) and (58) will hold and to apply the approximate formulae of Subsection **5** it remains to calculate σ^2. As we will see in the following section

$$\mathsf{D}(X(t)-X(0))=t\left(\frac{\mathsf{D}\tau^s}{\mathsf{M}\tau^e}+\frac{(\mathsf{M}\tau^s)^2\,\mathsf{D}\tau^e}{(\mathsf{M}\tau^e)^3}\right)+o(t)$$

so that σ^2 is equal to the coefficient of t on the right-hand side of this equation. Since in our example

$$\mathsf{D}\tau^e=\mathsf{D}\zeta+\mathsf{D}\nu=\tfrac{1}{12}+1=\tfrac{13}{12}\,,$$

in the first case we have

$$\sigma_1^2=(\tfrac{17}{20})^2\cdot\tfrac{2}{3}+(\tfrac{27}{20})^2\cdot\tfrac{13}{12}(\tfrac{2}{3})^3\approx 1.07$$

and in the second,

$$\sigma_2^2=(\tfrac{19}{20})^2\cdot\tfrac{2}{3}+(\tfrac{29}{20})^2\cdot\tfrac{13}{12}(\tfrac{2}{3})^3\approx 1.28\,.$$

In accordance with the main assertion of the previous Subsection we have in the first case

$$P_1(x)=\mathsf{P}(w_1^c(t)>x)\approx\exp\{-0.2x/1.07\}\approx\exp\{-0.187x\}$$

and in the second

$$P_2(x) = \mathsf{P}(w_2^c(t) > x) \approx \exp\{-0.2x/3.84\} \approx \exp\{-0.052x\}.$$

We now compare the derived estimates with the exact values of the functions $P_1(x)$ and $P_2(x)$. We know that the distribution of $w^c(t)$ coincides with that of $\sup_{v \leqslant 0}(X(0) - X(v))$, where $X(v)$ is a governing process with stationary increments. As we will see in the next section, we can assume that the process $Y(t) = X(0) - X(-t)$ is constructed as follows: $Y(t) + t$ is piecewise constant except at the points $\tau_0, \tau_0 + \tau_2^e, \tau_0 + \tau_2^e + \tau_3^e, \ldots$ at which it has jumps of sizes $\tau_2^s, \tau_3^s, \ldots$; the variable τ_0 does not depend on the terms of these sequences and is distributed according to

$$\mathsf{P}(\tau_0 > x) = \frac{1}{\mathsf{M}\tau^e} \int_x^\infty \mathsf{P}(\tau^e > x)\, dx.$$

To calculate the empirical distribution functions $\sup_{t \geqslant 0} Y(t)$ was computed in each of the two cases using 1000 trials. These distribution functions, which we denote by $F_1^*(x)$ and $F_2^*(x)$ were determined at the points $x_0 = 0, x_1, \ldots, x_{49}, x_{50} = \infty$ which were chosen in such a way that

$$\mathsf{P}(\xi \in (x_{j-1}, x_j)) = \tfrac{1}{50}, \quad j = 1, \ldots, 50,$$

where ξ is a random variable with d.f. $F_i(x) = 1 - \exp(2a_i x/\sigma_i^2)$ for $i = 1, 2$, resp.

We present below a table of the values of $F_i^*(x)$ at the points x_k (it is clear that in each of the cases under consideration $F_i(x_k) = k/50$).

Table 1. The values of $F_i^*(x_k)$, $i = 1, 2$

k	$F_1^*(x_k)$	$F_2^*(x_k)$	k	$F_1^*(x_k)$	$F_2^*(x_k)$	k	$F_1^*(x_k)$	$F_2^*(x_k)$
0	0.097	0.022	17	0.338	0.357	34	0.656	0.685
1	.104	.039	18	.350	.382	35	.682	.698
2	.109	.058	19	.369	.402	36	.701	.723
3	.122	.074	20	.383	.420	37	.720	.746
4	.134	.086	21	.396	.437	38	.737	.760
5	.157	.099	22	.422	.457	39	.755	.771
6	.174	.117	23	.442	.475	40	.775	.787
7	.197	.146	24	.458	.494	41	.791	.805
8	.213	.173	25	.484	.511	42	.811	.830
9	.228	.194	26	.505	.533	43	.834	.850
10	.246	.214	27	.528	.554	44	.849	.873
11	.256	.236	28	.548	.570	45	.868	.889
12	.270	.263	29	.571	.600	46	.900	.914
13	.285	.286	30	.580	.611	47	.920	.934
14	.300	.301	31	.605	.627	48	.944	.950
15	.310	.319	32	.622	.640	49	.980	.976
16	.323	.335	33	.644	.666			

The data in this table are plotted in Figs. 2 and 3. We have also plotted $F_i^*(x)$ and $F_i(x)$, $i = 1, 2$, in the usual (not logarithmic) scale (Figs. 4 and 5).

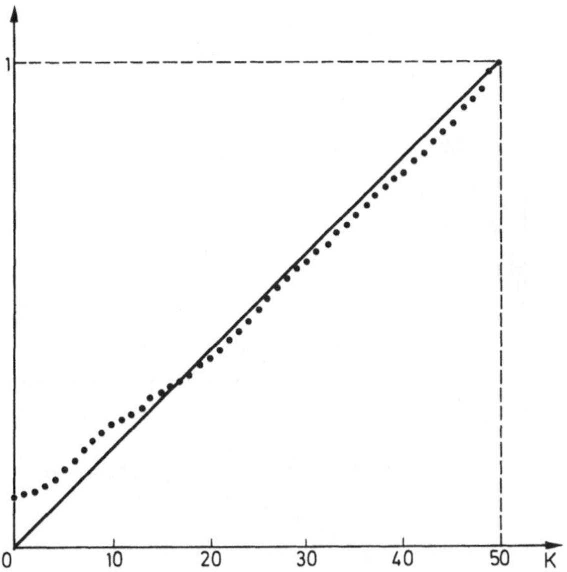

Fig. 2. $F_1^*(x_k)$

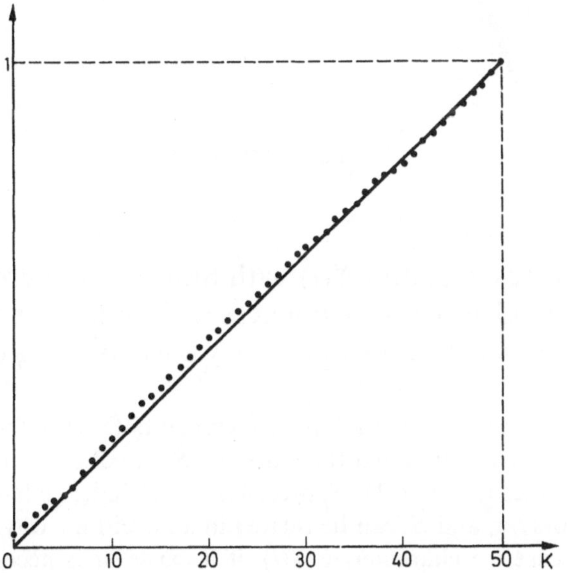

Fig. 3. $F_2^*(x_k)$

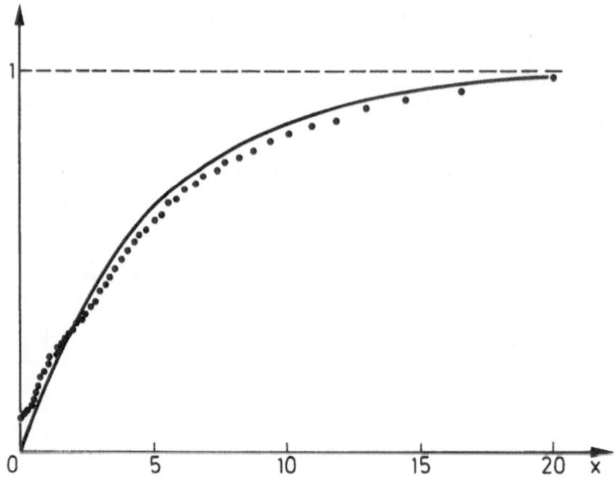

Fig. 4. $F_1^*(x_k)$

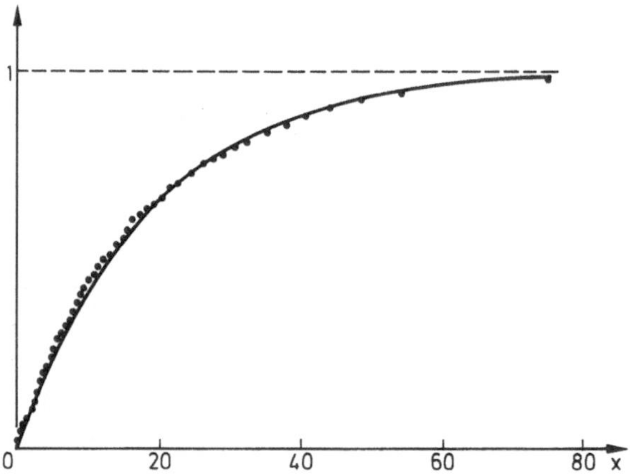

Fig. 5. $F_2^*(x_k)$

§9. The Processes $X(t)$ and $Y(t)$ with Stationary Increments Corresponding to Governing Sequences with Independent Terms. The Connection between the Stationary Distributions of $w^c(t)$ and w^k

1. How does the process with stationary increments $\{X(t);\ t \leqslant 0\}$ look when the governing sequences are such that the pairs (τ_1^e, S_j), $j=2, 3, \ldots$ (intervals between jumps and jump sizes (see Fig. 1); S_j precedes τ_j^e) are independent and identically distributed vectors? τ_j^e and S_j can be related in an arbitrary way. It's not hard to see that in this case, to characterize $X(t)$, if it exists, it is necessary to find the corresponding distribution of the time τ_1^e of the first jump of $X(t)$ after $t=0$. The subsequent behavior of $X(t)$ will be determined by the pairs (τ_j^e, S_j), $j=2, 3, \ldots$.

The situation is similar for the process with stationary increments $\{Y(t) = X(0) - X(-t);\ t \geqslant 0\}$. However, since $Y(t)$ is "inverse" to $X(t)$, its jump variables will depend, generally speaking, on *previous* time intervals (but not on the *subsequent* ones, as for $X(t)$). Hence, to characterize $Y(t)$ it is necessary to find the corresponding joint distribution of the time τ_0 of its first jump after $t=0$ and the size S_0 of this jump. The subsequent behavior of $Y(t)$ will be determined by the pairs (τ^e_{-1}, S_{-1}), (τ^e_{-2}, S_{-2}), ..., which are distributed like the initial (τ^e_j, S_j) for $j \geq 2$. However, as already mentioned, the interval τ^e_{-j} here precedes the jump S_{-j}.

Since the distribution of $w^c(t)$ coincides with that of $\sup_{t \geq 0} Y(t)$, we first construct the process $\{Y(t), t \geq 0\}$. In order to find the distribution of (τ_0, S_0), we proceed as follows: we consider the distribution of these r.v.'s for the process $\{Y_N(t); t \geq -N\}$ (denote them by $(\tau_{(N)}, S_{(N)})$), which is defined by: $Y_N(-N)=0$ and, as before

$$\frac{dY_N(t)}{dt} = -1$$

everywhere except for the points $-N+\tau^e_{-1}$, $-N+\tau^e_{-1}+\tau^e_{-2}$, ..., where Y_N has jumps of sizes $S_{-1}, S_{-2}, \ldots$. Roughly speaking, Y_N is, with accuracy up to translations of the ordinate axis, the "conditional" process $\{Y(t); -\infty < t < \infty\}$ for $t \geq -N$ under the condition that $Y(t)$ has a jump at the point $-N$. It is not hard to see that the limiting distribution of $(\tau_{(N)}, S_{(N)})$ for $N \to \infty$, if it exists, will possess the required properties.

Let us write

$$P(x, y) = \mathsf{P}(\tau^e_j \geq x, S_j < y), \quad j \geq 2.$$

By the formula of total probability we have

$$\begin{aligned} G_N(x, y) \equiv \mathsf{P}(\tau_{(N)} \geq x, S_{(N)} < y) &= P(N+x, y) + \\ &+ \int_0^N \mathsf{P}(\tau^e_{-1} \in dt) \mathsf{P}(N-t+x, y) \\ &+ \int_0^N \mathsf{P}(\tau^e_{-1} + \tau^e_{-2} \in dt) \mathsf{P}(N-t+x, y) + \cdots \\ &= \int_0^N dH(t) \mathsf{P}(N-t+x, y), \end{aligned} \tag{61}$$

where $H(t)$ is the renewal function of the sequence $\tau^e_2, \tau^e_3, \ldots$:

$$H(t) = E(t) + \mathsf{P}(\tau^e_2 < t) + \mathsf{P}(\tau^e_2 + \tau^e_3 < t) + \cdots,$$

$$E(t) = \begin{cases} 0, & t \leq 0 \\ 1, & t > 0. \end{cases}$$

If $\mathsf{M}\tau^e < \infty$ and τ^e is nonlattice (in the following formulae we will omit for brevity the subscript $j \geq 2$ on τ^e_j and S_j), then by the fundamental renewal theorem (Appendix 1) for $N \to \infty$

$$G_N(x, y) \to \frac{1}{\mathsf{M}\tau^e} \int_0^\infty \mathsf{P}(v+x, y)\, dv.$$

It is now easy to show directly that the *process* $\{Y(t); t \geq 0\}$ *with the joint distribution*

$$\mathsf{P}(\tau_0 \geq x, S_0 < y) = \frac{1}{\mathsf{M}\tau^e} \int_x^\infty P(v, y)\, dv \tag{62}$$

is a process with strictly stationary increments.

If τ_j^e and S_j, $j\geqslant 2$, are independent, then obviously

$$\mathsf{P}(\tau_0\geqslant x, S_0<y)=\frac{\mathsf{P}(S<y)}{\mathsf{M}\tau^e}\int_x^\infty \mathsf{P}(\tau^e\geqslant v)\,dv\,.$$

In a completely analogous way we can show that the joint distribution of the time τ_1^e of the first jump and its size S_1 for the process $\{X(t), t\geqslant 0\}$ with stationary increments will have exactly the same form.

The finiteness of $\mathsf{M}\tau^e$ *and "nonlatticeness" of* τ^e *are necessary and sufficient conditions for the existence of the processes* $Y(t)$ *and* $X(t)$ *with stationary increments.*

We also remark that in accordance with the fact that a homogeneous process with independent increments also has stationary increments, the right side of (62) takes the function

$$P(v, y)=e^{-\alpha v}P(y)$$

into itself. This is the only function possessing such a property.

If τ^e and S are lattice and $\mathsf{M}\tau^e<\infty$, then in place of the processes $Y(t)$ and $X(t)$ it is necessary to construct *sequences* $Y(n)$ and $X(n)$ with stationary increments (see § 7, Subsection 3). The result will be analogous to (62):

$$\mathsf{P}(\tau_0=k, S_0=j)=\frac{1}{\mathsf{M}\tau^e}\mathsf{P}(\tau^e>k, S=j)\,.$$

We will now find an expression for $\mathsf{M}(X(t)-X(0))=\mathsf{M}\,Y(t)$ in terms of the moments $\mathsf{M}\tau^e$ and $\mathsf{M}S$. We denote by $t+\tau(t)$ and $\eta(t)$ the time of appearance and the number of the first jump of the process $\{X(t); t\geqslant 0\}$ after time $t\geqslant 0$, so that $\tau_1^e=\tau(0)$ and $\eta(0)=1$. Then we can write

$$X(t)-X(0)=S_2+\dots+S_{\eta(t)}-t\,.$$

Since the event $\{\eta(t)\leqslant k\}$ is measurable w.r.t. the σ-algebra generated by the random variables τ_1^e, S_2, τ_2^e, ..., S_k, τ_k^e ($\eta(t)$ does not depend on the future), we have by Wald's identity

$$\mathsf{M}(X(t)-X(0))=-t+\mathsf{M}S\mathsf{M}(\eta(t)-1)\,.$$

But $\eta(t)-1$ is the number of jumps of $X(t)$ on $(0, t)$. In virtue of the stationarity of the increments of X and the integral renewal theorem, we have for $t\to\infty$

$$t\mathsf{M}(\eta(1)-1)=\mathsf{M}(\eta(t)-1)=\int_0^t \mathsf{P}(\tau(0)\in du)H(t-u)$$
$$\sim\int_0^t \mathsf{P}(\tau(0)\in du)\frac{t-u}{\mathsf{M}\tau^e}\sim\frac{t}{\mathsf{M}\tau^e}\,.$$

Thus, $\mathsf{M}(\eta(1)-1)$ must equal $1/\mathsf{M}\tau^e$.

From this we finally obtain

$$\mathsf{M}(X(t)-X(0))=t\left(\frac{\mathsf{M}S}{\mathsf{M}\tau^e}-1\right).\quad\square \tag{63}$$

If S_j and τ_j^e are independent, then it is not difficult to determine $\mathsf{D}(X(t)-X(0))$, assuming, of course, that $\mathsf{D}S$ and $\mathsf{D}\tau^e$ exist.

In fact,

$$\mathsf{D}(X(t)-X(0))=\mathsf{D}(S_2+\dots+S_{\eta(t)}),$$

where $\eta(t)$ does not depend on the sequence $S_2, S_3, \dots$. Hence,

$$\begin{aligned}\mathsf{D}(X(t)-X(0))&=\textstyle\sum_{k=0}^{\infty}\mathsf{P}(\eta(t)-1=k)\mathsf{M}[(S_2+\dots+S_{k+1}\\&\quad-k\mathsf{M}S)+\mathsf{M}S(k-\mathsf{M}(\eta(t)-1))]^2\\&=\mathsf{D}S\mathsf{M}(\eta(t)-1)+(\mathsf{M}S)^2\mathsf{D}(\eta(t)-1).\end{aligned}$$

Since (see Smith [19], and Cox [61])

$$\mathsf{D}(\eta(t)-1)=\frac{t\mathsf{D}\tau^e}{(\mathsf{M}\tau^e)^3}+o(t),$$

we obtain

$$\mathsf{D}(X(t)-X(0))=t\left(\frac{\mathsf{D}S}{\mathsf{M}\tau^e}+\frac{(\mathsf{M}S)^2\mathsf{D}\tau^e}{(\mathsf{M}\tau^e)^3}\right)+o(t).$$

(If $\mathsf{M}(\tau^e)^3<\infty$, then $o(t)$ can be replaced by

$$\frac{1}{6}+\frac{(\mathsf{D}\tau^e)^2}{2(\mathsf{M}\tau^e)^4}-\frac{\mathsf{M}(\tau^e)^3}{3(\mathsf{M}\tau^e)^3}+o(1),$$

where the rate of decrease $o(1)$ is determined by the rate of convergence to 0 of the "tails" $\mathsf{P}(\tau^e\geqslant x)$ for $x\to\infty$.)

It is clear that the derived formulae also hold for the process $Y(t)$.

2. We now find the *relation between the stationary distributions of* $w^c(t)$ *and* w^k (see §§ 3, 6). From the construction of the process with stationary increments $Y(t)$ it follows that

$$\bar{Y}=\max(0, S_0-\tau_0+Y),$$

where $S_0-\tau_0$ and Y are independent, and

$$Y=\sup(0, \xi_{-1}, \xi_{-1}+\xi_2, \dots).$$

Consequently,[8]

$$w^c(0)\underset{d}{=}\max(0, S_0-\tau_0+w^0),$$

where $S_0-\tau_0$ and w^0 are independent ($w^0\underset{d}{=}Y$).

If τ^e is *exponentially-distributed* and is independent of S, then, as already mentioned, the distribution of (τ_0, S_0) coincides with that of (τ^e, S) and

$$w^c(0)\underset{d}{=}\max(0, \xi_0+Y)\underset{d}{=}Y\underset{d}{=}w^0.$$

Hence, in this case the stationary distributions of the virtual waiting time and the waiting time at the instant of arrival of a request coincide.[9] This means, in

[8] We will write $\xi\underset{d}{=}\eta$ if $\mathsf{P}(\xi\geqslant x)=\mathsf{P}(\eta\geqslant x)$ and $\xi\underset{d}{\geqslant}\eta$ if $\mathsf{P}(\xi\geqslant x)\geqslant\mathsf{P}(\eta\geqslant x)$ for all x.

[9] A paradox seems to be contained here: the sequence w_n is generated by the *lowest* points of the trajectory $w(t)$ and it would seem that $w_n\underset{d}{<}w(t)$. However, the fact is that the larger values of t usually fall within the *longer* intervals between jumps of $w(t)$, so that the distribution of $w_{\eta(t)}$, where $\eta(t)$ is the number of the first jump after time t, turns out to be displaced in the direction of smaller values.

particular, that all of the assertions of Subsection 2 of § 7 as well as the corresponding formulae of 2 § 8 carry over to the distribution of w^k.

In the general case, the distribution of $S_0 - \tau_0$ is defined by the density

$$\mathsf{P}(S_0 - \tau_0 \in dx) = \frac{1}{\mathsf{M}\tau^e} \mathsf{P}(S > x, S - \tau^e < x)\, dx . \tag{64}$$

This formula follows from Eq. (62).

§ 10. Estimates of the Rate of Convergence of the Distributions of w_n and $w(t)$ to Stationarity. Connection with the Queue Length

1. An estimate of the rate of convergence of the distributions of $\{w_{n+k}; k \geqslant 0\}$ for $n \to \infty$ to that of $\{w^k; k \geqslant 0\}$ was given by Theorems 2 and 4. In order to estimate the probabilities appearing in those propositions, it is necessary to impose additional conditions on the sequence $\{\xi_k\}$. The essence of these conditions will become clear in the sequel, although the form in which they are given may differ.

Set

$$\mathsf{M}\xi_k = a \quad \text{and} \quad \zeta_k = \xi_k - a$$

and assume that the conditions of Theorems 2 and 4 hold, and that the $(2+\gamma)$-th moment of ζ_k exists and is finite:

$$\mathsf{M}|\xi_k - a|^{2+\gamma} = m_{2+\gamma} < \infty, \quad \gamma > 0 . \tag{65}$$

Moreover, we must assume that some weak dependence condition holds for terms of $\{\xi_k\}$ separated from one another by long time intervals. A condition on the decrease of the correlation function will not be enough here.

Theorem 19. *Assume Condition* (65) *holds for* $\gamma = 2$ *and that*

$$|\mathsf{M}\zeta_{j-b}\zeta_j\zeta_k\zeta_{k+d}| \leqslant m_4 \min(\varphi(b), \varphi(d)); \quad j \leqslant k, b \geqslant 0, d \geqslant 0 , \tag{66}$$

where $\varphi(n) \leqslant 1$ *decreases in such a way that*

$$\textstyle\sum_1^\infty n\varphi(n) = \varphi < \infty .$$

Then for an arbitrary fixed (*nonrandom*) *initial value* w_1

$$-\frac{cn^2}{\left(n - \dfrac{x - w_1}{a}\right)^4} \leqslant \mathsf{P}(w^0 < x) - \mathsf{P}(w_{n+1} < x) \leqslant \frac{c}{n},$$

where

$$c = 3m_4 \frac{8\varphi + 1}{a^4} .$$

If $\mathbf{w}_n$ *and* $\mathbf{w}$ *are realizations of the random sequences*

$$\{w_{n,k}=w_{n+k};\, k\geqslant 0\} \quad \textit{and} \quad \{w^k;\, k\geqslant 0\}$$

(*see Theorem* 4) *and B is an arbitrary measurable set in the sample space, then*

$$|\mathsf{P}(\mathbf{w}_{n+1}\in B)-\mathsf{P}(\mathbf{w}\in B)|\leqslant\frac{c}{n}+\frac{cn^2}{(n+w_1/a)^4}\,.$$

Proof. Using the equality

$$\begin{aligned}(\textstyle\sum_{j=1}^n \zeta_j)^4 &= \textstyle\sum \zeta_j^4+4\sum_{i<j}\zeta_i^3\zeta_j+4\sum_{i<j}\zeta_i\zeta_j^3+6\sum_{i<j}\zeta_i^2\zeta_j^2\\ &\quad+12\textstyle\sum_{i<j<k}(\zeta_i^2\zeta_j\zeta_k+\zeta_i\zeta_j^2\zeta_k+\zeta_i\zeta_j\zeta_k^2)+24\sum_{i<j<k<l}\zeta_i\zeta_j\zeta_k\zeta_l\,,\end{aligned}$$

the stationarity of the sequence $\{\zeta_j\}$ and (66), we get

$$\begin{aligned}\mathsf{M}(\textstyle\sum_{j=1}^n \zeta_j)^4 &\leqslant m_4\{n+8\textstyle\sum_{i<j}\varphi(j-i)+6\sum_{i<j}1\\ &\quad+12\textstyle\sum_{i<j<k}[\varphi(k-j)+\varphi(j-i)+\min(\varphi(k-j),\varphi(j-i))]\\ &\quad+24\textstyle\sum_{i<j<k<l}\min(\varphi(l-k),\varphi(j-i))\}\leqslant m_4 3n^2(8\varphi+1)\,.\end{aligned}$$

Put $A_n=\{Y_n>x\}$. Then by Čebyšev's inequality

$$\mathsf{P}(A_n)=\frac{3m_4n^2(8\varphi+1)}{(x-an)^4}\leqslant\frac{3m_4(8\varphi+1)}{a^4n^2}\,.$$

Since

$$\mathsf{P}(\overline{Y}_n\leqslant x,\, Y>x)\leqslant\mathsf{P}(\textstyle\bigcup_{k=n+1}^\infty A_k)\leqslant\dfrac{3m_4(8\varphi+1)}{a^4n}\,,$$

to prove the first assertion of the theorem we need only use Theorem 2.

The proof of the second claim proceeds quite analogously with the help of Theorem 4 if we notice that

$$\mathsf{P}(X_n>-\max(w_1,\xi_0+Y))\leqslant\mathsf{P}(X_n>-w_1)+\textstyle\sum_{k=n+1}^\infty \mathsf{P}(X_k>0)\,. \quad \square$$

It is clear that by imposing other conditions on the sequence $\{\xi_k\}$, we can obtain better estimates of the rate of convergence. For example, we can assume that the sequence $\{\xi_k\}$ satisfies (instead of (66)) a uniform strong mixing condition with sufficiently rapid rate of convergence to 0. This condition goes as follows:

If $\mathfrak{M}_k^l$ is the σ-algebra generated by the random variables $\xi_{k+1},\dots,\xi_{k+l}$, then for arbitrary $A\in\mathsf{M}_{k+n}^\infty$

$$|\mathsf{P}_{\mathfrak{M}_{-\infty}^k}(A)-\mathsf{P}(A)|<\alpha_1(n)\downarrow 0 \quad \text{as } n\to\infty\,. \tag{67}$$

This condition can be written in terms of moments: if ξ is $\mathfrak{M}_{-\infty}^k$-measurable and η is $\mathfrak{M}_{k+n}^\infty$-measurable, and if $\mathsf{M}|\xi|^p<\infty$, $\mathsf{M}|\eta|^q<\infty$; $p,q>1$, $1/p+1/q=1$, then

$$|\mathsf{M}\xi\eta-\mathsf{M}\xi\cdot\mathsf{M}\eta|\leqslant\mathsf{M}^{1/p}|\xi|^p\cdot\mathsf{M}^{1/q}|\eta|^q\alpha_2(n)\downarrow 0 \quad \text{as } n\to\infty \tag{68}$$

(see Ibragimov and Linnik [34], p. 392 [Russ. ed.]. (67) implies (68) for $\alpha_2<2\alpha_1^{1/p}$ and (68) implies (67) when $\alpha_1\leqslant\alpha_2$).

The assumption on the existence of $\mathsf{M}e^{\lambda\xi_k}$ for some $\lambda>0$ and on the exponential decrease of the function $\alpha_1(n)$ in Condition (67) will evidently allow us to obtain an exponential estimate for the difference

$$\mathsf{P}(\mathbf{w}_n \in B) - \mathsf{P}(\mathbf{w} \in B) . \tag{69}$$

On the other hand, if Condition (65) is satisfied only for some $0<\gamma<1$, then Condition (67) (or (68)) allows us to obtain an interesting estimate for the difference (69) in this case as well. As an example we give

Theorem 20. *Assume that for some* $0<\gamma\leqslant 1$ *and* $\alpha_1(n)$ *Conditions* (65) *and* (67) *are satisfied. Moreover, let*

$$r(n) = \mathsf{M}\zeta_j\zeta_{n+j} \quad \text{and} \quad r = \sum_{n=1}^{\infty} r(n) < \infty , \quad r(0) + 2r > 0 .$$

Then for some c_1 *(depending, in general, on* w_1*) we have in the notation of Theorem* 19

$$|\mathsf{P}(\mathbf{w}_n \in B) - \mathsf{P}(\mathbf{w} \in B)| < c_1 n^{-\gamma/2} .$$

Proof. The following well-known fact will be used: if, in addition to (65) and (67),

$$d_n^2 = \mathsf{M}(Y_n - an)^2 \to \infty \tag{70}$$

as $n\to\infty$, then there exists a constant c_2 for which

$$\mathsf{M}|Y_n - an|^{2+\gamma} \leqslant c_2 d_n^{2+\gamma} \tag{71}$$

(see Ibragimov and Linnik [34], p. 432 [Russ. ed.]; also Doob [23]).

We need only verify the validity of (70). One has

$$\begin{aligned} d_n^2 &= \mathsf{M}\Big(\sum_{j=1}^{n} \zeta_j\Big)^2 = n\mathsf{M}\zeta_1^2 + 2\sum_{i<j} \mathsf{M}\zeta_i\zeta_j \\ &= nr(0) + 2\sum_{j=1}^{n-1}(n-j)r(j) = n\Big(r(0) + 2\sum_{j=1}^{n-1} r(j)\Big) - 2\sum_{j=1}^{n-1} jr(j) . \end{aligned}$$

Since $R(n) = \sum_{j=n}^{\infty} r(j) \to 0$ as $n\to\infty$, we get

$$\sum_{j=1}^{n-1} jr(j) = \sum_{j=1}^{n-1} j(R(j) - R(j+1)) = \sum_{j=1}^{n-1} R(j) - (n-1)R(n) = o(n) .$$

Consequently, $d_n^2 = n(r(0)+2r) + o(n)$ and (71) holds. Now we need only set up estimates completely analogous to those in the proof of the previous theorem. □

More precise estimates of the rate of convergence for independent ξ_k will be given in Chapter 4.

2. Theorems 19 and 20 allow us to construct without difficulty similar estimates for the convergence of the distributions of $\{w_t(u) = w(t+u), u\geqslant 0\}$ to that of $\{w^c(u), u\geqslant 0\}$ by imposing corresponding restrictions on the sequence $\{X(n) - X(n-1)\}$. For example, a probability of the form $\mathsf{P}(\bigcup_{u>t}\{Y(u)>x\})$ appearing in the estimate (35), does not exceed

$$\mathsf{P}\Big(\bigcup_{k\geqslant[t]}\{Y(k) > x-1\}\Big) \leqslant \sum_{k\geqslant[t]} \mathsf{P}(Y(k) > x-1) ,$$

where the last sum can be estimated as in the proofs of Theorems 19 and 20. The situation is the same for the remaining formulae in the estimates of Theorem 11 and in (35) and (36).

3. *The Relationship between the Distributions of the Waiting Time and Queue Length.*

Up to now we have considered two basic characteristics of the systems $\langle G, G, G, 1\rangle$: the waiting time w_n until service of the first customer in the n-th lot and the virtual waiting time $w(t)$ which a customer arriving at time t would have to wait until service. (We recall that the systems $\langle G, G, G, 1\rangle$ are at the same time models for the study of other systems, see § 2.) Equally widespread in queueing theory are two other characteristics which were mentioned in § 2: q_n—the length of the queue at the moment of arrival of the n-th lot (not counting the customers of this lot) and $q(t)$—the length of the queue at time t. In both cases, the queue length includes a request undergoing service so that $w(t)=0$ and $q(t)=0$ are equivalent.

We first consider the distribution of q_n for the systems $\langle G, 1, G, 1\rangle$ ($v_j^e \equiv v_j^s \equiv 1$). It is clear that the event $\{q_n > k\}$ occurs iff the waiting time of the $(n-k)$-th customer satisfies

$$w_{n-k} > \tau_{n-k}^e + \dots + \tau_{n-1}^e .$$

We now assume that the sequence $\{\tau_j^e, \tau_j^s\}$ is *stationary* and *metrically transitive* and $\mathsf{M}\xi_j < 0$. We will also assume that the *initial condition* w_1 is *measurable* w.r.t. the *σ-algebra* $\mathfrak{M}$ generated by the governing sequence $\{\tau_j^e, \tau_j^s\}$ and extended if necessary as described in § 3. Then performing the random variable shift $U^{-(n-k)}$ we get

$$\mathsf{P}(q_n > k) = \mathsf{P}(w_{n-k} > \tau_{n-k}^e + \dots + \tau_{n-1}^e) = \mathsf{P}(U^{-(n-k)} w_{n-k} > \tau_0^e + \dots + \tau_{k-1}^e) .$$

By Theorem 1

$$\begin{aligned} U^{-(n-k)} w_{n-k} &= U^{-(n-k)} \max(w_1 + X_{n-k-1}, X_{n-k-1} - X_1, \dots, X_{n-k-1} - X_{n-k-1}) \\ &= \max(0, Y_1, Y_2, \dots, Y_{n-k-2}, U^{-(n-k)} w_1 + Y_{n-k-1}) . \end{aligned}$$

Since

$$\max_{j \ge 0} Y_{n-k-1+j} \xrightarrow[\text{a.s.}]{} -\infty \quad \text{for } n \to \infty$$

it follows that

$$\lim_{n\to\infty} \mathsf{P}(q_n > k) = \mathsf{P}(Y > \tau_0^e + \dots + \tau_{k-1}^e) = \mathsf{P}(w^{-1} > \tau_0^e + \dots + \tau_{k-1}^e) \quad \text{exists.}$$

We therefore have

Theorem 21. *Under the formulated conditions the limiting (stationary) distribution of q_n (the random variable corresponding to this distribution is denoted by q^0) exists and is such that*

$$\lim_{n\to\infty} \mathsf{P}(q_n > k) = \mathsf{P}(q^0 > k) = \mathsf{P}(w^0 > \tau_1^e + \dots + \tau_k^e) .$$

(Recall that $w^k = U^k w^0 = U^{k+1} Y$, $Y = w^{-1} = \sup(0, \xi_{-1}, \xi_{-1} + \xi_{-2}, \dots)$.)

If we denote by $\eta(t)=\min\{k: \sum_{j=1}^{k} \tau_j^e \geq t\}$ the first passage time through the level t in a random walk with jumps $\tau_1^e, \tau_2^e, \ldots$, then this theorem implies, in particular, that for the systems $\langle G_I, 1, G, 1\rangle$ we have a representation

$$q^0 \underset{d}{=} \eta(w^0)$$

where w^0 is independent of $\{\tau_j^e\}$.

If we consider the systems $\langle G, G, E, 1\rangle$, i.e., the case in which $\mathsf{P}(\tau^s > x) = e^{-\alpha x}$, then assuming that the limiting distribution of q_n exists, we can write what is in a certain sense the inverse relation for q^0 and w^0:

$$w^0 \underset{d}{=} \tau_1^s + \cdots + \tau_{q^0}^s,$$

which is a consequence of the equality

$$w_n \underset{d}{=} \tau_1^s + \cdots + \tau_{q_n}^s$$

since in this case the system is vacated according to a Poisson process which does not depend on the past history.

For the systems $\langle G, G, E, 1\rangle$ it is clear that analogous relations will also hold for the virtual waiting time and the queue length

$$w(t) = \tau_1^s + \cdots + \tau_{q(t)}^s$$

so that for the limiting distribution we have

$$w^0 \underset{d}{=} \tau_1^s + \cdots + \tau_{q^0}^s, \qquad w^c \underset{d}{=} \tau_1^s + \cdots + \tau_{q^c}^s,$$

where $\{\tau_j^s\}$ does not depend on q^0 and q^c.

Several other connections between the waiting time and queue length are established in §§ 26, 27 (Subsection 6) and § 28.

§ 11. Theorems on the Stability of the Stationary Waiting Time under a Change of the Governing Sequences

1. The main question we will consider in this section is the following: under what conditions will a small change in the finite-dimensional distributions of the governing sequence lead to a small change in the distribution of the stationary waiting time? The importance of this question can be illustrated as follows. In the consideration of the stationary sequence $\{w^k\}$ in actual problems it is customary to make some assumptions on the nature of the governing sequence (say, that the ξ_k are independent or that τ^e is exponentially distributed, etc.) which are actually only approximately true. For example, the assumption that the duration τ_j of messages which can be transmitted by a communication line is exponentially distributed (this is the service time of customers in the corresponding service system) obviously cannot hold in the neighbourhood of 0 because of the existence of official words which must introduce messages (thus $\mathsf{P}(\tau_j^s > \varepsilon)$ equals one for small ε and not $e^{-\alpha\varepsilon}$); nonetheless, this assumption is customarily made and is often justified.

The question to be treated here is whether or not the solution in this way of "idealized" problems is close to that of the actual problem. The answer is generally speaking no. An example of this is given in the remark following Theorem 22.

We will present below some very wide sufficient conditions which guarantee the required closeness and which are nearly necessary. The discussion of these conditions in the case where the elements of the governing sequence $\{\xi_n\}$ are independent will be continued in § 21.

2. In order to state the problem more precisely we must introduce a family of stationary sequences $\xi^{(n)}=\{\xi_k^{(n)}; -\infty<k<\infty\}$ and a stationary sequence $\xi=\{\xi_k; -\infty<k<\infty\}$. Put $Y^{(n)}=\sup_{k\geq 0} Y_k^{(n)}$, where $Y_k^{(n)}=\sum_{j=1}^{k}\xi_{-j}^{(n)}$. The question of convergence of the distribution of the stationary waiting time in answered by

Theorem 22.[10] *Assume that the finite dimensional distributions of the sequences $\xi^{(n)}$ converge weakly to the corresponding distributions of the sequence ξ, which is assumed to be ergodic with $\mathsf{M}\xi_k<0$. Then for the convergence*

$$\mathsf{P}(Y^{(n)}<t)\Rightarrow \mathsf{P}(Y<t)$$

it is sufficient that

$$\mathsf{M}(\xi_1^{(n)};\xi_1^{(n)}\geq 0)\rightarrow \mathsf{M}(\xi_1;\xi_1\geq 0)\,. \tag{72}$$

Remark. Without Condition (72) it is easy to see that the theorem will no longer hold. The condition

$$\mathsf{M}\xi^{(n)}\rightarrow \mathsf{M}\xi$$

is also not sufficient for the convergence of the distributions of $Y^{(n)}$. The following example shows this. Let the sequences $\{\xi_k^{(n)}\}$, $n=1,2,\ldots$, be given on a single probability space and be formed from the independent random variables $\xi_k^{(n)}$ which assume the four values $-2n$, -1, 1 and $2n$ with probabilities $1/n$, $\frac{3}{4}-1/n$, $\frac{1}{4}-1/n$ and $1/n$. In this case

$$\mathsf{M}\xi^{(n)}=-\tfrac{1}{2}=\mathsf{M}\xi,\qquad \mathsf{P}(Y<\infty)=1\,.$$

At the same time, $\liminf_{n\to\infty}\mathsf{P}(\overline{Y}^{(n)}>n)>0$. Indeed, let A_n be the event that the value $-2n$ does not occur among the first n trials but the value $2n$ occurs at least once. Then

$$A_n\subset\{\overline{Y}^{(n)}\geq n\}$$

and

$$\mathsf{P}(A_n)=\left(1-\frac{1}{n}\right)^n\left(1-\left(1-\frac{1}{n-1}\right)^n\right)\rightarrow q_1=e^{-1}(1-e^{-1})>0.\quad\square$$

We proceed now to the proof of the theorem. We will need two lemmas.

[10] In essence, this assertion is related to limit theorems on the convergence of the distributions of functionals of a sequence of random processes. It can also be described in a natural way as a theorem on the continuous dependence of the distribution of the stationary waiting time (or the distribution of $\overline{Y}$) on the distribution of the governing sequences.

Lemma 1. *Let ξ be an arbitrary stationary sequence. Then for arbitrary $c>0$*

$$\mathsf{P}(Y>0)\leqslant\frac{1}{c}\mathsf{M}(\xi_1;\xi_1\geqslant 0)+\mathsf{P}(\xi_1>-c).$$

Proof. We will use the inequality (Doob [23])

$$\mathsf{M}\ (\xi_1;\ Y>0)\geqslant 0 \tag{73}$$

which holds under the hypotheses of the theorem and for $\mathsf{M}|\xi_1|<\infty$.

We consider first the narrower class of sequences for which $\mathsf{M}|\xi_1|<\infty$ and $\mathsf{P}(\xi_1\in(-c,0))=0$ for some $c>0$. Then by (73)

$$\mathsf{M}(\xi_1;\xi_1\geqslant 0,\ Y>0)\geqslant -\mathsf{M}(\xi_1;\xi_1\leqslant -c,\ Y>0)\geqslant c\mathsf{P}(\xi_1\leqslant -c,\ Y>0).$$

Consequently,

$$\mathsf{P}(Y>0)=\mathsf{P}(\xi_1\geqslant 0,\ Y>0)+\mathsf{P}(\xi_1\leqslant -c,\ Y>0)\leqslant\mathsf{P}(\xi_1\geqslant 0)+\frac{1}{c}\mathsf{M}(\xi_1;\xi_1\geqslant 0). \tag{74}$$

Now assume that the probability of falling in the interval $(-c,0)$ is positive. Introduce the random variables

$$\xi_j^*=\begin{cases}\xi_j & \text{if } \xi_j\notin(-c,0)\\ 0 & \text{if } \xi_j\in(-c,0).\end{cases}$$

Then, using (67) and putting $Y_k^*=\sum_{j=1}^k\xi_j^*$, $Y_j^*=\sup_{k\geqslant 0}Y_k^*$, we can write

$$\mathsf{P}(Y>0)\leqslant\mathsf{P}(Y^*>0)\leqslant\mathsf{P}(\xi_1^*\geqslant 0)+\frac{1}{c}\mathsf{M}(\xi_1^*;\xi_1^*\geqslant 0)$$

$$=\mathsf{P}(\xi_1>-c)+\frac{1}{c}\mathsf{M}(\xi_1;\xi_1\geqslant 0).$$

It remains to remove the restriction $\mathsf{M}\xi_1\neq-\infty$. This can be done with the help of a procedure completely analogous to that which we just used. One must introduce the random variables $\xi_j^{**}=\max(\xi,-A)$ for $A>c$ and employ the inequality $\mathsf{P}(Y>0)\leqslant\mathsf{P}(Y^{**}>0)$ (with the obvious agreement on the notation Y^{**}). The lemma is proved. □

Lemma 2. *If $(\xi_1^{(n)},\dots,\xi_L^{(n)})$ is a vector whose distribution converges weakly to that of $(\xi_1,\dots,\xi_L)$ for $n\to\infty$ and*

$$\mathsf{M}(\xi_j^{(n)};\xi_j^{(n)}\geqslant 0)\to\mathsf{M}(\xi_j;\xi_j\geqslant 0),\quad j=1,\dots,L, \tag{75}$$

then the sum $Y_L^{(n)}=\sum_{j=1}^L\xi_j^{(n)}$ satisfies

$$\mathsf{M}(Y_L^{(n)};Y_L^{(n)}\geqslant 0)\to\mathsf{M}(Y_L;Y_L\geqslant 0).$$

Proof. It is obviously enough to prove the assertion for $L=2$. Since always

$$\liminf_{n\to\infty}\mathsf{M}(Y_2^{(n)};Y_2^{(n)}\geqslant 0)\geqslant\mathsf{M}(Y_2;Y_2\geqslant 0),$$

we must show that

$$\limsup_{n\to\infty}\mathsf{M}(Y_2^{(n)};Y_2^{(n)}\geqslant 0)\leqslant\mathsf{M}(Y_2;Y_2\geqslant 0). \tag{76}$$

Let $F_n(x, y)$ and $F(x, y)$ be the distribution functions of $(\xi_1^{(n)}, \xi_2^{(n)})$ and (ξ_1, ξ_2). Then

$$\mathsf{M}(Y_2^{(n)}; Y_2^{(n)} \geqslant 0) - \mathsf{M}(Y_2; Y_2 \geqslant 0) = \int_{x+y>0} (x+y)\, d(F_n(x, y) - F(x, y)).$$

Using the convergence $F_n(x, y) \Rightarrow F(x, y)$ we get for arbitrary fixed M

$$\begin{aligned}
&\limsup_{n\to\infty} \int_{x+y>0} x\, d(F_n(x, y) - F(x, y)) \\
&\leqslant \limsup_{n\to\infty} \int_{0\leqslant x<M,\, x+y>0} + \limsup_{n\to\infty} \int_{x\geqslant M,\, x+y>0} \\
&\leqslant \limsup_{n\to\infty} \int_{x\geqslant M} x\, dF_n(x, y) \\
&= \limsup_{n\to\infty} \mathsf{M}(\xi_1^{(n)}; \xi_1^{(n)} \geqslant M) \equiv \varepsilon(M).
\end{aligned}$$

But it is easy to see that the uniform (w.r.t. n) convergence $\mathsf{M}(\xi_1^{(n)}; \xi_1^{(n)} \geqslant M) \to 0$ as $M \to \infty$ is necessary and sufficient for (75) to hold. Hence, $\varepsilon(M) \to 0$ as $M \to \infty$ and so

$$\limsup_{n\to\infty} \int_{x+y>0} x\, d(F_n(x, y) - F(x, y)) \leqslant 0;$$

$$\limsup_{n\to\infty} \int_{x+y>0} y\, d(F_n(x, y) - F(x, y)) \leqslant 0.$$

Inequality (76), and along with it the lemma, are proved. □

Proof of the theorem. Putting $\psi(N) = \mathsf{P}(\sup_{k\geqslant N} Y_k \geqslant 0)$ we assume that

$$\varphi(N) \equiv \limsup_{n\to\infty} \mathsf{P}(\sup_{k\geqslant N} Y_k^{(n)} \geqslant 0) \to 0 \tag{77}$$

as $N \to \infty$. Then by the strong law of large numbers, for arbitrary $\varepsilon > 0$ we can choose an N such that $\max(\psi(N), \varphi(N)) < \varepsilon$. If $x \geqslant 0$ is a point of continuity of the distribution of $\max_{0\leqslant k\leqslant N} Y_k$ for the chosen value of N, we then find that

$$\begin{aligned}
\limsup_{n\to\infty} \mathsf{P}(Y^{(n)} \geqslant x) &\leqslant \varepsilon + \limsup_{n\to\infty} \mathsf{P}(Y^{(n)} \geqslant x; \sup_{k\geqslant N} Y_k^{(n)} < 0) \\
&= \varepsilon + \limsup_{n\to\infty} \mathsf{P}(\max_{0\leqslant k<N} Y_k^{(n)} \geqslant x) \\
&= \varepsilon + \mathsf{P}(\max_{0\leqslant k<N} Y_k \geqslant x) \leqslant \varepsilon + \mathsf{P}(Y \geqslant x).
\end{aligned}$$

Obtaining in an analogous way the reversed inequality for lim inf (one must use the relation $\psi(N) < \varepsilon$) we likewise establish (since ε is arbitrary) that for almost all x

$$\lim_{n\to\infty} \mathsf{P}(Y^{(n)} \geqslant x) = \mathsf{P}(Y \geqslant x)$$

exists.

Hence, to prove the theorem, we must show that (77) holds.

For arbitrary $\beta > 0$ and integer $L > 0$ we have

$$\begin{aligned}
\mathsf{P}(\sup_{k\geqslant N} Y_k^{(n)} \geqslant 0) &\leqslant \mathsf{P}(Y_N^{(n)} \geqslant -\beta N) + \mathsf{P}(Y_N^{(n)} < -\beta N; \sup_{k\geqslant N} Y_k^{(n)} > 0) \\
&\leqslant \mathsf{P}(Y_N^{(n)} \geqslant -\beta N) + \mathsf{P}(Y_N^{(n)} \leqslant -\beta N; \\
&\qquad \textstyle\bigcap_{k=0}^{\infty} \{Y_{N+kL}^{(n)} - Y_N^{(n)} \leqslant -\beta k L\}; \sup_{k\geqslant N} Y_k^{(n)} \geqslant 0) \\
&\quad + \mathsf{P}(\textstyle\bigcup_{k=0}^{\infty} \{Y_{N+kL}^{(n)} - Y_N^{(n)} > -\beta k L\}).
\end{aligned}$$

Denote the terms on the right-hand side of this inequality, in the order in which they occur, by I_1, I_2 and I_3. Then by the stationarity of $\xi^{(n)}$ and Lemma 1 (with $c=\beta L$)

$$I_3=\mathsf{P}(\sup_{k\geqslant 0}(Y_{kL}^{(n)}+\beta kL)>0)$$

$$\leqslant \mathsf{P}\left(\frac{Y_L^{(n)}}{L}>-2\beta\right)+\frac{1}{\beta}\mathsf{M}\left(\frac{Y_L^{(n)}+\beta L}{L};\frac{Y_L^{(n)}+\beta L}{L}>0\right);$$

and

$$I_2=\mathsf{P}\left(\bigcup_{j=0}^{\infty}\bigcup_{k=jL+1}^{(j+1)L}\left\{\xi_k^{(n)}>\frac{N\beta+jL\beta}{L}\right\}\right)$$

$$\leqslant L\sum_{j=0}^{\infty}\mathsf{P}\left(\xi_1^{(n)}>j\beta+\frac{NB}{L}\right)\leqslant\frac{L}{\beta}\int_{N\beta/L-\beta}^{\infty}P_n(t)\,dt,$$

where $P_n(t)=\mathsf{P}(\xi_1^{(n)}>t)$. The convergence $\mathsf{M}(\xi_1^{(n)};\xi_1^{(n)}>0)\to\mathsf{M}(\xi_1;\xi_1>0)$ implies, as we have already noted, the uniform (w.r.t. n) convergence $\varphi_n(u)=\int_u^\infty P_n(t)\,dt\to 0$ for $u\to\infty$. Thus, there exists a function $\varphi(u)\to\infty$ for $u\to\infty$ such that

$$I_2\leqslant\frac{L}{\beta}\varphi\left(\frac{N\beta}{L}\right).$$

Since for each L and for $N\to\infty$, $L\varphi(N/L)\to 0$, we can find a function $M=M(L)$ increasing without bound for $L\to\infty$ and such that $L\varphi(M(L)/L)\to 0$ for $L\to\infty$. Henceforth we will assume the numbers L and N are related by $N=(1/\beta)M(L)$.

Thus, for arbitrary N (or L) and β we get the estimate

$$\mathsf{P}(\sup_{k\geqslant N}Y_k^{(n)}\geqslant 0)\leqslant\mathsf{P}\left(\frac{Y_N^{(n)}}{N}>-\beta\right)+\mathsf{P}\left(\frac{Y_L^{(n)}}{L}>-2\beta\right)$$

$$+\frac{1}{\beta}\mathsf{M}\left(\frac{Y_L^{(n)}+\beta L}{L};\frac{Y_L^{(n)}+\beta L}{L}>0\right)+\frac{L}{\beta}\varphi\left(\frac{N\beta}{L}\right).$$

Consequently, for the values of β (forming an everywhere dense set) which are points of continuity of the distributions of $-Y_N/N$ and $-Y_L/2L$ we obtain by Lemma 2

$$\limsup_{n\to\infty}\mathsf{P}(\sup_{k\geqslant N}Y_k^{(n)}\geqslant 0)\leqslant A(N,\beta)+\frac{L}{\beta}\varphi\left(\frac{N\beta}{L}\right),\tag{78}$$

where

$$A(N,\beta)=\mathsf{P}\left(\frac{Y_N}{N}>-\beta\right)+\mathsf{P}\left(\frac{Y_L}{L}>-2\beta\right)+\frac{1}{\beta}\mathsf{M}\left(\frac{Y_L+\beta L}{L};\frac{Y_L+\beta L}{L}>0\right).$$

We now prove that by choosing N and β appropriately, the right side of (78) can be made as small as we like. Take an arbitrary $\varepsilon>0$ and consider the quantity $A(N,\beta)$.

For $\beta=-a/3$, where $a=\mathsf{M}\xi_k<0$, the terms of $A(N,\beta)$ satisfy the relations

$$\frac{1}{\beta}\mathsf{M}\left(\frac{Y_L+\beta L}{L};\frac{Y_L+\beta L}{L}>0\right)\leqslant\frac{1}{\beta}\mathsf{M}\left(\frac{Y_L}{L}-a;\frac{Y_L}{L}>\frac{a}{2}\right);$$

$$\mathsf{P}\left(\frac{Y_L}{L}>-2\beta\right)\leqslant\mathsf{P}\left(\frac{Y_L}{L}>\tfrac{2}{3}a\right),$$

so that

$$A(N,\beta)\leqslant 2\mathsf{P}\left(\frac{Y_L}{L}>\tfrac{2}{3}a\right)+\frac{1}{\beta}\mathsf{M}\left|\frac{Y_L}{L}-a\right|$$

Since by the strong law of large numbers $\mathsf{M}|Y_L/L-a|\to 0$ for $L\to\infty$ (L_1-convergence of Y_L/L to a) we can now choose N (and hence also L) so large that

$$A(N,\beta)\leqslant\varepsilon/2 \quad\text{and}\quad \frac{L}{\beta}\varphi\left(\frac{N\beta}{L}\right)\leqslant\varepsilon/2.$$

Hence, we have shown that the left side of (78) can be made as small as we like by appropriate choice of N. Thus, (77) and the theorem are proved. □

3. We now give an example which shows that (72) is not necessary for the convergence of the distributions of $Y^{(n)}$ (convergence of the finite-dimensional distributions of $\xi^{(n)}$ is assumed along with the condition $\mathsf{M}\xi_k<0$).

Let $\xi_k^{(n)}$ be independent, with

$$\xi_k^{(n)}=\begin{cases} n^2-1 & \text{with probability } n^{-2},\\ -1 & \text{with probability } 1-n^{-2}-n^{-1},\\ -n^2-1 & \text{with probability } n^{-1}.\end{cases}$$

The finite-dimensional distributions of the constructed sequences $\xi^{(n)}$ obviously converge to the distributions of the sequence $\xi\equiv(-1,-1,\ldots)$. The distribution of $Y^{(n)}$ also converges. In fact, since $\mathsf{P}(Y=0)=1$ it is enough to show that $\mathsf{P}(Y^{(n)}>0)\to 0$ for $n\to\infty$. We have

$$\mathsf{P}(Y^{(n)}>0)\leqslant\mathsf{P}(Y_*^{(n)}>0)=\mathsf{P}(Y_{**}^{(n)}>0)\,, \tag{79}$$

where $Y_*^{(n)}$ and $Y_{**}^{(n)}$ are constructed w.r.t. the sums of random variables $\xi_k^{(n)*}=\xi_k^{(n)}+1$ and

$$\xi_k^{(n)**}=\begin{cases} 1 & \text{with probability } \dfrac{n^{-2}}{n^{-1}+n^{-2}}=(n+1)^{-1},\\[2ex] -1 & \text{with probability } \dfrac{n^{-1}}{n^{-1}+n^{-2}}=n(n+1)^{-1}\end{cases}$$

(the sequence $\xi^{(n)**}$ is obtained from the sequence $\xi^{(n)*}$ by deleting zero terms). By Lemma 1 appearing in the proof of Theorem 22, the right side of (79) does not exceed

$$\mathsf{P}(\xi_1^{(n)**}>-1)+\mathsf{M}(\xi_1^{(n)**};\,\xi_1^{(n)**}\geqslant 0)=(n+1)^{-1}+(n+1)^{-1}\to 0$$

for $n\to\infty$. Hence $\mathsf{P}(Y^{(n)}=0)\to\mathsf{P}(Y=0)=1$, although (72) of Theorem 22 is not satisfied since

$$\mathsf{M}(\xi_1^{(n)};\,\xi_1^{(n)}\geqslant 0)=1-n^{-2}\to 1\neq\mathsf{M}(\xi_1;\,\xi_1\geqslant 0)=0\,. \quad □$$

Later, in Chapter 4 (§ 21) we will explain the remark made at the beginning of this section to the effect that (72), although not necessary, is nearly so.

We will also prove that if $\xi_k^{(n)}$ is constructed with the aid of the governing sequence $\{\tau_j^{(n)e}, \nu_j^{(n)e}, \tau_j^{(n)s}\}$ and the formula $\xi_k^{(n)} = S_k^{(n)} - \tau_k^{(n)e}$ (compare with (3), § 2), where $S_k^{(n)}$ is the time spent serving the k-th lot of requests to arrive (if $\nu_j^{(n)e} \equiv 1$ then $S_k^{(n)} = \tau_k^{(n)s}$) and if $S_k^{(n)}$ and $\tau_k^{(n)e}$ are independent, then it is sufficient for (72) to hold that

$$\mathsf{M} S_k^{(n)} \to \mathsf{M} S_k.$$

This implies, in particular, the following

Corollary. *For the systems $\langle G_I, 1, G_I, 1\rangle$ the distributions of the suprema $Y^{(n)}$ (or what is the same, the stationary waiting times) will converge if the distributions of $\tau_j^{(n)e}$ and $\tau_j^{(n)s}$ converge and if for $n \to \infty$*

$$\mathsf{M}\tau_j^{(n)s} \to \mathsf{M}\tau_j^s. \tag{80}$$

4. To conclude this subsection we give a numerical example which shows that (80) is essential.

Let τ_j^e and τ_j^s be exponentially distributed:

$$\mathsf{P}(\tau_j^e > x) = e^{-x}, \qquad \mathsf{P}(\tau_j^s > x) = e^{-2x}.$$

We consider three ways of constructing the sequences $\{\tau_j^{(n)e}, \tau_j^{(n)s}\}$:

1. $\tau_j^{(n)e} = \tau_j^e$, $\tau_j^{(n)s} = \tau_j^s + 2^{-n}$, $n = 1, 2, \dots$;

2. $\tau_j^{(n)e} = \tau_j^e$,

$$\tau_j^{(n)s} = \begin{cases} \tau_j^s & \text{with probability } 1 - 2^{-n} \\ \tau_j^s + 2^{n-3} & \text{with probability } 2^{-n}; \end{cases}$$

3. $\tau_j^{(n)s} = \tau_j^s$,

$$\tau_j^{(n)e} = \begin{cases} \tau_j^e & \text{with probability } 1 - 2^{-n} \\ \tau_j^e + 2^{n-3} & \text{with probability } 2^{-n}, \end{cases}$$

where the selection with probabilities 2^{-n} and $1 - 2^{-n}$ is carried out independently for each j.

It is clear that in the second case (80) is not satisfied, since

$$\mathsf{M}\tau_j^{(n)s} = \mathsf{M}\tau_j^s + \tfrac{1}{8}.$$

By Subsection **2** § 10, the stationary distributions of w^k and $w^c(t)$ coincide in our case. Hence, in the search for w^k (equivalently, for Y) we can use the stationary solution of Beneš' equation. By the formula of Hinčin (Cor. 3 § 8), we get

$$\mathsf{M}\, e^{i\lambda Y} = \frac{1 - \alpha \mathsf{M} S}{1 - \alpha \mathsf{M} S \dfrac{f(\lambda) - 1}{i\lambda \mathsf{M} S}},$$

where, under our conditions

$$\alpha = \frac{1}{\mathsf{M}\tau_j^e} = 1, \qquad \mathsf{M} S = \mathsf{M}\tau^s = \tfrac{1}{2}$$

and

$$f(\lambda) = \mathsf{M}\, e^{i\lambda\tau_j^s} = 2\int_0^\infty e^{i\lambda x - 2x}\, dx = \frac{2}{2 - i\lambda}.$$

Thus,

$$\mathsf{M}\, e^{i\lambda Y} = \frac{2-i\lambda}{2(1-i\lambda)}, \qquad \mathsf{P}(Y=0)=\tfrac{1}{2} \quad \text{and} \quad \mathsf{P}(Y>x)=\tfrac{1}{2}e^{-x} \quad \text{for } x \geqslant 0.$$

These formulae also follow immediately from the results of Chapters 2 and 4.

To obtain the empirical distribution of $Y^{(n)}$ for each of the 5 values $n=2, 3, \ldots, 6$ in the three cases under consideration, we carried out 1000 trials. To estimate $\sup_x |\mathsf{P}(Y<x)-\mathsf{P}(Y^{(n)}<x)|$, the points $0=x_0, x_1, \ldots, x_{199}, x_{200}=\infty$ were chosen such that

$$\mathsf{P}(Y \in (x_{j-1}, x_j)) = \tfrac{1}{400}, \quad j=1, \ldots, 200.$$

For the indicated values of n we obtained the following sample values A_n^* of the variable

$$A_n = \sup_{1 \leqslant i \leqslant 200} |\mathsf{P}(Y<x_i)-\mathsf{P}(Y^{(n)}<x_i)|:$$

In the first case

n	2	3	4	5	6
A_n^*	.326	.145	.082	.026	.013

In the second

n	2	3	4	5	6
A_n^*	.146	.167	.185	.200	.178

In the third

n	2	3	4	5	6
A_n^*	.084	.074	.058	.050	.030

The values $n>6$ were not considered since with the chosen method of producing trials, for $n>6$ A_n^* becomes an unsatisfactory estimate for A_n in the first and third cases.

In these two cases, A_n converges to 0 in agreement with Theorem 22; in the second case, A_n converges with growing n to a limit near 0.2, although

$$\mathsf{P}(\xi_k^{(n)} \neq \xi_k) = 2^{-n} \to 0 \quad \text{for } n \to \infty.$$

We present below the graphs we obtained for the empirical distribution functions $F_n^*(x)$ corresponding to the functions $F_n(x)=\mathsf{P}(Y^{(n)}<x)$ for $n=3$ and 6. The values of the functions $F_n^*(x)$ and $F(x)=\mathsf{P}(Y<x)$ were calculated at the points x_{4k}, $k=0, \ldots, 50$ (in the logarithmic scale) for the latter two methods of constructing the approximating sequences.

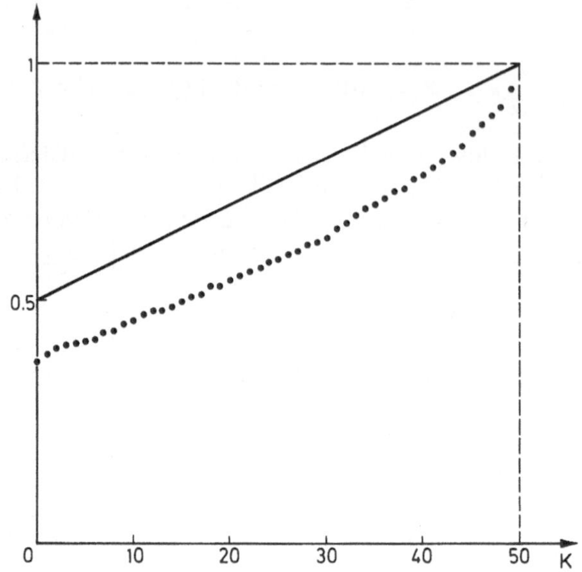

Fig. 6. $F(x_{4k})$ and $F_3^*(x_{4k})$. (Second method: $n=3$)

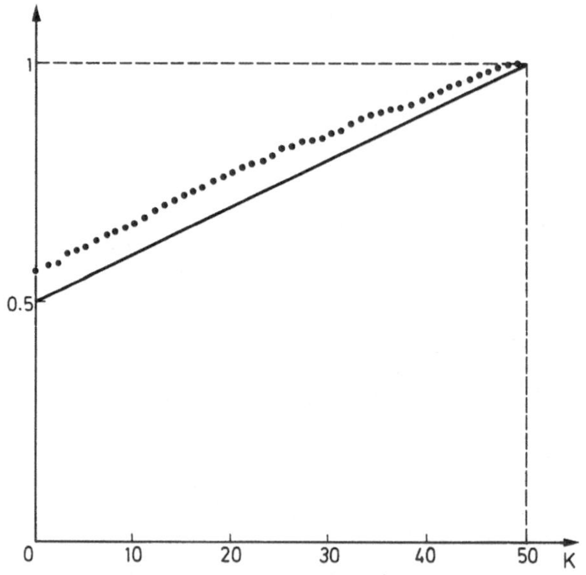

Fig. 7. $F(x_{4k})$ and $F_3^*(x_{4k})$. (Third method: $n=3$)

5. It is not hard to see that an assertion completely analogous to Theorem 22 will also hold for the stationary virtual waiting time $w^c(t)$ or, equivalently, for $Y=\sup_{t\geq 0} Y(t)$, where $Y(t)$ is a process with stationary increments "inverse" to $X(t)$: $Y(t)=X(0)-X(-t)$.

We will consider along with $Y(t)$ a sequence of governing processes $\{Y^{(n)}(t); t\geq 0\}$, $n=1, 2, \ldots$, with stationary increments and set $Y^{(n)}=\sup_{t\geq 0} Y^{(n)}(t)$, $\eta_k^{(n)}=Y^{(n)}(k+1)-Y^{(n)}(k)$, $k=0, 1, \ldots$.

Table 2. The values of $F_n^*(x_{4k})$ for the graphs on p. 60 and 62

	Second Method		Third Method	
$4k$	$n=3$	$n=6$	$n=3$	$n=6$
0	0.374	0.401	0.565	0.527
4	.388	.413	.569	.533
8	.398	.424	.578	.546
12	.407	.432	.597	.554
16	.410	.439	.608	.567
20	.416	.445	.613	.576
24	.422	.453	.626	.584
28	.433	.461	.636	.592
32	.440	.472	.644	.605
36	.450	.484	.655	.620
40	.460	.493	.663	.624
44	.471	.504	.673	.633
48	.476	.515	.689	.638
52	.481	.522	.704	.651
56	.488	.532	.710	.663
60	.501	.541	.721	.668
64	.508	.545	.730	.681
68	.522	.552	.641	.692
72	.529	.557	.751	.701
76	.533	.566	.761	.712
80	.545	.572	.771	.720
84	.553	.585	.778	.729
88	.563	.593	.786	.738
92	.570	.603	.794	.745
96	.578	.612	.807	.756
100	.587	.616	.819	.763
104	.597	.630	.824	.773
108	.606	.640	.832	.783
112	.615	.649	.837	.789
116	.624	.660	.841	.802
120	.633	.664	.849	.807
124	.649	.668	.862	.817
128	.663	.678	.875	.829
132	.678	.685	.883	.840
136	.694	.692	.891	.850
140	.701	.701	.894	.858
144	.715	.707	.899	.869
148	.729	.711	.906	.875
152	.737	.726	.910	.884
156	.754	.729	.919	.895
160	.763	.733	.933	.906
164	.780	.739	.939	.914
168	.792	.744	.948	.922
172	.811	.758	.956	.935
176	.827	.768	.967	.946
180	.849	.777	.975	.957
184	.869	.782	.984	.962
188	.888	.792	.989	.969
192	.911	.806	.996	.980
196	.945	.830	.999	.987

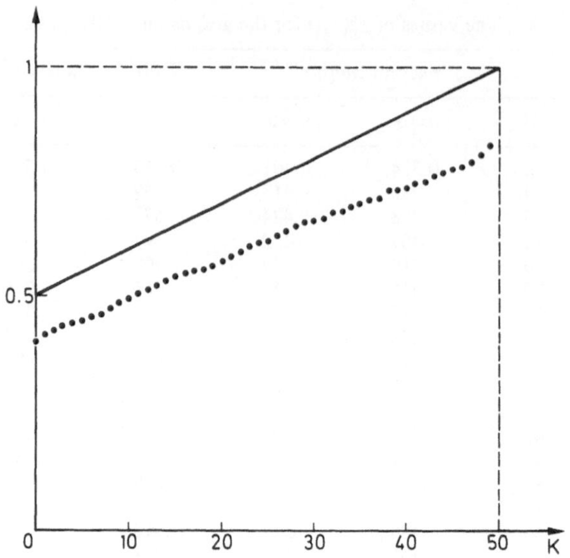

Fig. 8. $F(x_{4k})$ and $F_6^*(x_{4k})$. (Second method: $n=6$)

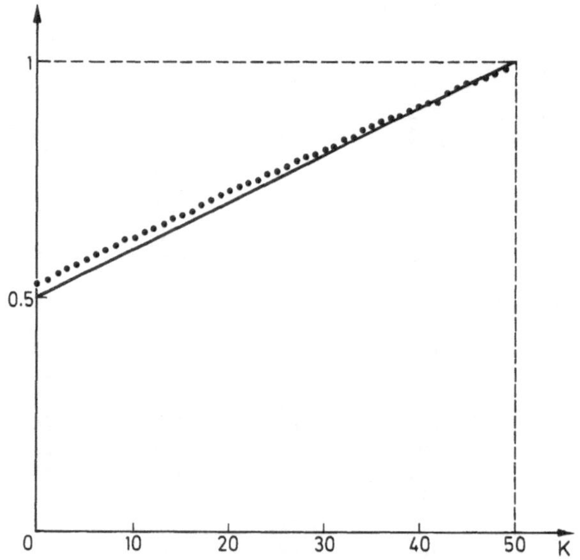

Fig. 9. $F(x_{4k})$ and $F_6^*(x_{4k})$. (Third method: $n=6$)

Theorem 23. *Assume that the finite-dimensional distributions of the processes* $Y^{(n)}(t)$ *converge weakly to the corresponding distributions of the process* $Y(t)$, $\mathsf{M}(Y(t+1)-Y(t))=a<0$. *Then if the sequence* $\eta_k=Y(k+1)-Y(k)$, $k=0, 1, \ldots$, *is ergodic it is sufficient for the convergence*

$$\mathsf{P}(\overline{Y}^{(n)}<t)\Rightarrow\mathsf{P}(\overline{Y}<t)$$

that

$$\mathsf{M}(\eta_1^{(n)};\eta_1^{(n)}\geqslant 0)\rightarrow\mathsf{M}(\eta_1;\eta_1\geqslant 0)\,.$$

The proof of this theorem causes no difficulty since it follows at once from Theorem 22 that

$$\mathsf{P}\left(\sup_{t \geqslant N} Y^{(n)}(t) \geqslant 0\right) \leqslant \mathsf{P}\left(\sup_{k \geqslant N} \sum_{j=0}^{k-1} \eta_j^{(n)} \geqslant -1\right) \rightarrow 0$$

for $N \rightarrow \infty$ uniformly in n. This means that the convergence of the distributions of the $\bar{Y}^{(n)}$ is guaranteed by the convergence of the distributions of the $Y^{(n)}(t)$ on the finite interval $(0, N)$ or, equivalently, by that of the finite-dimensional distributions of the $Y^{(n)}(t)$.

Chapter 2

Some Boundary Problems for Processes Continuous from below with Independent Increments. Their Connection with the Distribution of $w(t)$

We have already seen that the assumption that the governing process $X(t)$ (see § 6) be one with independent increments immediately simplifies the form of Beneš' equation and leads to compact formulae, such as, for example, the equality

$$\mathsf{P}(w^c(t) \geqslant x) = -ah(x),$$

where $h(x)$ is the renewal density of the process.

However, for such governing processes the distribution of $w(t)$ can be investigated much more completely with the result that one gets explicit formulae for all of the basic characteristics of the system.

The fact is that processes with independent increments which are *continuous from above* or *below* (see § 12), such as the processes $X(t)$ and $Y(t)$ of Chapter 1, occupy a special place. They are distinguished by the fact that the distribution of all their basic boundary functionals (i.e., functionals of the sample paths connected with the crossing by these paths of certain boundaries) can be expressed explicitly by means of the distribution of the values of the process at fixed moments of time.

The exposition of another approach to the investigation of boundary functionals which is more general (and applicable to sequences) is contained in Chapters 3 and 4.

§ 12. Boundary Problems for Processes Continuous from below with Independent Increments

1. A left-continuous homogeneous process with independent increments $\{Y(t), t \geqslant 0\}$, $Y(0)=0$, will be called *continuous from below* if

$$\mathsf{P}(Y(t+0) \geqslant Y(t) \text{ } \textit{for all } t) = 1 .$$

Because of the independence and stationarity of the increments,

$$\mathsf{M}\, e^{i\lambda Y(t)} = e^{t\psi(\lambda)},$$

where the Lévy-Hinčin representation holds for $\psi(\lambda)$:

$$\psi(\lambda)=im\lambda-\sigma^2\lambda^2+\int_{-\infty}^{\infty}\left(e^{i\lambda u}-1-\frac{i\lambda u}{1+u^2}\right)\frac{1+u^2}{u^2}G(du). \tag{1}$$

Here $G(B)$ is a finite measure on the real line ($G(0)=0$) and m and σ^2 are the coefficients of drift and diffusion, resp. When $\int u^{-2}G(du)<\infty$, the integral in (1) corresponds to the component which is a so-called compound Poisson process.

$$F(B)=\int_B \frac{1+u^2}{u^2}G(du)$$

defines (with accuracy up to the constant $\alpha=F(-\infty,\infty)$, called the *intensity* of the process) the jump distribution of this process.

If $\int u^{-1}G(du)<\infty$ and $\sigma^2=0$, then the sample paths have bounded variation w.p.1. In this case the representation at (1) can be written as

$$\psi(\lambda)=im_1\lambda+\int(e^{i\lambda u}-1)\frac{1+u^2}{u^2}G(du), \tag{2}$$

where $m_1=m-\int u^{-1}G(du)$.

It is not difficult to see that *for the continuity of $Y(t)$ from below it is necessary and sufficient that the process have no negative jumps. That is, that in* (1)

$$G((-\infty,0))=0.$$

In this case the integral in (1) converges uniformly in any closed domain in the half-plane $\operatorname{Im}\lambda>0$. Hence, the function $\psi(\lambda)$ is analytic in the domain $\operatorname{Im}\lambda>0$ and is also continuous on the boundary $\operatorname{Im}\lambda=0$. The function $\psi_1(\mu)=\psi(-i\mu)=\ln \mathsf{M}\, e^{\mu Y(1)}$, is easily seen to be convex on the segment $(-\infty,\mu_+]$ of the real axis ($0\leqslant\mu_+=\sup\{\mu\colon \psi_1(\mu)<\infty\}$), and

$$\psi_1'(0)=\mathsf{M}Y(1)=a, \qquad \psi_1''(0)=\mathsf{D}Y(1).$$

The Laplace transformation of the characteristic function of $Y(t)$ (the double transformation) is

$$\int_0^\infty e^{-pt}\mathsf{M}\, e^{i\lambda Y(t)}\,dt=\int_0^\infty e^{t\psi(\lambda)-pt}\,dt=\frac{1}{p-\psi(\lambda)},$$

$$\operatorname{Im}\lambda\geqslant 0, \qquad \operatorname{Re}p>\operatorname{Re}\psi(\lambda).$$

From Fig. 10 it is clear that the function $p-\psi(\lambda)$ has no more than two zeròs on the imaginary axis. The zero with greatest imaginary part is defined for all $p\geqslant\inf_\mu \psi_1(\mu)$. Denote it by $-i\mu(p)$ so that $\psi_1(\mu(p))\equiv p$, $\mu(p)<0$ for $p>0$.

2. We will now find the distribution of $\underline{Y}(t)=\inf_{0\leqslant u\leqslant t} Y(u)$ for processes which are continuous from below. Set

$$H(x,t)=\mathsf{P}(\underline{Y}(t)\leqslant x), \quad x<0,$$

and

$$F(x,t)=\mathsf{P}(Y(t)<x), \quad -\infty<x<\infty.$$

The function $H(x, t)$, as function of t, can also be interpreted as the distribution function $H(x, t)=\mathsf{P}(\eta(x)\leqslant t)$ of the first passage time $\eta(x)$ of the trajectories $Y(t)$ through the level $x\leqslant 0$:

$$\eta(x)=\sup\{t\colon \underline{Y}(t)>x\}=\inf\{t\colon \underline{Y}(t)\leqslant x\}.$$

As is clear from the definition, $\eta(x)$ is an improper random variable equal to ∞ on trajectories for which $\underline{Y}(\infty)>x$.

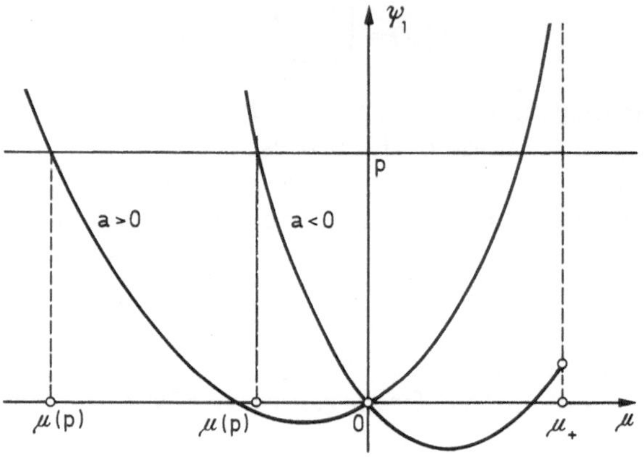

Fig. 10. The function ψ_1

Theorem 1. *For $x<0$ and $t>0$*

$$-\int_{-\infty}^{x}\frac{1}{y}H(y, t)\,dy=\int_{0}^{t}\frac{1}{u}F(x, u)\,du \tag{3}$$

and

$$\mathsf{M}\,(e^{-p\eta(x)};\,\eta(x)<\infty)=e^{-x\mu(p)}.$$

If the distribution of $Y(t)$ has a density $f(x, t)$ at the point x, then the distribution of $\eta(x)$ has a density $h(x, t)$ at the point t and (3) *can be written as*

$$h(x, t)=-\frac{x}{t}f(x, t). \tag{4}$$

Proof. In virtue of the continuity from below, an arbitrary path of $Y(t)$ attaining the level x also intersects any intermediate level between 0 and x w.p.1. Hence, for arbitrary x_1, $0\geqslant x_1\geqslant x$, the total probability formula holds in the form

$$H(x, t)=\textstyle\int_0^t d_uH(x_1, u)H(x-x_1, t-u). \tag{5}$$

Analogously,

$$F(x, t)=\textstyle\int_0^t d_uH(x_1, u)F(x-x_1, t-u).$$

Hence, setting for $\operatorname{Re} p>0$

$$\tilde{h}(x, p)=\textstyle\int_0^\infty e^{-pt}\,d_tH(x, t)\quad\text{and}\quad \tilde{F}(x, p)=\int_0^\infty F(x, t)\,e^{-pt}\,dt,$$

we find[1] that

$$\tilde{h}(x,p)=\tilde{h}(x_1,p)\tilde{h}(x-x_1,p)\,,\qquad \tilde{F}(x,p)=\tilde{h}(x_1,p)\tilde{F}(x-x_1,p)\,. \tag{7}$$

The first of these relations together with the monotonicity in x and boundedness of $\tilde{h}(x,p)$ for $p>0$ yields

$$\tilde{h}(x,p)=e^{xa(p)},\quad x\leqslant 0\,, \tag{8}$$

where $a(p)$ is some function regular for $\operatorname{Re} p>0$ and satisfying the inequality $a(p)>0$ for $p>0$. The second relation at (7) leads to

$$\tilde{F}(x,p)=\tilde{F}(0,p)\,e^{xa(p)},\quad x\leqslant 0\,.$$

We now consider for $p>0$ and $0\leqslant \operatorname{Im}\lambda\leqslant \operatorname{Re}[a(p)]$ the double transformation of $F(x,t)$:

$$\frac{1}{p-\psi(\lambda)}=\int_{-\infty}^{\infty} e^{i\lambda x}\,d_x\tilde{F}(x,p)=\int_{-\infty}^{0} e^{i\lambda x}a(p)\tilde{F}(0,p)\,e^{xa(p)}\,dx+\int_0^{\infty} e^{i\lambda x}\,d_x\tilde{F}(x,p)$$

$$=\frac{\tilde{F}(0,p)a(p)}{i\lambda+a(p)}+\int_0^{\infty} e^{i\lambda x}\,d_x\tilde{F}(x,p)\,. \tag{9}$$

The last integral is obviously an analytic function in the entire half-plane $\operatorname{Im}\lambda>0$. Hence, the right side of (9) is regular in the domain $\operatorname{Im}\lambda>0$ except at the point $\lambda=ia(p)$, where it has a simple pole with principal part $-i\tilde{F}(0,p)a(p)$. From Eq. (9) it follows that the function $(p-\psi(\lambda))^{-1}$ also enjoys this property. This means that $p-\psi(\lambda)$ has a single zero $\lambda=ia(p)=-i\mu(p)$ in the upper half-plane. The principal part of the pole of $(p-\psi(\lambda))^{-1}$ is

$$\lim_{\lambda\to -i\mu(p)}\frac{\lambda+i\mu(p)}{p-\psi(\lambda)}=-\frac{1}{\psi'(-i\mu(p))}=i\tilde{F}(0,p)\mu(p)\,.$$

Differentiating the identity $p-\psi(-i\mu(p))\equiv 0$, we get

$$1=-\psi'(-i\mu(p))i\mu'(p)\,,$$

so that

$$\tilde{F}(0,p)=\frac{i}{\psi'(-i\mu(p)\mu(p))}=\frac{\mu'(p)}{\mu(p)}\,,$$

[1] Concerning the existence of integrals of the functions $F(\cdot,t)$, we remark that $F(x,t)$, as a function of t, is always integrable in the sense of Riemann. Indeed, the following alternatives hold: (∗) *either $Y(t)$ is a compound Poisson process with drift or $F(x,t)$ is continuous in x.* In the first case the integrability follows from the explicit formula

$$F(x,t)=\sum_{k=0}^{\infty} e^{-\alpha t}\frac{(\alpha t)^k}{k!}\,\mathsf{P}(S_k<x-mt)\,,$$

where α is the intensity of the process and m is the drift. In the second it follows from the continuity of $F(x,t)$ in t which in turn follows from the stochastic continuity of $Y(t)$ ($Y(\Delta)\underset{P}{\to}0$ for $\Delta\to 0$) and the equation

$$F(x,t+\Delta)-F(x,t)=\int \mathsf{P}(Y(\Delta)\in du)(F(x-u,t)-F(x,t))\,. \tag{6}$$

The notation $\xi\in du$ used here is understood as inclusion in the interval $[u,u+du)$. This ambiguous notation for du is convenient and should lead to no misunderstanding.

and moreover,

$$\tilde{F}(x,p)=\frac{\mu'(p)}{\mu(p)}e^{-x\mu(p)}=-\int_{-\infty}^{x}\mu'(p)\,e^{-y\mu(p)}\,dy$$
$$=\int_{-\infty}^{x}\frac{1}{y}\frac{\partial}{\partial p}e^{-y\mu(p)}\,dy=\int_{-\infty}^{x}\frac{1}{y}\frac{\partial}{\partial p}\int_{0}^{\infty}e^{-pt}\,d_tH(y,t)\,dy$$
$$=-\int_{-\infty}^{x}\int_{0}^{\infty}\frac{t}{y}e^{-pt}\,d_tH(y,t)\,dy=-\int_{0}^{\infty}te^{-pt}\,d_t\int_{-\infty}^{x}\frac{H(y,t)}{y}\,dy. \tag{10}$$

The last integral is always finite. For example, from the equality

$$\int_0^\infty e^{-pt}\mathsf{M}\,e^{i\lambda \underline{Y}(t)}\,dt=-\frac{\mu(p)}{p(i\lambda-\mu(p))},$$

which results from (8), it follows that all moments of $\underline{Y}(t)$ are finite. In particular, $\int_{-\infty}^{0}H(y,t)\,dy<\infty$. The change of order of integration in (10) is legitimate because of the positivity of the integrand.

Hence, as is obvious from (10),

$$-t\frac{\partial}{\partial t}\int_{-\infty}^{x}\frac{H(y,t)}{y}\,dy=F(x,t)\quad\text{exists a.e.}$$

and this is equivalent to (3). □

Remark. The first statement of Theorem 1 is also necessary for the continuity from below of $Y(t)$ (see [54]).

Corollary. *For $x<0$*

$$\mathsf{P}(\eta(x)<\infty)=e^{-x\mu(0)},$$

$$\mathsf{M}(\eta(x)\mid\eta(x)<\infty)=\frac{x}{\psi_1'(\mu(0))}$$

and

$$\mathsf{M}(\eta^2(x)\mid\eta(x)<\infty)=\frac{x\psi_1''(\mu(0))}{[\psi_1'(\mu(0))]^3}+\left[\frac{x}{\psi_1'(\mu(0))}\right]^2.$$

If $\mathsf{M}\,Y(1)=a<0$ exists, then these relations reduce to

$$\mathsf{P}(\eta(x)<\infty)=1,\qquad \mathsf{M}\eta(x)=\frac{x}{a}\quad\textit{and}\quad \mathsf{D}\eta(x)=\frac{x\mathsf{D}\,Y(1)}{a^3}.$$

3. Equation (5) says that $\{\eta_1(y)=\eta(-y);\,y\geqslant 0\}$ is also a *homogeneous process with independent increments.* However, in contrast to $Y(t)$, it can be improper, since

$$\mathsf{P}(\eta_1(y)<\infty)=e^{y\mu(0)}<1$$

for $\mu(0)<0$ (this always holds if $a>0$). Considering the increments $\Delta\eta_1(y)$ of the process $\eta_1(y)$ during the "time" interval Δy, we see that either $\eta_1(y)$ goes to ∞

during this interval with probability

$$-\mu(0)\,\Delta y+o(\mu(0)\,\Delta y)\,,$$

or with the complementary probability the increment $\Delta\eta_1(y)$ is a random variable independent of the previous history with the distribution $\mathsf{P}(\eta_1(\Delta y)<x \mid \eta_1(\Delta y)<\infty)$. This says that $\eta_1(y)$ can be represented as the sum

$$\eta_1(y)=Z(y)+B(y)$$

of two independent processes: an ordinary homogeneous process with independent increments $\{Z(y); y\geqslant 0\}$ with

$$\mathsf{M}\, e^{i\lambda Z(y)}=\mathsf{M}(e^{i\lambda\eta_1(y)} \mid \eta_1(y)<\infty)\,, \tag{11}$$

and a Poisson process $B(y)$ with an infinite jump and intensity $-\mu(0)$ (so that $\mathsf{P}(B(y)=0)=e^{y\mu(0)}$), which realizes an exit to ∞ or a "break" of the process $\eta_1(y)$.

The paths of the process $Z(t)$ are obviously monotone and consequently, they have bounded variation w.p.1. This means that the characteristic function of $Z(t)$ has a representation of the form (compare with (2))

$$\mathsf{M}\, e^{i\lambda Z(y)}=\exp\left\{yi\lambda m_Z+y\int_0^\infty (e^{iu\lambda}-1)\frac{1}{u}\,G_Z(du)\right\}, \tag{12}$$

where $G_Z(du)$ is a measure on $(0,\infty)$ for which

$$\int\frac{G_Z(du)}{1+u^2}<\infty \quad\text{and}\quad \int\frac{u}{1+u^2}\,G_Z(du)<\infty\,.$$

On the other hand, by (11) and Theorem 1

$$\mathsf{M}\, e^{-pZ(y)}=e^{y(\mu(p)-\mu(0))}. \tag{13}$$

This allows us to find the values of m_Z and G_Z in two important special cases. Let $Y(t)$ first have a continuous density

$$f(x,t)=\frac{\partial}{\partial x}F(x,t)\,.$$

Then $f(x,t)$ is also continuous in t (see (6)) and

$$\tilde f(x,p)=\int_0^\infty f(x,t)\,e^{-pt}\,dt=\frac{\partial}{\partial x}\tilde F(x,p)=-\mu'(p)\,e^{-x\mu(p)}\,,$$

whence $\mu'(p)=-\tilde f(0,p)$. This yields

$$\mu(p)-\mu(0)=-\int_0^p \tilde f(0,v)\,dv=-\int_0^p\int_0^\infty e^{-vt}f(0,t)\,dt\,dv=\int_0^\infty (e^{-pt}-1)\frac{f(0,t)}{t}\,dt\,.$$

Comparing this with (12) and (13), we find that in this case

$$m_Z=0 \quad\text{and}\quad G_Z(du)=f(0,u)\,du \tag{14}$$

with

$$\int_0^1 f(0,u)\,du<\infty \quad\text{and}\quad \int_1^\infty\frac{f(0,u)}{u}\,du<\infty\,.$$

Therefore, $Z(y)$ is a pure jump process (without drift or diffusion) with spectral density $f(0, u)/u$. We necessarily have $\int_0^1 f(0, u)/u\,du = \infty$ since otherwise $Z(y)$ would be a Poisson process without drift, which is impossible because of the continuity from below of $Y(t)$.

Formulae (14) make it natural to assume that in the general case in (12)

$$m_Z = 0 \quad \text{and} \quad G_Z(du) = \mathsf{P}(Y(u) \in [0, du)) . \tag{15}$$

But here the question arises of how to interpret the expression on the right in (15) and whether or not it is a measure.

We now consider the first of the alternatives (*) in the footnote on p. 67 in which $Y(t)$ is a compound Poisson process with drift. Here, in the representation

$$\psi(\lambda) = im\lambda + \alpha \int_0^\infty (e^{i\lambda u} - 1) F(du) , \tag{16}$$

$m < 0$, $F(B)$ is a probability measure (the jump distribution of the process) and α is the intensity of the process. It is evident that $(1/u)G_Z(du)$ in (12) will be a finite measure since the number of jumps of $Z(y)$ during a finite interval is finite w.p.1.

We will assume without loss of generality that $Y(t)$ has drift $m = -1$. We have

$$\mu'(p) = -\frac{\partial}{\partial x} \tilde{F}(x, p) \,|_{x=0} ,$$

$$\begin{aligned} \mu(p) - \mu(0) &= -\int_0^p \frac{\partial}{\partial x} \tilde{F}(x, v) \,|_{x=0}\, dv \\ &= -\frac{\partial}{\partial x} \int_0^p \int_0^\infty F(x, t)\, e^{-vt}\, dt\, dv \,|_{x=0} = \frac{\partial}{\partial x} \int_0^\infty F(x, t) \frac{e^{-pt} - 1}{t}\, dt \,|_{x=0} . \end{aligned} \tag{17}$$

By the formula of total probability

$$F(x, t) = \sum_{k=0}^\infty e^{-\alpha t} \frac{(\alpha t)^k}{k!} \mathsf{P}(\Sigma_k < x + t) ,$$

where Σ_k is the sum of the first k jumps of $Y(t)$. Using the representation

$$\mathsf{P}(\Sigma_k < x + t) = \int \mathsf{P}(\Sigma_k \in du)\, \delta(x + t - u) ,$$

where

$$\delta(x) = \begin{cases} 1, & x > 0 , \\ 0, & x \leqslant 0 , \end{cases}$$

we can write (17) as

$$\begin{aligned} &\frac{\partial}{\partial x} \int \sum_{k=0}^\infty \mathsf{P}(\Sigma_k \in du) \int_0^\infty \frac{e^{-pt} - 1}{t} \frac{e^{-\alpha t}(\alpha t)^k}{k!} \delta(x + t - u)\, dt \,|_{x=0} \\ &\quad = \frac{\partial}{\partial x} \int \sum_{k=0}^\infty \mathsf{P}(\Sigma_k \in du) \int_{u-x}^\infty \frac{e^{-pt} - 1}{t} \frac{e^{-\alpha t}(\alpha t)^k}{k!}\, dt \,|_{x=0} . \end{aligned} \tag{18}$$

The interchange of the order of integration and summation in these formulae is legitimate due to the nonnegativity of the integrands. But in (18) the last integral is differentiable w.r.t. x and the result of differentiation is integrable uniformly w.r.t. x. Hence,

$$\mu(p)-\mu(0)=\int \sum_{k=0}^{\infty} \mathsf{P}(\Sigma_k \in du)\frac{e^{-pu}-1}{u}\frac{e^{-\alpha u}(\alpha u)^k}{k!}$$

$$=-p+\int (e^{-pu}-1)\frac{1}{u}\sum_{k=1}^{\infty}\frac{e^{-\alpha x}(\alpha u)^k}{k!}\mathsf{P}(\Sigma_k \in du). \quad (19)$$

It is obviously natural to interpret the sum appearing in (19) as $\mathsf{P}(Y(v)\in[0, dv)$; *the number of jumps on* $[0, v)\geqslant 1)$ and the entire right side of (19) can be understood as

$$\mu(p)-\mu(0)=\ln \mathsf{M}\, e^{-pZ(1)}=\int_{[0,\infty)}(e^{-pv}-1)\frac{1}{v}\mathsf{P}(Y(v)\in[0, dv)).$$

In this sense, the hypothesis of (15) turns out to be true. More precisely, by (19) $Z(y)$ has here (in the notation of (12)) drift $m_Z=1$ and spectral measure, concentrated on $(0, \infty)$, equal to

$$G_Z(dv)=\mathsf{P}(Y(v)\in[0, dv)),$$

where the right side is to be understood as the sum on the right in (19).

We can now summarize what has been said in

Theorem 2. *The process* $\{\eta_1(y)=\eta(-y), y\geqslant 0\}$ *can be represented as the sum of two independent processes:*

$$\eta_1(y)=Z(y)+B(y),$$

where $B(y)$ *is a Poisson process exiting to* ∞ *with intensity* $-\mu(0)$:

$$\mathsf{M}\, e^{i\lambda B(y)}=e^{\mu(0)y}, \qquad \operatorname{Im}\lambda>0.$$

If $\mu(0)=0$, *then* $B(y)\equiv 0$. $Z(y)$ *is a homogeneous, monotone process with independent increments and characteristic function*

$$\mathsf{M}\, e^{i\lambda Z(y)}=\exp\left\{i\lambda y m_Z+y\int_{+0}^{\infty}(e^{i\lambda u}-1)\frac{1}{u}G_Z(du)\right\}, \quad (20)$$

where $G_Z(du)/(1+u)$ *is a finite measure.*

If $Y(t)$ *has a continuous density* $f(x, t)$, *then*

$$m_Z=0,\ G_Z(du)=f(0, u)\,du \quad \text{and} \quad \int\frac{f(0, u)}{u}du=\infty.$$

If $Y(t)$ *is a compound Poisson process with drift, which can be assumed equal to* -1 *without loss of generality, then in* (20) *we have*

$$m_Z=1, \qquad G_Z(du)=\mathsf{P}(Y(u)\in(0, du)); \qquad \alpha_Z=\int_{+0}^{\infty}\frac{G_Z(du)}{u}<\infty. \quad (21)$$

Hence, in this case, $Z(t)$ is also a compound Poisson process with drift. The right side of the middle equation at (21) is to be understood as the sum on the right side of (19). If $Y(u)$ has a density $f(x, u)$ in the segment $x \in (-u, \infty)$ (i.e., if the jumps of $Y(u)$ have a density), then

$$G_Z(du) = f(0, u)\, du\,.$$

We remark that for the Poisson process under consideration the measure $\mathsf{P}(Y(t) \in [x, x+dt))$ can also be understood as in (19) for arbitrary x. Then, as is easily seen, the assertion of Theorem 1 can be written as

$$\mathsf{P}(\eta(x) \in dt) = -\frac{x}{t}\,\mathsf{P}(Y(t) \in [x, x+dt))\,.$$

We also mention a useful relation for the spectral function of the process $Z(t)$.

Corollary. *If $Y(t)$ is the compound Poisson process at* (16), *where $m=-1$ and*

$$f_1(\mu) = \int e^{\mu u}\, dF(u)\,,$$

then the Laplace transformations for the jump variables ξ_η and ξ_Z of the processes $\eta_1(y)$ and $Z(y)$ are equal to

$$\mathsf{M}(e^{-p\xi_\eta}; \xi_\eta < \infty) = f_1(\mu(p))$$

and

$$\mathsf{M}\, e^{-p\xi_Z} = \frac{1}{\alpha_Z}\int_{+0}^{\infty} e^{-pu}\frac{G_Z(du)}{u} = \frac{f_1(\mu(p))}{f_1(\mu(0))}\,, \tag{22}$$

$$\alpha_Z = \alpha + \mu(0)\,.$$

Proof. If $\tilde{y}$ is the instant of a jump of the process $\eta_1(u)$, then $\tilde{t} = \eta_1(\tilde{y}-0)$ is the instant of the jump of $Y(t)$. But $\eta_1(\tilde{y}+0) - \eta_1(\tilde{y}-0)$ is the time of return of the process $Y(t)$, $t > \tilde{t}$, to the position $Y(\tilde{t}-0)$. This means that $\eta_1(\tilde{y}+0) - \eta_1(\tilde{y}-0)$ is distributed like the time $\eta(-\xi_Y)$, where ξ_Y is a variable independent of $\{Y(t)\}$ with the same distribution $F(B)$ as the jump distribution of the process $Y(t)$:

$$\begin{gathered}\mathsf{P}(\xi_\eta < x) = \int F(du)\,\mathsf{P}(\eta(-u) < x)\,,\\ \mathsf{M}(e^{-p\xi_\eta}; \xi_\eta < \infty) = \int F(du)\, e^{u\mu(p)} = f_1(\mu(p))\,.\end{gathered} \tag{23}$$

The conditional distribution of ξ_η, given that $\xi_\eta < \infty$, obviously coincides with the distribution of ξ_Z:

$$\mathsf{M}\, e^{-p\xi_Z} = \frac{f_1(\mu(p))}{f_1(\mu(0))}\,.$$

By (13) we obtain

$$\ln \mathsf{M}\, e^{-pZ(1)} = -p + \alpha_Z\left(\frac{f_1(\mu(p))}{f_1(\mu(0))} - 1\right) = \mu(p) - \mu(0)\,.$$

Taking into account the fact that

$$\psi_1(\mu) = -\mu + \alpha(f_1(\mu) - 1), \psi_1(\mu(p)) = p, \tag{24}$$

we find that $\alpha_Z = \alpha + \mu(0)$.

The assertion of the corollary can also be obtained using only (13). But it seems to us that (23) is more intuitive. □

We mention that in the important special case $\mu(0) = 0$ the intensity α_Z of the process $Z(y) = \eta(-y)$ coincides with the intensity α of the process $Y(t)$ (which is easy to establish directly), and

$$\mathsf{M}\, e^{-p\xi_Z} = f_1(\mu(p)) = \frac{1}{\alpha} \int_0^\infty e^{-pu} \frac{\mathsf{P}(Y(u) \in [0, du))}{u}. \tag{25}$$

We also remark that by (24) $z(p) \equiv f_1(\mu(p))$ satisfies the equations

$$z = 1 + \frac{\mu(p) + p}{\alpha} \quad \text{and} \quad z = f_1(\alpha(z-1) - p). \tag{26}$$

Using the results obtained, we can also find the joint distribution of $(\underline{Y}(t), \theta_t^*)$, $(\underline{Y}, \theta^*)$, etc., where $\underline{Y} = \underline{Y}(\infty)$ and θ_t^* and θ^* are the first times $Y(u)$ attains inf $Y(u)$ in the intervals $(0, t)$ and $(0, \infty)$, resp. For example, if $Y(t)$ is defined by (16) and has the density $f(x, t)$ at the point x, then (see (22), (26))

$$\begin{aligned}\mathsf{P}\{\underline{Y} \in dx, \theta^* \in dt\} &= \mathsf{P}(\eta(x) \in dt\} \alpha\, dx (1 - f_1(\mu(0)))\\ &= -\frac{x}{t} f(x, t)\, dt\, dx \alpha \left(-\frac{\mu(0)}{\alpha}\right) = \mu(0) \frac{x}{t} f(x. t)\, dt\, dx.\end{aligned} \tag{27}$$

The first equality at (27) allows us for $\mu(0) < 0, p \geqslant 0$ and $\operatorname{Im} \lambda < -\mu(p)$ to write

$$\mathsf{M}\, e^{i\lambda \underline{Y} - p\theta^*} = -\int_{-\infty}^0 e^{i\lambda x} \mu(0)\, e^{-x\mu(p)}\, dx = \frac{-\mu(0)}{i\lambda - \mu(p)}.$$

We mention another identity which can be useful. Let $m = -1$ and suppose $F(u)$ has a density. Then, using the form of the spectral function of the process Z and the total probability formula, we get

$$\frac{\mathsf{P}(\xi_Z \in dt)}{dt} = \frac{f(0, t)}{t(\alpha + \mu(0))} = \frac{1}{f_1(\mu(0))} \left[\int_0^t F(du) \frac{u}{t} f(-u, t) + \frac{F(dt)}{dt} e^{-\alpha t}\right].$$

Since

$$f_1(\mu(0)) = 1 + \frac{\mu(0)}{\alpha} = \frac{\alpha + \mu(0)}{\alpha},$$

we find that

$$\alpha^{-1} f(0, t) = \int_0^t dF(u) u f(-u, t) + \frac{dF(t)}{dt} t e^{-\alpha t}. \quad □$$

4. We determine finally *the joint distribution of the random variables* $\overline{Y}(t) = \sup_{0 \leqslant u \leqslant t} Y(u)$ *and* $Y(t)$. Let

$$P\{x, y, t\} = \mathsf{P}\{\overline{Y}(t) \geqslant x, Y(t) < x - y\}$$

and

$$\tilde{P}(x, y, p) = \int_0^\infty e^{-pt} P(x, y, t)\, dt.$$

Theorem 3. *If $Y(t)$ is continuous from below, then*

$$P(x,y,t)=-\frac{\partial}{\partial x}\int_0^t \mathsf{M}\left(\frac{Y(u)}{u};\ Y(u)<-y\right)F(x,t-u)\,du\,, \tag{28}$$

or, what is the same,

$$\tilde{P}(x,y,p)=-\frac{\partial}{\partial x}\frac{\tilde{F}(p,x)\,e^{y\mu(p)}}{\mu(p)}, \quad x>0,\ y\geqslant 0,\ \operatorname{Re}p>0\,.$$

Proof. We assume initially that $\sigma^2>0$ in the representation (1) so that for each $t>0$ there exists a smooth density $f(x,t)$. Since $Y(t)$ is continuous from below, any of its trajectories passing into the region $Y\geqslant x$ must pass at least once through the level x on the way to the point $z<x$. Denote by $\rho(x)$ the time of the first passage "from above" of the level $x>0$ (obviously, this random variable will be defined on not all of Ω). Then by the total probability formula

$$P(x,y,t)=\textstyle\int_0^t d\mathsf{P}(\rho(x)<u)\mathsf{P}(Y(t-u)<-y)\,,$$

$$f(x,t)=\textstyle\int_0^t d\mathsf{P}(\rho(x)<u)f(0,t-u)\,.$$

Going to Laplace transformations, we find

$$\tilde{P}(x,y,p)=\frac{\tilde{f}(x,p)}{\tilde{f}(0,p)}\textstyle\int_{-\infty}^{-y}\tilde{f}(z,p)\,dz=\tilde{f}(x,p)\int_{-\infty}^{-y}e^{-z\mu(p)}\,dz=-\dfrac{\tilde{f}(x,p)}{\mu(p)}e^{y\mu(p)}\,.$$

This expression can also be written as

$$\tilde{f}(x,p)\int_{-\infty}^{-y}\int_0^{\infty}-\frac{z}{t}f(z,t)\,e^{-pt}\,dt\,dz=-\tilde{f}(x,p)\int_0^{\infty}\mathsf{M}\left(\frac{Y(t)}{t};\ Y(t)<-y\right)e^{-pt}\,dt\,, \tag{29}$$

which is equivalent to the first assertion of the theorem. Now let $\sigma^2=0$ in (1). Taking the symbol σ as some other positive number, we denote by $\omega_\sigma(t)$ a Wiener process with parameters

$$\mathsf{M}\omega_\sigma(t)=0 \quad\text{and}\quad \mathsf{M}\omega_\sigma^2(t)=\sigma^2 t\,.$$

From the equality ($\eta(\varepsilon)=\inf\{t\colon \omega_\sigma(t)\geqslant\varepsilon\}$)

$$\mathsf{P}(\omega_\sigma(t)>\varepsilon)=\textstyle\int_0^t d\mathsf{P}(\eta(\varepsilon)<u)\frac{1}{2}=\frac{1}{2}\mathsf{P}\big(\sup\limits_{0\leqslant u\leqslant\varepsilon}\omega_\sigma(u)>\varepsilon\big)$$

it follows that for some $\varepsilon>0$

$$\mathsf{P}(|\omega_\sigma(t)|>\varepsilon)\to 0 \quad\text{and}\quad \mathsf{P}\big(\sup_{u\leqslant t}|\omega_\sigma(u)|>\varepsilon\big)\to 0$$

for $\sigma\to 0$. Consequently, if P_σ, M_σ and F_σ denote the functions appearing in (28) and corresponding to the process $Y_\sigma(t)=Y(t)+\omega_\sigma(t)$, where ω_σ and Y are independent, we find that at each point of continuity (x,y) of $P(x,y,t)$

$$P_\sigma(x,y,t)\to P(x,y,t)\,.$$

as $\sigma\to 0$.

Analogous relations will hold for M_σ and F_σ.

Using now the equality

$$\int_{x_0}^{x} P_\sigma(x, y, t)\,dx = \int_0^t \mathsf{M}\left(\frac{Y_\sigma(u)}{u};\ Y_\sigma(u) < -y\right)(F_\sigma(x_0, t-u) - F_\sigma(x, t-u))\,du$$

and well-known theorems on the convergence of integrals, we obtain (28). □

It is not difficult to find the double transformation of $P(x, y, t)$:

$$\begin{aligned}\int_0^\infty e^{i\lambda x}\tilde{P}(x, y, p)\,dx &= \frac{-e^{y\mu(p)}}{\mu(p)}\int_0^\infty e^{i\lambda x}\tilde{f}(x, p)\,dx\\ &= -\frac{e^{y\mu(p)}}{\mu(p)}\left[\frac{1}{p-\psi(\lambda)} + \int_{-\infty}^{0} e^{i\lambda x}\mu'(p)\,e^{-x\mu(p)}\,dx\right]\\ &= \frac{-e^{y\mu(p)}}{\mu(p)}\left[\frac{1}{p-\psi(\lambda)} + \frac{\mu'(p)}{i\lambda-\mu(p)}\right]\\ &= -\frac{e^{-y\mu(p)}}{\mu(p)}\left[\frac{1}{p-\psi(\lambda)} + \frac{1}{\psi'(-i\mu(p))(\lambda+i\mu(p))}\right].\end{aligned}$$

In the square brackets stands the function $1/(p-\psi(\lambda))$, from which is subtracted the principal part of its only singularity in the upper half-plane.

Formulae similar to (28) can also be derived for other distributions. For example, using the fact that the "inverse" process

$$Y_t(v) = Y(t) - Y(t-v)$$

is distributed on $[0, t]$ like $Y(v)$, we can write in the case of (16) ($\overline{Y} = \overline{Y}(\infty)$ and $\theta = \theta_\infty$, where $\theta_t = \inf(u\colon \overline{Y}(u) = \overline{Y}(t))$)

$$\mathsf{P}(\overline{Y} \in dx,\ \theta \in dt) = \alpha\,dt\,\mathsf{P}(\overline{Y} = 0)\int_0^\infty dF(y)\,\mathsf{P}(\underline{Y}(t) > -y,\ Y(t) \in dx - y),$$

where the last factor can easily be expressed in terms of the function $F(x, t)$ using Theorem 1. Obviously, analogous formulae can also be obtained for the joint distribution of $(\overline{Y}(t), \theta_t)$ for finite t.

§ 13. Properties of the Distribution of $w(t)$. The Busy Period

1. We turn now to the study of the distribution of the virtual waiting time $w(t)$. In § 6 of Chapter 1 we proved that

$$w(t) = \sup_{0 \leqslant u \leqslant t}(X(t) - X(0) + w(0),\ X(t) - X(u)),$$

where $X(t)$ is the governing process, $X(0) = w(0)$. As already remarked, for $X(t)$ to be a homogeneous process with independent increments it is necessary and sufficient that the time intervals $\tau_1^e, \tau_2^e, \ldots$ be independent and exponentially distributed according to the law $\mathsf{P}(\tau_j^e \geqslant x) = e^{-\alpha x}$ and that the random variables S_i (see (3), § 2) be independent and identically distributed. This will be the case, for example, in the system $\langle E, G_I, G_I, 1\rangle$. Suppose that

$$\mathsf{P}(S_i<x)=F(x) \quad \text{and} \quad f(\lambda)=\mathsf{M}\exp(i\lambda S_i)\,.$$

Then, if $w(0)$ is independent of $\{X(t)-X(0)\}$ we have

$$\mathsf{M}\,e^{i\lambda X(t)}=\mathsf{M}\,e^{i\lambda w(0)}\exp\{t[-i\lambda+\alpha(f(\lambda)-1)]\}\,,$$

so that the process $X(t)-X(0)$ is a *continuous from below, compound Poisson process with drift* -1, *intensity* α *and jump distribution* $F(x)$.

Obviously, the process $Y(t)=X(0)-X(-t)$ will also have these properties. In terms of this process we can write

$$w(t)\underset{d}{=}\sup_{0\leqslant u\leqslant t}(Y(t)+w(0),\, Y(u))\,. \tag{30}$$

In the sequel we will assume for simplicity that $w(0)\geqslant 0$ is nonrandom. However, all of the results formulated above remain valid (with the obvious changes) for random $w(0)$ not depending on $\{Y(t)\}$.

Theorem 4.

$$\mathsf{P}(w(t)<x)=F(x-w(0),t)+\frac{\partial}{\partial x}\int_0^t \mathsf{M}\left(\frac{Y(u)}{u};\, Y(u)<-w(0)\right)F(x,t-u)\,du\,, \tag{31}$$

$$\mathsf{P}(w(t)=0)=-\mathsf{M}\left(\frac{Y(t)}{t},\, Y(t)<-w(0)\right),$$

$$\textstyle\int_0^\infty e^{-pt}\,\mathsf{P}(w(t)=0)\,dt=-\dfrac{e^{w(0)\mu(p)}}{\mu(p)}\,.$$

These relations evidently yield the solution of Beneš' equation. In particular, the last equalities determine the solution of Eq. (48) §7 for the function $p(u)$ in §7, Chapter 1.

Proof. The first assertion follows immediately from a comparison of the equality

$$\mathsf{P}(w(t)<x)=\mathsf{P}(Y(t)+w(0)<x)-\mathsf{P}(\overline{Y}(t)\geqslant x,\, Y(t)+w(0)<x)\,,$$

which is a consequence of (30), with Theorem 3.

Comparing (31) with Beneš' equation we obtain

$$-\int_0^t \mathsf{M}\left(\frac{Y(u)}{u};\, Y(u)<-w(0)\right)F(x,t-u)\,du=\textstyle\int_0^t \mathsf{P}(w(u)=0)F(x,t-u)\,du\,.$$

Since $\tilde{F}(x,p)\neq 0$ for $p>0$, we obtain from this the second claim of the theorem. The proof of the third follows by comparing Eq. (29) in the proof of Theorem 3 with the relations preceding that equality. $\square$

2. From Theorem 4 we can easily obtain the following formulae (see Corollary 3 § 8), already known in the case $\mathsf{M}Y(1)=a<0$:

$$\mathsf{P}(w^c(t)=0)=\lim_{t\to\infty}\mathsf{P}(w(t)=0)=-a\,,$$

$$\mathsf{P}(w^c(t)\geqslant x)=-a\frac{\partial}{\partial x}\int_0^\infty \mathsf{P}(0\leqslant Y(t)\leqslant x)\,dt=-ah(x)\,,$$

where $h(x)$ is the renewal density for the process $Y(t)$, and

$$\mathsf{M}\exp(i\lambda w^c(t))=\frac{1-\alpha\mathsf{M}S}{1-\alpha\mathsf{M}S\dfrac{f(\lambda)-1}{i\lambda\mathsf{M}S}}=-a\sum_{k=0}^\infty(\alpha\mathsf{M}S)^k\left(\frac{f(\lambda)-1}{i\lambda\mathsf{M}S}\right)^k.$$

Since $(f(\lambda)-1)/i\lambda\mathsf{M}S$ is the characteristic function of the distribution with density $\varphi(x)=\mathsf{P}(S\geqslant x)/\mathsf{M}S$, the last formula implies, in particular, that the distribution of $w^c(t)$ has a discontinuity at the point $x=0$ equal to $-a$ and is continuously differentiable in the interval $(0,\infty)$, with

$$\frac{\partial\mathsf{P}(w^c(t)<x)}{\partial x}=-a\sum_{k=1}^\infty(1+a)^k\varphi^{(k)}(x)\,, \tag{32}$$

where $\varphi^{(k)}$ is the k-fold convolution of φ. If $(1+a)$ is small, then (32) provides an effective means of calculating the distribution of $w^c(t)$. For light traffic, when $a\to-1$, the density of $w^c(t)$ is equivalent to $(a+1)\varphi(x)$.

From (32) it also follows that

$$\kappa(w^c(t))=\kappa(S)-1\,, \tag{33}$$

where $\kappa(\xi)=\sup\{\gamma\colon \mathsf{M}\xi^\gamma<\infty\}$. Indeed, by Minkowski's inequality for identically distributed variables and $\gamma\geqslant1$

$$\mathsf{M}\,|\xi_1+\cdots+\xi_k|^\gamma\leqslant k^\gamma\mathsf{M}\,|\xi_1|^\gamma\,.$$

Thus,

$$\mathsf{M}(w^c(t))^\gamma\leqslant-\frac{a}{\mathsf{M}S}\sum_1^\infty(1+a)^k k^\gamma\int\mathsf{P}(S\geqslant x)x^\gamma\,dx=-\frac{a\mathsf{M}S^{1+\gamma}}{\mathsf{M}S(1+\gamma)}\sum_1^\infty(1+a)^k k^\gamma\,;$$

$$\mathsf{M}(w^c(t))^\gamma\geqslant-\frac{a\mathsf{M}S^{1+\gamma}(1+a)}{\mathsf{M}S(1+\gamma)}\,.$$

For $\gamma<1$ an analogous relation can be obtained by using the fact that in this case $\mathsf{M}\,|\xi_1+\cdots+\xi_k|^\gamma\leqslant k\mathsf{M}\,|\xi_1|^\gamma$.

We now calculate formulae for the first two moments, which follow from (32):

$$\mathsf{M}w^c(t)=-\frac{1+a}{a}\frac{\mathsf{M}S^2}{2\mathsf{M}S}\,;\qquad \mathsf{D}w^c(t)=-\frac{1+a}{a}\left(\frac{\mathsf{M}S^3}{3\mathsf{M}S}-\frac{1+a}{a}\left(\frac{\mathsf{M}S^2}{2\mathsf{M}S}\right)^2\right).$$

3. Theorem 4 with well-known limit theorem allows us to obtain without great difficulty estimates of the asymptotic behavior of $\mathsf{P}(w(t)<x)$ as $t\to\infty$. In this connection an important rôle is played by asymptotic estimates of the probability $\mathsf{P}(w(t)=0)$.

Consider, for example, the case $a=0$, $\mathsf{D}Y(1)=\sigma^2<\infty$ ($\mathsf{M}S^2<\infty$). We have

$$-\mathsf{M}(Y(t),\, Y(t)<-y)=yF(t,-y)-\int_{-y}^{0}F(t,x)\,dx+\int_{-\infty}^{0}F(t,x)\,dx\,.$$

Denote the sum of the first two terms on the right by $m(t,y)$. Since $F(t,x\sigma\sqrt{t})$ converges to $\Phi(x)$ (the normal law) along with its first two moments by the central limit theorem, we have for $y=o(\sqrt{t})$ (in particular, for each fixed y)

$$m(t,y)=o(y)\quad\text{and}\quad \int_{-\infty}^{0}F(t,x)\,dx\sim\sigma\sqrt{t}\int_{-\infty}^{0}\Phi(x)\,dx=\frac{\sigma\sqrt{t}}{\sqrt{2\pi}}\,.$$

Hence, for $w(0)=o(\sqrt{t})$ we get

$$\mathsf{P}(w(t)=0)=\frac{\sigma}{\sqrt{2\pi t}}+o\left(\frac{1}{\sqrt{t}}\right)+o\left(\frac{w(0)}{t}\right).$$

If $\mathsf{M}\,|Y(1)|^k<\infty$ for $k>2$, we can obtain more precise results, for example, in the form of asymptotic expansions. Thus, if $\mathsf{M}\,|Y(1)|^3<\infty$, then using well-known refinements of the central limit theorem (Feller [25] Ch. 16; Ibragimov and Linnik [34], p. 146) we get for $w(0)<c\sqrt{t}$

$$\begin{aligned}\mathsf{P}(w(t)=0)&=\frac{\sigma}{\sqrt{2\pi t}}+o\left(\frac{1}{t}\right)+\frac{\sigma}{\sqrt{t}}\int_{-w(0)/\sigma\sqrt{t}}^{0}\left[\Phi\left(-\frac{w(0)}{\sigma\sqrt{t}}\right)-\Phi(u)\right]du+O\left(\frac{w^2(0)}{t^2}\right)\\&=\frac{\sigma}{\sqrt{2\pi t}}+o\left(\frac{1}{t}\right)+O\left(\frac{w^2(0)}{t^{3/2}}\right).\end{aligned}$$

Using these representations we can also study the behavior of $\mathsf{P}(w(t)<x)$ in a way analogous to our investigation of the distribution of $w(t)$ for the systems with heavy traffic in §8.

If a double sequence is introduced we can also consider the "transient regime" in which $\mathsf{M}Y(1)=c/\sqrt{t}$ depends on t. In this case the asymptotic behavior of $\mathsf{P}(w(t)=0)$ will, generally speaking, be different ($w(0)=o(\sqrt{t})$):

$$\mathsf{P}(w(t)=0)\sim\frac{\sigma\sqrt{t}}{t}\int_{-\infty}^{c/\sigma}\Phi(x)\,dx\,.$$

We will return later to the study of the transient regime in our investigation of the properties of the sequence w_n (recall that the distributions of $w^c(t)$ and w^k coincide under the conditions of this section since τ_j^e is exponentially distributed).

The estimation of the rate of convergence of $\mathsf{P}(w(t)=0)$ to its limit in the case where $a\neq 0$ is fixed obviously leads to the estimation of the probability of large deviations for $Y(t)$. If $\mathsf{M}\,e^{\mu S}<\infty$ for some $\mu>0$, then it is easy to see that the rate will be exponential; if $\mathsf{M}S^k<\infty$, $k>1$, then according to some power (in this connection see also § 10 and Appendix 4).

4. *The Busy Period.* Let $t_1, t_2, \ldots$ be the times the system goes from the idle to the busy state, so that $w(t_i-0)=0$ and $w(t_i+0)>0$. Suppose $t_i'>t_i$ is the first time after t_i that the system is again idle:

$$w(t_i'+0)=0\quad\text{and}\quad w(t)>0\quad\text{for } t\in(t_i,t_i')\,.$$

The time intervals $T_1=t'_1-t_1$, $T_2=t'_2-t_2$, etc. are called *busy periods* of the system. Obviously, in our case the T_i are i.i.d. variables. The t_i are the jump times of the governing process $X(t)$.

Considering the graph of $X(u)$ and recalling that $w(t)=X(t)-\inf_{0\leqslant u\leqslant t}(0, X(u))$, it is not difficult to show that the T_i coincide with the jumps ξ_η of the process $\eta_1(y)=\eta(-y)$, $y\geqslant 0$ (see § 12). Hence, from Theorem 2 and its corolllary we have

Theorem 5. *If* $f_1(\mu)=\mathsf{M}\exp\mu S_j$, *then for* $p>0$

$$\mathsf{M}(\exp(-pT_1))=f_1(\mu(p)) \tag{34}$$

(*this implies* $\mathsf{P}(T_1<\infty)=f_1(\mu(0))$) *and*

$$\mathsf{P}(T_1\in du\mid T_1<\infty)=\frac{\mathsf{P}(X(u)-X(0)\in(0,du))}{u(\alpha+\mu(0))}.$$

If $a<0$, *then* $\mu(0)=0$ *and* T_i *is a proper random variable,*

$$\mathsf{P}(T_i\in du)=\frac{\mathsf{P}(X(u)-X(0)\in(0,du))}{\alpha u}.$$

If $X(u)$ *has a density* $f(0,u)$ *for* $x=0$, *then* T_1 *also has a density, equal to* $f(0,u)/\alpha u$.

Differentiating (34) for $a<0$, we obtain

$$\mathsf{M}T_1=-\frac{\mathsf{M}S}{a},\qquad \mathsf{M}T_1^2=-\frac{\mathsf{M}S\cdot\mathsf{D}X(1)}{a^3}+\frac{\mathsf{M}S^2}{a^2}, \tag{35}$$

and so forth, so that $\mathsf{M}T_1^k$ is a function of the first k moments of S.

5. In addition to the busy period, we are also interested in the distribution of the time

$$v(t)=\inf\{u\colon w(t-u)=0\},$$

during which the system continuously occupies the busy state before time t. In Theorem 14, Chapter 1, we showed that the distribution of $(w(t), v(t))$ converges for $t\to\infty$ to that of $(\overline{Y},\theta)$.

We will find the limiting distribution of $v(t)$ (and thereby that of θ; see Subsection 4 § 12) directly with the aid of Theorem 5.

Let the time t_1 be fixed (formally, one ought to carry out the exposition in the language of conditional expectations). Then for $t-t_1>x$

$$\mathsf{P}(v(t)>x)=\mathsf{P}(T_1>t-t_1)+\int_0^{t-t_1-x}d\mathsf{P}(T_1+\tau_1<u)\mathsf{P}(T_2>t-t_1-u)$$
$$+\int_0^{t-t_1-x}d\mathsf{P}(T_1+\tau_1+T_2+\tau_2<u)\mathsf{P}(T_3>t-t_1-u)+\cdots,$$

where $\tau_1,\tau_2,\ldots$ are the time intervals during which the system is idle. Because of the properties of the exponential distribution, $\{\tau_j\}$ does not depend on $\{T_j\}$ and $\mathsf{P}(\tau_j>x)=e^{-\alpha x}$.

Let $H(u)$ denote the renewal function of the random variable $T_1+\tau_1$ and set $\mathsf{P}(T_1>x)=P(x)$; we then have

$$\mathsf{P}(v(t)>x)=\int_0^{t-t_1-x}dH(u)P(t-t_1-u),$$

and by the fundamental renewal theorem we find that as $t \to \infty$

$$P(v(t) > x) \to \frac{1}{MT_1 + \frac{1}{\alpha}} \int_x^\infty P(v)\,dv\,,$$

so that by (35)

$$\lim_{t \to \infty} P(v(t) > x) = P(\theta > x) = -a\alpha \int_x^\infty P(v)\,dv\,.$$

This means that $\theta = 0$ with probability $1 + a\alpha MT_1 = 1 - \alpha MS = -a$ (which we already know), and for $x > 0$ the distribution of θ has the density (see also Subsection 4 of Appendix 1)

$$-a\alpha P(x) = -a \int_x^\infty \frac{P(X(u) - X(0) \in (0, du))}{u}\,.$$

A whole series of other properties of the random variables $(\bar{Y}, \theta)$ will be found in Chapters 3 and 4, where we consider boundary problems for arbitrary sequences with independent increments (see also § 9).

§ 14. Discrete Time

1. A homogeneous random sequence $\{Y_n; n \geqslant 0\}$, $Y_0 = 0$, with independent increments will be called *continuous from below* if the random variable Y_1 is an integer, $Y_1 \geqslant -1$. Obviously, Y_n can be represented as the sum of independent random variables $\xi_1, \xi_2, \ldots$, with the same distribution as Y_1. If $Mz^{\xi_1} = d(z)$, then

$$Mz^{Y_n} = (Mz^{Y_1})^n = d^n(z)\,.$$

The double transformation of Y_n equals

$$\sum_0^\infty p^n Mz^{Y_n} = \frac{1}{1 - p\,d(z)}, \quad 0 < |z| \leqslant 1,\ |p| < |d^{-1}(z)|\,.$$

When $p \leqslant 1/\inf_{z \geqslant 0} d(z)$, the function $1 - p\,d(z)$ has no more than two real zeros. The smallest of these always exists. Denote it by $z(p)$, so that $z(p) < 1$ for $p < 1$ (Fig. 11).

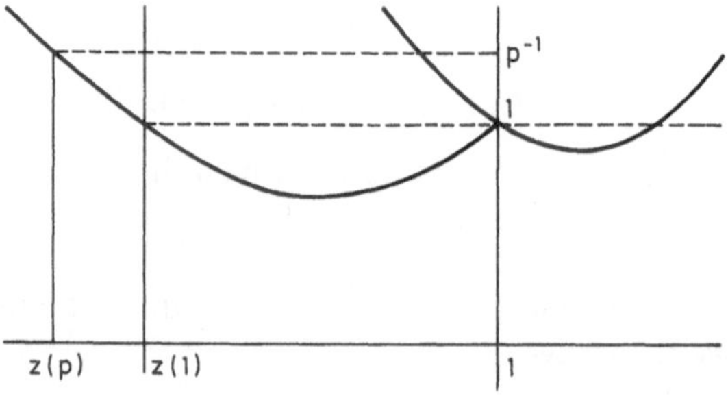

Fig. 11. The zeros of $1 - pd(z)$

The formulae just derived, as well as the rest of the exposition of this section, are completely analogous to the contents of § 12. We will therefore present many of the arguments and calculations in less detail. For $x<0$, we set

$$\eta(x)=\inf\{k\colon Y_k\leqslant x\},$$
$$h(x,n)=\mathsf{P}(\eta(x)=n)=\mathsf{P}\big(\min_{k<n} Y_k>x,\ Y_n=x\big),$$

and $p(x,n)=\mathsf{P}(Y_n=x)$. We have

Theorem 6. *For $x<0$ and $n>0$*

$$h(x,n)=-\frac{x}{n}p(x,n),\qquad \mathsf{M}(p^{\eta(x)};\eta(x)<\infty)=z^{-x}(p).$$

Proof. Since the trajectory of $Y_n, n=0,1,\dots$, on paths from 0 to x passes through all intermediate levels, by the total probability formula

$$h(x,n)=\textstyle\sum_{k=0}^{n} h(-1,k)h(x+1,n-k)$$

and

$$p(x,n)=\textstyle\sum_{k=0}^{n} h(x,k)p(0,n-k),$$

where we assume that

$$h(x,0)=\begin{cases}0, & \text{if } x<0,\\ 1, & \text{if } x=0,\end{cases}$$

$$h(0,k)=\begin{cases}0, & \text{if } k>0,\\ 1, & \text{if } k=0\end{cases}$$

and

$$p(0,0)=1.$$

Setting

$$\tilde{h}(x,p)=\textstyle\sum_{n=0}^{\infty} h(x,n)p^n \quad\text{and}\quad \tilde{p}(x,p)=\sum_{n=0}^{\infty} p(x,n)p^n,$$

we get

$$\tilde{h}(x,p)=\tilde{h}(-1,p)\tilde{h}(x+1,p)=\tilde{h}^{-x}(-1,p)\equiv\tilde{h}^{-x}(p)$$

and

$$\tilde{p}(x,p)=\tilde{h}^{-x}(p)\tilde{p}(0,p).$$

Thus, for $p<1$ and $1\geqslant|z|\geqslant\tilde{h}(p)$

$$\frac{1}{1-pd(z)}=\textstyle\sum_{x=-\infty}^{\infty} z^x\tilde{p}(x,p)=\sum_{-\infty}^{0} z^x\tilde{p}(0,p)\tilde{h}^{-x}(p)+\sum_{1}^{\infty} z^x\tilde{p}(x,p)$$

$$=\frac{\tilde{p}(0,p)}{1-\dfrac{\tilde{h}(p)}{z}}+\textstyle\sum_{1}^{\infty} z^x\tilde{p}(x,p).$$

The right side of this equation is analytic in the domain $|z|<1$ excluding the point $z=\tilde{h}(p)$, where it has a simple pole. This means that the function $1-pd(z)$ has a single zero of multiplicity one: $z=z(p)=\tilde{h}(p)$ in the domain $|z|<1$. We have

$$\lim_{z\to z(p)} \frac{\tilde{p}(0,p)}{1-z(p)/z}(1-pd(z)) = 1 = \lim_{z\to z(p)} z\tilde{p}(0,p)\frac{[1-pd(z(p))-p(z-z(p))\,d'(z(p))]}{z-z(p)},$$

so that $\tilde{p}(0,p)=-1/pz(p)\,d'(z(p))$.

On the other hand, from the identity $1-pd(z(p))\equiv 0$ it follows that

$$d(z(p))+pd'(z(p))z'(p)=0, \qquad \tilde{p}(0,p)=\frac{pz'(p)}{z(p)},$$

and

$$\tilde{p}(x,p)=pz'(p)z^{-x-1}(p)=-\frac{p}{x}\frac{\partial}{\partial p}z^{-x}(p)$$

$$=-\frac{p}{x}\frac{\partial}{\partial p}\tilde{h}(x,p)=-\sum_{n=0}^{\infty}\frac{n}{x}h(x,n)p^n .$$

The theorem is proved. □

Corollary. *For $p<1$ the function $z(p)$ is the generating function of some random variable η_1 which with probability*

$$1-z(1)=1-\sum_{n=1}^{\infty}\frac{p(-1,n)}{n}\geqslant 0$$

can assume the value ∞ ($z(1)=1$ if $\mathsf{M}\xi_1=a\leqslant 0$). For $x<0$ $\eta(x)$ can be represented as

$$\eta(x)=\eta_1+\eta_2+\cdots+\eta_{-x},$$

where the η_k are independent and distributed like η_1:

$$\mathsf{P}(\eta_k=n)=\mathsf{P}(\eta(-1)=n)=\frac{p(-1,n)}{n}, \quad n=1,2,\ldots .$$

From the theorem it also follows that for $a<0$

$$\mathsf{M}\eta_1=-\frac{1}{a} \quad\text{and}\quad \mathsf{D}\eta_1=-\frac{\mathsf{D}\xi_1}{a^3}.$$

The analogue of Theorem 3 here is the following assertion. First let

$$\underline{Y}_n=\min_{k\leqslant n} Y_k \quad\text{and}\quad \overline{Y}_n=\max_{k\leqslant n} Y_k .$$

Theorem 7. *For $x\geqslant 0$, $y\leqslant x$ and $B\subset(-\infty,x-1]$*

$$\mathsf{P}(\overline{Y}_n\geqslant x,\ Y_n=y)=\mathsf{P}(\underline{Y}_n\leqslant -x+y,\ Y_n=y)$$

and

$$\mathsf{P}(\overline{Y}_n\geqslant x,\ Y_n\in B)=-\sum_{k=1}^{n}\mathsf{M}\left(\frac{Y_k}{k},\ Y_k\in B-x\right)p(x,n-k).$$

Proof. The first assertion is obvious when it becomes clear what the event $\{\overline{Y}_n \geqslant x, Y_n = y\}$ is for the "backward" trajectory $Y_n - Y_n, Y_{n-1} - Y_n, \ldots, -Y_n$, which is obviously also a sequence with independent increments. Moreover,

$$\begin{aligned}
\mathsf{P}(\overline{Y}_n \geqslant x, Y_n \in B) &= \textstyle\sum_{y \in B} \mathsf{P}(\underline{Y}_n \leqslant -x+y, Y_n = y) \\
&= \textstyle\sum_{y \in B} \sum_{k=1}^{n} \mathsf{P}(\eta(-x+y) = k) p(x, n-k) \\
&= \textstyle\sum_{y \in B} \sum_{k=1}^{n} \frac{x-y}{k} p(-x+y, k) p(x, n-k) \\
&= -\textstyle\sum_{k=1}^{n} \mathsf{M}\left(\frac{Y_k}{k}, Y_k \in B - x\right) p(x, n-k). \quad \square
\end{aligned}$$

Remark 1. Writing out the total probability formula w.r.t. passage of the level $-x+y+1$ for $B \ni y < x-1$, we obtain quite analogously

$$\mathsf{P}(\overline{Y}_{n-1} \geqslant x-1, Y_n \in B) = -\textstyle\sum_{k=1}^{n-1} \mathsf{M}\left(\frac{Y_k}{k}; Y_k \in B - x + 1\right) p(x-1, n-k). \quad (36)$$

We also have (compare with § 12) ($y \geqslant 0$)

$$\tilde{P}(x, y, p) = \textstyle\sum_{n=0}^{\infty} p^n \mathsf{P}(\overline{Y}_n \geqslant x, Y_n = x - y) = \tilde{p}(x, p) z^y(p),$$

$$\tilde{P}(z, y, p) = \textstyle\sum_{x=0}^{\infty} z^x \tilde{P}(x, y, p) = z^y(p)\left[\frac{1}{1 - pd(z)} + \frac{1}{(z - z(p)) pd'(z(p))}\right].$$

Remark 2. The assertion of Theorem 6 is also *necessary* for the continuity from below of the sequence $\{Y_n; n \geqslant 0\}$. Indeed, suppose for $x < 0$ that

$$h(x, n) = -\frac{x}{n} p(x, n).$$

Setting $n = 1$, we find that

$$h(x, 1) = \mathsf{P}(Y_1 \leqslant x) = \textstyle\sum_{k=-x}^{\infty} p(-k, 1) = -xp(x, 1).$$

If we put $p(z) = \sum_{k=1}^{\infty} z^k p(-k, 1)$, we get

$$\frac{p(z) - p(1)}{z - 1} = p'(z).$$

Since the solution of this equation has the form $p(z) = a + bz$, $p(-k, 1) = 0$ for $k \geqslant 2$.

2. *The Distribution of* $w(n)$. We have seen that the sequences $X(n)$ and $Y(n)$ in Chapter 1 will be continuous from below, homogeneous and have independent increments iff τ_j^e and S_j, $j = 1, 2, \ldots$, are independent (of one another and as sequences) and

$$\mathsf{P}(\tau_j^e = k) = q_1 p_1^{k-1}; \quad k = 1, 2, \ldots, p_1 \geqslant 0, p_1 + q_1 = 1.$$

Here we have

$$\mathsf{M}\, e^{i\lambda X(n)} = \mathsf{M}\, e^{i\lambda w(0)} (q_1 f(\lambda) + p_1)^n e^{-i\lambda n}.$$

We proved in Chapter 1 that

$$w(n) = \max_{k \leqslant n-1} (X(n), X(n) - X(k) + 1),$$

so that ($w(0) \geqslant S_1 > 0$ in this case)

$$\mathsf{P}(w(n) < x) = \mathsf{P}(Y(n) < x - w(0)) - \mathsf{P}(\overline{Y}(n) \geqslant x - 1,\ Y(n) < x - w(0)).$$

Comparing this with Theorem 15A and (36), we obtain

Theorem 8. *If $x \geqslant 2$ and $w(0)$ is nonrandom, then*

$$\mathsf{P}(w(n) < x) = \mathsf{P}(Y(n) < x - w(0)) + \textstyle\sum_{k=1}^{n-1} \mathsf{M}\left(\frac{Y_k}{k}, Y_k < -w(0) + 1\right) p(x-1, n-k),$$

$$\mathsf{P}(w(n) = 0) = -\mathsf{M}\left(\frac{Y_n}{n}, Y_n < -w(0) + 1\right),$$

$$\textstyle\sum_{n=0}^{\infty} p^n \mathsf{P}(w(n) = 0) = \frac{z^{w(0)}(p)}{1 - z(p)}.$$

With the obvious modifications, these assertions remain valid for random $w(0)$ not depending on $\{X(n) - X(0);\ n \geqslant 1\}$. From this it is not difficult to show again that for $a < 0$,

$$\lim_{n \to \infty} \mathsf{P}(w(n) = 0) = -a \quad \text{and} \quad \lim_{n \to \infty} \mathsf{P}(w(n) > x) = -ah(x), \quad x \geqslant 0,$$

where $h(x)$ is the renewal density of the random variable $Y(1)$. It is also easy to find a formula for $\mathsf{M}z^{w^c(n)}$ from which, as in the case of continuous time, it follows (compare with (33)) that

$$\kappa(w^c(n)) = \kappa(S) - 1.$$

The asymptotic behavior of $\mathsf{P}(w(n) = 0)$ can be estimated just as in Subsection 3 § 13. The distribution of the busy periods $T_1, T_2, \ldots$ (they are independent as before) coincides with that of η_1 (see Theorem 6 and its corollary):

$$\mathsf{P}(T_j = k) = \frac{p(-1, k)}{k}, \qquad \mathsf{M}T_1 = -\frac{1}{a} \quad \text{and} \quad \mathsf{D}T_1 = -\frac{\mathsf{D}Y(1)}{a^3}.$$

Using these formulae, we can find, as in the continuous case, the limiting distribution of the time $v(n)$ continuously spent by the system in the busy state up to time n.

Chapter 3

Boundary Problems for Sequences with Independent Increments and Factorization Identities

We have already seen that the study of the basic characteristics of queueing systems (and of many other service systems as well; see Chapters 5–8) leads to certain functionals of the trajectories generated by the governing sequences. It is natural to call these *boundary functionals*, since all of them are connected with attainment by these trajectories of certain boundaries. In this regard, it seems fitting to present here the basic methods of investigation and the corresponding results relating to the problem of finding the distributions of the most important and (for us) interesting boundary functionals (we will call these *boundary problems*). We will consider functionals of the sequences $\{Y_n; n \geqslant 0\}$ formed by sums of i.i.d. variables (see also §§ 5, 9).

Many of the results in Chapters 3 and 4 can be generalized to governing sequences defined on the states of a finite Markov chain. Significant progress in this direction has been made in [46] and [50]. An extension of these results is doubtless of interest.

Our attention in this chapter is devoted mainly to analytic methods of obtaining so-called multi-parametric factorizations of identities. These methods are reasonably simple and apparently of wider applicability than direct probabilistic approaches. The derived identities will allow us to investigate many properties of the trajectory Y_n, $n=0, 1, 2, \ldots$, and to obtain as corollaries almost all of the known results on the questions under consideration.

§ 15. Preliminary Remarks

Let $\xi, \xi_1, \xi_2, \ldots$ be a sequence of i.i.d. variables and

$$Y_n = \sum_{k=1}^{n} \xi_k, \qquad f(\lambda) = \mathsf{M}\, e^{i\lambda\xi}.$$

If $|z|<1$ and $\operatorname{Im} \lambda = 0$, then $|zf(\lambda)|<1$ and the function $\mathfrak{w}_z(\lambda) = 1 - zf(\lambda)$ can be represented as

$$\mathfrak{w}_z(\lambda) = \phi(z) \frac{\mathfrak{w}_{z+}(\lambda)}{\mathfrak{w}_{z-}(\lambda)}, \tag{1}$$

where

$$\mathfrak{w}_{z+}(\lambda)=\exp\left\{-\int_{+0}^{\infty} e^{i\lambda x}\, d_x\left[\sum_{k=1}^{\infty}\frac{z^k}{k}\mathsf{P}(Y_k<x)\right]\right\},$$

$$\mathfrak{w}_{z-}(\lambda)=\exp\left\{\int_{-\infty}^{-0} e^{i\lambda x}\, d_x\left[\sum_{k=1}^{\infty}\frac{z^k}{k}\mathsf{P}(Y_k<x)\right]\right\}$$

and

$$\phi(z)=\exp\left\{-\sum_{k=1}^{\infty}\frac{z^k}{k}\mathsf{P}(Y_k=0)\right\}.$$

This representation can be obtained easily using the fact that for $|zf(\lambda)|<1$

$$1-zf(\lambda)=\exp\left\{-\sum_{k=1}^{\infty}\frac{z^k}{k}f^k(\lambda)\right\}=\exp\left\{-\int_{-\infty}^{\infty} e^{i\lambda x}\, d_x\left[\sum_{1}^{\infty}\frac{z^k}{k}\mathsf{P}(Y_k<x)\right]\right\}.$$

The representation (1) effects a so-called *factorization* of the function $\mathfrak{w}_z(\lambda)$ since $\mathfrak{w}_{z+}(\lambda)$ is analytic in λ in the domain $\operatorname{Im}\lambda>0$ and is continuous on the boundary $\operatorname{Im}\lambda=0$. Analogous properties are enjoyed by $\mathfrak{w}_{z-}(\lambda)$ in the lower half-plane. The representation (1) itself, with factors possessing the above properties, will be referred to as a *factorization*. Furthermore, in (1)

$$\inf_{\operatorname{Im}\lambda\gtrless 0}|\mathfrak{w}_{z\pm}(\lambda)|>0 \quad\text{and}\quad \mathfrak{w}_{z+}(i\infty)=1\,.$$

We have seen that the characteristic functions of the joint distributions of a number of boundary functionals of the trajectory

$$(0,0),(1,Y_1),(2,Y_2),\ldots \tag{2}$$

can be expressed directly in terms of the components $\mathfrak{w}_{z\pm}(\lambda)$ and $\phi(z)$.

For this trajectory, which we will consider as the path of a particle moving in the (n,t)-plane, we introduce the boundaries (levels) $n=\text{const}>0$ and $t=\text{const}>0$. Connected with the passage of these levels are the following random variables, which we will call *upper functionals* (*characteristics*).

For the level n:

$$\overline{Y}_n=\max(0,Y_1,\ldots,Y_n),\quad \overline{Y}_0=0,\ \overline{Y}_\infty=\overline{Y};$$

$$\bar{y}_n=\max(Y_1,\ldots,Y_n),\quad \bar{y}_\infty=\bar{y};$$

$$\theta_n=\min\{k\colon Y_k=\overline{Y}_n\},\quad \theta_0=0;$$

$$\theta^n=\max\{k\leqslant n\colon Y_k=\overline{Y}_n\},\quad \theta^0=0\,.$$

For the level t:

$$\eta(t)=\min\{k\geqslant 1\colon Y_k\geqslant t\},\qquad \chi(t)=Y_{\eta(t)}-t\,.$$

If $\bar{y}<t$, we will set $\eta(t)=\infty$. The variable $\chi(t)$ will remain undefined on the set $\{\bar{y}<t\}$. The *lower characteristics* for the levels n and $t\leqslant 0$ are defined symmetric-

ally. We will denote them by the same symbols supplied with an asterisk. For example,

$$\overline{Y}_n^* = \min(0, Y_1, \ldots, Y_n),$$

$$\theta_*^n = \max\{k \leqslant n : Y_k = \overline{Y}_n^*\},$$

$$\eta^*(t) = \min\{k \geqslant 1 : Y_k \leqslant t\}, \qquad \chi^*(t) = Y_{\eta^*(t)} - t.$$

The random variable Y_n will be understood as being both an upper and lower characteristic.

A remarkable fact is that *the characteristic function of the joint distribution of all upper or all lower characteristics can be determined explicitly in terms of the functions* $\mathfrak{w}_{z\pm}(\lambda)$. In this connection it is sufficient to know the characteristic function of only *two* joint distributions:

$$1)\ \mathsf{P}(\overline{Y}_n \in dx, \theta_n = k)$$

and

$$2)\ \mathsf{P}(\chi(t) \in dy, \eta(t) = l). \tag{3}$$

If $F(x) = \mathsf{P}(\xi < x)$ is not continuous, then along with (1) it is also necessary to know the characteristic function of the distribution $\mathsf{P}(\overline{Y}_n \in dx, \theta^n = k)$.

Indeed, for the joint distribution

$$\mu_{n,t}(dx, dy, dz, k, l) = \mathsf{P}(\theta_n = k, \overline{Y}_n \in dx, \eta(t) = l, \chi(t) \in dy, Y_n \in dz)$$

we have for $l \leqslant n$

$$\mu_{n,t} = \mathsf{P}(\eta(t) = l, \chi(t) \in dy)\mathsf{P}(\theta_{k-l} = k - l, \overline{Y}_{k-l} \in dx - t - y)$$
$$\times \mathsf{P}(\overline{Y}_{n-k} = 0, Y_{n-k} \in dz - x), \tag{4$_1$}$$

and for $l > n$

$$\mu_{n,t} = \mathsf{P}(\theta_k = k, \overline{Y}_k \in dx)\mathsf{P}(\overline{Y}_{n-k} = 0, Y_{n-k} \in dz - x)$$
$$\times \mathsf{P}(\eta(t - z) = l - n, \chi(t - z) \in dy). \tag{4$_2$}$$

But probabilities of the form $\mathsf{P}(\overline{Y}_n = 0, Y_n \in dz)$, which enter into these equalities, coincide with $\mathsf{P}^*(\overline{Y}_n \in dz, \theta^n = n)$ for "backward" trajectories from the point (n, S_n) to $(0, 0)$ (i.e., for trajectories generated by the random variables $-\xi_n$, $-\xi_{n-1}, \ldots, -\xi_1$). On the other hand, the representations (4) of the measure $\mu_{n,t}$ by means of the distributions (3) are obviously such that by proceeding to characteristic functions and transformations w.r.t. n and t we obtain expressions of the convolution type. This leads us to the product of transformations of (3) for distinct values of the arguments. As is easy to verify, the situation does not change if the random variables θ^n and $\bar{y}_n$ are inserted into the distribution $\mu_{n,t}$.

We will obtain below explicit formulae for transforms of the distributions (3). Transforms of other joint distributions of boundary characteristics can be obtained without difficulty by the reader using formulae of the form (4) and the remarks already made.

The components $\mathfrak{w}_{z\pm}$ of the factorization are associated with transforms of other random variables related in a natural way to boundary functionals. For example, the number $K_n(t)$ of intersections of the trajectories (2) of the level $t \geqslant 0$ and the time $T_n(t)$ spent by these trajectories above the level t (i.e., the number of indices $k, n \geqslant k \geqslant 1$, for which $Y_k \geqslant t$). These are functionals depending on both of the parameters n and t. The relations between transforms of $T_n(t)$ and $K_n(t)$ and the components $\mathfrak{w}_{z\pm}$ are much more complicated (see, for example, [38]) and we will not concern ourselves with them here. We mention merely one consequence of these relations by virtue of which the distributions of $T_n(+0)$ and θ_n coincide. This also holds for the distributions of $T_n(0)$ and θ^n.

In a number of special cases one can also express by means of the factorization components the transforms of certain boundary functionals connected with the random walk (2) in a *strip*, i.e., when there exist both upper and lower boundaries ([37]). These special cases are furnished, for example, by random walks for which $\chi(t) = Y_{\eta(t)} - t$ does not depend on the previous history $Y_1, \ldots, Y_{\eta(t)-1}$ (compare with Corollary 12 of this chapter and with the remarks concerning it).

§ 16. The First Factorization Identity and Its Consequences

Theorem 1. *For* $|z| < 1$, $|\rho| < 1/|z|$, $\operatorname{Im} \lambda = 0$ *we have*

$$(1-z) \sum_{n=0}^{\infty} z^n \mathsf{M}(\rho^{\theta_n} \exp(i\lambda \overline{Y}_n)) = \frac{\mathfrak{w}_{z+}(0)}{\mathfrak{w}_{z\rho+}(\lambda)} . \tag{5}$$

The sum on the left is obviously the triple transform of the first of the distributions at (3).

Proof. Set

$$F(x) = \mathsf{P}(\xi < x) \quad \text{and} \quad u_{n,k}^x = \mathsf{P}(\theta_n = k, \overline{Y}_n < x) .$$

We have $u_{n,k}^0 = u_{n,n+l}^x = 0$ for all $k \geqslant 1$ and $l \geqslant 1$. By the total probability formula

$$u_{n+1,k+1}^x = \int_{-\infty}^x dF(t) u_{n,k}^{x-t}, \quad n \geqslant k \geqslant 1 ;$$
$$u_{n,1}^x = (F(x) - F(+0)) \mathsf{P}(\overline{Y}_{n-1} = 0) .$$

This implies that for $|z| < 1$ and $k \geqslant 2$

$$u_k^x(z) \equiv \sum_{n=1}^{\infty} z^n u_{n,k}^x = \sum_{n=k}^{\infty} u_{n,k}^x z^n = z \int_{-\infty}^x dF(t) u_{k-1}^{x-t}(z)$$

and

$$u_1^x(z) = (F(x) - F(+0)) z \varphi(z), \; \varphi(z) = \sum_{n=0}^{\infty} z^n \mathsf{P}(\overline{Y}_n = 0) .$$

Moreover, for $|\rho| < |z|^{-1}$

$$u^x(z, \rho) \equiv \sum_{k=1}^{\infty} \rho^k u_k^x(z) = z\rho \int_{-\infty}^x dF(t) u^{x-t}(z, \rho) + z\rho\varphi(z)(F(x) - F(+0)) .$$

Finally, noting that $u^x(z, \rho)$ along with $u^x_{n,k}$ is a function of bounded variation in x, we get for $\operatorname{Im} \lambda = 0$

$$\begin{aligned} u(\lambda, z, \rho) &\equiv \int_{+0}^{\infty} e^{i\lambda x}\, d_x u^x(z, \rho) \\ &= z\rho f(\lambda) u(\lambda. z, \rho) + z\rho\varphi(z) f(\lambda) \\ &\quad - z\rho \int_{-\infty}^{+0} e^{i\lambda x}\, d_x\Big(\int_{-\infty}^{x} dF(t) u^{x-t}(z, \rho) + \varphi(z) F(x)\Big). \end{aligned}$$

Hence,

$$u(\lambda, z, \rho)(1 - z\rho f(\lambda)) = -\varphi(z)(1 - z\rho f(\lambda)) + \varphi(z) + h_-,$$

where $h_- = z\rho \int_{-\infty}^{+0} e^{i\lambda x}\, dH_{z,\rho}(x)$ and $H_{z,\rho}(x)$ is of bounded variation for the chosen values of z and ρ. Using (1) we can now write the equality

$$u(\lambda, z, \rho)\mathfrak{w}_{z\rho+}(\lambda) = -\varphi(z)\mathfrak{w}_{z\rho+}(\lambda) + (\varphi(z) + h_-)\mathfrak{w}_{z\rho-}(\lambda)\phi^{-1}(z\rho).$$

Because of the properties of the factorization components, the last addend is an analytic function of λ in the lower half-plane and is continuous on the boundary $\operatorname{Im} \lambda = 0$. As the derivation of the above equality shows, this function can be extended to a function analytic in the upper half-plane. Consequently, it is entire and bounded and is necessarily equal to a constant. This means that

$$(u(\lambda, z, \rho) + \varphi(z))\mathfrak{w}_{z\rho+}(\lambda) = c(z, \rho)$$

does not depend on λ. Setting $\operatorname{Im} \lambda = \infty$, we obtain

$$\varphi(z) = c(z, \rho).$$

In addition, it is easy to verify that

$$u(\lambda, z, \rho) + \varphi(z) = \sum_{n=0}^{\infty} z^n \mathsf{M}(\rho^{\theta_n} \exp(i\lambda \bar{Y}_n)).$$

For $\rho = 1$ and $\lambda = 0$ this yields the equality

$$\frac{1}{1-z} = \frac{\varphi(z)}{\mathfrak{w}_{z+}(0)} \tag{6}$$

which concludes the proof. □

We have used here the Wiener–Hopf method for solving integral equations on the half-line.[1]

It is clear that an equation symmetric to (5) also holds for the random variables θ_n^* and $\bar{Y}_n^*$.

In an entirely analogous way one can prove

Theorem 2. *For* $|z| < 1$, $|\rho| < 1/|z|$ *and* $\operatorname{Im} \lambda \geq 0$

$$(1 - z) \sum_{n=0}^{\infty} z^n \mathsf{M}(\rho^{\theta^n} \exp(i\lambda \bar{Y}_n)) = \frac{\mathfrak{w}_{z+}(0)\phi(z)}{\mathfrak{w}_{z\rho+}(\lambda)\phi(z\rho)}.$$

[1] This procedure was used by the author in [8] to investigate the properties of the function $u(\lambda, z, \rho)$ for $|z| \geq 1$ and $\operatorname{Im} \lambda < 0$. This made it possible to establish limit theorems for the joint distribution of $\bar{Y}_n$ and Y_n and a number of other quantities.

To prove this theorem we need a consequence of Theorem 1 given by

Corollary 1. *For* $|u|<1$ *and* $\operatorname{Im}\lambda\geqslant 0$

$$\textstyle\sum_{n=0}^{\infty} u^n \mathsf{M}(\exp(i\lambda \overline{Y}_n);\, \theta_n=n)=\frac{1}{\mathfrak{w}_{u+}(\lambda)}\cdot$$

This equality results from Theorem 1 by setting $z\rho=u$, $|u|<1$, and letting z tend to 0.

Proof of Theorem 2. We start with the following system of equations for $u_{n,k}^x=\mathsf{P}(\theta^n=k,\, \bar{y}_n<x)$:

$$u_{n+1,k+1}^x=\textstyle\int_{-\infty}^{x} dF(t)u_{n,k}^{x-t}, \quad n\geqslant k\geqslant 1\,;$$

$$u_{n,1}^x=(F(x)-F(0))\mathsf{P}(\bar{y}_{n-1}<0)\,.$$

Setting $\varphi(z)=\sum_{n=0}^{\infty} z^n \mathsf{P}(\bar{y}_n<0)$ (here $\mathsf{P}(\bar{y}_0<0)=1$), we find as before that the triple transform

$$u(\lambda, z, \rho)=\textstyle\int_0^{\infty} e^{i\lambda x}\, d_x\big(\sum_{n=1}^{\infty} z^n \sum_{k=0}^{n} \rho^k u_{n,k}^x\big)$$

has the property that the function

$$(u(\lambda, z, \rho)+\varphi(z))\mathfrak{w}_{z\rho+}(\lambda)=c(z,\rho)$$

is independent of λ. Or, what is the same ($\max(0, \bar{y}_n)=\overline{Y}_n$),

$$\textstyle\sum_{n=0}^{\infty} z^n \mathsf{M}(\rho^{\theta^n}\exp(i\lambda\overline{Y}_n))\mathfrak{w}_{z\rho+}(\lambda)=c(z,\rho)\,, \tag{7}$$

from which it follows that

$$c(z,\rho)=\textstyle\sum_0^{\infty} z^n \mathsf{M}(\rho^{\theta n};\, \overline{Y}_n=0)\,.$$

To find $c(z,\rho)$ we note that

$$\mathsf{P}(\theta^n=k)=\mathsf{P}(Y_k=\overline{Y}_k)\mathsf{P}(\bar{y}_{n-k}<0)=\mathsf{P}^*(\overline{Y}_k=0)\mathsf{P}^*(\theta_{n-k}=n-k)\,,$$

where P^* denotes the probability "induced" by the sequence $-\xi_1, -\xi_2, \ldots$. This implies immediately that

$$\textstyle\sum_0^{\infty} z^n \mathsf{M}(\rho^{\theta n})=\sum_{n=0}^{\infty} (\rho z)^n \mathsf{P}^*(\overline{Y}_n=0)\sum_{n=0}^{\infty} z^n \mathsf{P}^*(\theta_n=n)\,.$$

By Corollary 1 and Eqs. (6) and (7), we find that

$$\textstyle\sum_0^{\infty} z^n \mathsf{M}\rho^{\theta n}=\frac{c(z,\rho)}{\mathfrak{w}_{z\rho+}(0)}=\frac{\mathfrak{w}_{z-}(0)}{1-z\rho}\frac{1}{\mathfrak{w}_{z\rho-}(0)}\,,$$

so that

$$c(z,\rho)=\frac{\mathfrak{w}_{z-}(0)}{\phi(z\rho)}=\frac{\mathfrak{w}_{z+}(0)\phi(z)}{(1-z)\phi(z\rho)}\cdot$$

The theorem is proved. □

We now consider some of the most interesting consequences of Theorems 1 and 2. (Regarding many of these, see also [2], [62]–[65].)

Corollary 2. *One obviously has*

$$[\sum_0^\infty z^n \mathsf{M}(\rho^{\theta^n} \exp(i\lambda \overline{Y}_n))]\,[\sum_0^\infty z^n \mathsf{M}(\rho^{\theta_*^n} \exp(i\lambda \overline{Y}_n^*))] = \frac{1}{(1-z)(1-z\rho f(\lambda))}.$$

In the sequel we will not dwell on results of this type, in which transformations are chosen whose products do not depend on the components of the factorization.

Corollary 3 (Spitzer's Identity). *Setting $\rho=1$ in* (5), *we get*

$$\sum_0^\infty z^n \mathsf{M} \exp(i\lambda \overline{Y}_n) = \frac{\mathfrak{w}_{z+}(0)}{(1-z)\mathfrak{w}_{z+}(\lambda)}$$
$$= \exp\left\{\sum_1^\infty \frac{z^k}{k}\left[1 + \int_{+0}^\infty (e^{i\lambda x}-1)\, d_x \mathsf{P}(Y_k<x)\right]\right\}$$
$$= \exp\left\{\sum_1^\infty \frac{z^k}{k} \mathsf{M} \exp i\lambda \max(0, Y_k)\right\}. \quad \square$$

In order to avoid trivial reservations in the assertions which follow, we will assume that ξ takes values of both signs with positive probability.

Corollary 4.

1) $$\mathsf{M}(z^{\eta(+0)}; \overline{Y}>0) = 1 - \exp\left\{-\sum_1^\infty \frac{z^k}{k} \mathsf{P}(Y_k>0)\right\},$$

$$\mathsf{M}(z^{\eta(0)}; \overline{y}\geq 0) = 1 - \phi(z)\mathfrak{w}_{z+}(0) = 1 - \exp\left\{-\sum_1^\infty \frac{z^k}{k} \mathsf{P}(Y_k \geq 0)\right\}. \tag{8}$$

2) *The following three assertions are equivalent:*

$$\mathsf{P}(\overline{Y}<\infty)=1, \qquad \mathsf{P}(\overline{Y}=0)>0 \quad \text{and} \quad \sum_1^\infty \frac{\mathsf{P}(Y_k>0)}{k} < \infty. \tag{9}$$

3) *If* $\mathsf{P}(\overline{Y}=\infty)=1$), *then*

$$\mathsf{M}\eta(0) = \frac{1}{\mathsf{P}(\overline{Y}^*=0)} \quad \text{and} \quad \mathsf{M}\eta(+0) = \frac{1}{\mathsf{P}(\overline{y}^*>0)}.$$

Proof. Since $\mathsf{P}(\eta(+0)=n) = \mathsf{P}(\overline{Y}_{n-1}=0) - \mathsf{P}(\overline{Y}_n=0)$, the first equality at (8) follows from the relations

$$\mathfrak{w}_{z+}(0) = (1-z)\sum_0^\infty z^n \mathsf{P}(\overline{Y}_n=0) = 1 - \sum_1^\infty z^n \mathsf{P}(\eta(+0)=n).$$

The second relation is obtained similarly using the fact that by Corollary 1

$$(1-z)\sum_0^\infty z^n \mathsf{P}(\overline{y}_n<0) = (1-z)\sum z^n \mathsf{P}^*(\theta_n=n)$$
$$= (1-z)\mathfrak{w}_{z-}(0) = \phi(z)\mathfrak{w}_{z+}(0).$$

The equivalence of the two last expressions at (9) follows immediately from (8). Moreover, if, for example, $\mathsf{P}(\overline{Y}=0)>0$, then $\mathsf{P}(\overline{Y}<\infty)>0$ and by the 0-1-law we have $\mathsf{P}(\overline{Y}<\infty)=1$. Conversely, suppose that $\mathsf{P}(\overline{Y}<\infty)=1$. Then there exist $x_0>0$ and $z_0>0$ such that $\mathsf{P}(\overline{Y}<x_0)>0$ and $\mathsf{P}(\xi<-z_0)>0$. Thus, for $k>x_0/z_0$

$$\begin{aligned}\mathsf{P}(\overline{Y}=0)&\geqslant \mathsf{P}(\xi_1<-z_0,\dots,\xi_k<-z_0)\mathsf{P}(\overline{Y}<kz_0)\\&\geqslant \mathsf{P}^k(\xi_1<-z_0)\mathsf{P}(\overline{Y}<x_0)>0\,.\end{aligned}$$

The third assertion follows from the fact that $\eta(0)$ is a proper random variable when $\mathsf{P}(\overline{Y}=\infty)=1)$ by (8) and

$$\frac{1-\mathsf{M}z^{\eta(0)}}{1-z}=\frac{\phi(z)\mathfrak{w}_{z+}(0)}{1-z}=\mathfrak{w}_{z-}(0)=\frac{1}{1-\mathsf{M}(z^{\eta^*(-0)},\ \overline{Y}^*<0)}\,.$$

The variable $\eta(+0)$ is treated analogously. □

We mention that the relations

$$\mathsf{P}(\overline{y}<0)>0,\qquad \sum_1^\infty\frac{\mathsf{P}(Y_k\geqslant 0)}{k}<\infty\quad\text{and}\quad \sum_0^\infty \mathsf{P}(\overline{Y}_n^*=0)<\infty\,.$$

will also be equivalent to (9). Here it is sufficient to note merely that one always has

$$\sum\frac{\mathsf{P}(Y_k=0)}{k}<\infty\,,$$

which follows from

Corollary 5.

$$\phi(z)=1-\mathsf{M}(z^{\eta(0)};\,\eta(0)<\infty,\,\chi(0)=0)\,.$$

Proof. We have

$$u_n=\mathsf{P}(\theta^n=n,\ \overline{Y}_n=0)=\sum_{k=1}^n \mathsf{P}(\eta(0)=k,\ \chi(0)=0)u_{n-k}\,.$$

Consequently, by Theorem 2 (as $z\to 0$, $z\rho=u$, $\lambda=i\infty$)

$$u(z)=\sum_{n=0}^\infty u_n z^n=\frac{1}{\phi(z)}=1+u(z)\sum_{k=1}^\infty z^k\mathsf{P}(\eta(0)=k,\ \chi(0)=0)\,.$$

The corollary is proved. □

If $\mathsf{M}|\xi|<\infty$, then by the strong law of large numbers, the conditions (9) are obviously equivalent to the condition $\mathsf{M}\xi<0$. On the other hand, from the relation

$$\{\mathsf{M}\xi<0\}\subset\left\{\sum_1^\infty\frac{\mathsf{P}(Y_k>0)}{k}<\infty\right\},$$

say, and using Corollary 4 (see (9)), it is easy to obtain the strong law of large numbers.

Corollary 6. *Condition (9) is necessary and sufficient to ensure that $\theta_\infty \leqslant \theta^\infty$ be proper random variables. We have*

$$\mathsf{M}\rho^{\theta_\infty} \exp(i\lambda \overline{Y}) = \frac{\mathfrak{w}_{1+}(0)}{\mathfrak{w}_{\rho+}(\lambda)}, \qquad \mathsf{M}\rho^{\theta^\infty} \exp(i\lambda \overline{Y}) = \frac{\phi(1)\mathfrak{w}_{1+}(0)}{\phi(\rho)\mathfrak{w}_{\rho+}(\lambda)}$$

and the joint distributions of $(\theta_\infty, \overline{Y})$ and $(\theta^\infty, \overline{Y})$ are infinitely divisible.

Proof. The first assertion follows immediately from relations of the form

$$\mathsf{P}(\theta^\infty = k) = \mathsf{P}(\theta^k = k)\mathsf{P}(\bar{y} < 0)$$

and the fact that

$$\textstyle\sum_0^\infty z^k \mathsf{P}(\theta^k = k) = \dfrac{1}{\phi(z)\mathfrak{w}_{z+}(0)}\cdot$$

Moreover,

$$\mathsf{P}(\theta_\infty = n,\ \overline{Y} \in dx) = \mathsf{P}(\theta_n = n,\ \overline{Y}_n \in dx)\mathsf{P}(\overline{Y} = 0)\,.$$

Consequently (see Corollaries 1 and 4),

$$\mathsf{M}\rho^{\theta_\infty} \exp(i\lambda \overline{Y}) = \frac{\mathsf{P}(\overline{Y}=0)}{\mathfrak{w}_{\rho+}(\lambda)} = \frac{\mathfrak{w}_{1+}(0)}{\mathfrak{w}_{\rho+}(\lambda)} = \exp\left\{\textstyle\sum_{k=1}^\infty \int_{+0}^\infty (\rho^k e^{i\lambda x} - 1)\dfrac{d\mathsf{P}(Y_k < x)}{k}\right\}. \tag{10}$$

The infinite divisibility of $(\theta^\infty, \overline{Y})$ is shown analogously. □

It is not hard to show that the joint distribution of $(\theta_\infty, \theta^\infty, \overline{Y})$ is also infinitely divisible. Moreover, θ_∞ and $\theta^\infty - \theta_\infty$ are independent, and

$$\mathsf{M}z^{\theta^\infty - \theta_\infty} = \frac{\phi(1)}{\phi(z)} = \exp\left\{\textstyle\sum_1^\infty (z^k - 1)\dfrac{\mathsf{P}(Y_k = 0)}{k}\right\}.$$

This is another probabilistic interpretation of the function $\phi(z)$ by which $\phi(1) = \mathsf{P}(\theta_\infty = \theta^\infty)$ is the probability that the trajectory (2) has a single absolute peak.

Corollary 7. *If the random variables ξ_k are symmetric and $F(x)$ is continuous, then the distribution of the random variables $\eta(0)$, $\eta(+0)$, θ_n and θ^n is defined for arbitrary n by the explicit formulae*

$$\mathsf{P}(\theta_n = n) = \mathsf{P}(\overline{Y}_n = 0) = \frac{(2n)!}{2^{2n}(n!)^2},$$

$$\mathsf{P}(\eta(0) = n) = \frac{\mathsf{P}(\overline{Y}_n = 0)}{2n-1} \tag{11}$$

and

$$\mathsf{P}(\theta_n = k) = \mathsf{P}(\theta_k = k)\mathsf{P}(\overline{Y}_{n-k} = 0)$$

and does not depend on $F(x)$.

Proof. In the case under consideration $\mathsf{P}(Y_k>0)=\mathsf{P}(Y_k\geqslant 0)=\frac{1}{2}$ and $\eta(0)=\eta(+0)$, $\theta_n=\theta^n$ w.p.1. Moreover,

$$\mathfrak{w}_{z+}(0)=\mathfrak{w}_{z+}(0)\phi(z)=\exp\left\{-\tfrac{1}{2}\textstyle\sum_1^\infty \frac{z^k}{k}\right\}=\sqrt{1-z}\,,$$

$$\mathsf{M}z^{\eta(0)}=1-\sqrt{1-z}\,,$$

$$\textstyle\sum_0^\infty z^n\mathsf{P}(\overline{Y}_n=0)=\frac{\mathfrak{w}_{z+}(0)}{1-z}=\frac{1}{\sqrt{1-z}}\cdot$$

From these we obtain

$$\mathsf{P}(\theta_n=n)=\mathsf{P}(\overline{Y}_n=0)=\frac{(2n)!}{2^{2n}(n!)^2}\,;$$

$$\mathsf{P}(\eta(0)=n)=\frac{\mathsf{P}(\overline{Y}_n=0)}{2n-1}\,,$$

$$\mathsf{P}(\theta_n=k)=\mathsf{P}(\theta_k=k)\mathsf{P}(\overline{Y}_{n-k}=0)\,.\quad\square$$

With the help of Stirling's formula we find that as $k\to\infty$ and $n-k\to\infty$

$$\mathsf{P}(\overline{Y}_n=0)\sim\sqrt{\frac{1}{\pi n}}\quad\text{and}\quad\mathsf{P}(\theta_n=k)\sim\frac{1}{\pi\sqrt{k(n-k)}}\cdot\tag{12}$$

These relations can be made more precise by using the fact that

$$\mathsf{P}(\overline{Y}_n=0)=\sqrt{\frac{1}{\pi n}}\left(1-\frac{1}{8n}+O\left(\frac{1}{n^2}\right)\right)\cdot$$

It is natural to expect that if $\mathsf{P}(Y_n>0)\neq\frac{1}{2}$, but is near $\frac{1}{2}$, then the approximate formulae (12) remain valid. A decisive rôle begins to be played here by the convergence of the series of differences

$$r=\textstyle\sum_1^\infty \frac{1}{k}\left(\frac{1}{2}-\mathsf{P}(Y_k>0)\right).$$

Sufficient for the finiteness of r is, for example, that $\mathsf{M}\xi=0$ and $\mathsf{D}\xi<\infty$ (see Corollary 10; also [55] and [65]).

If $|r|<\infty$, then we obviously have as $z\to 1$

$$\textstyle\sum z^n\mathsf{P}(\overline{Y}_n=0)\sim\frac{e^r}{\sqrt{1-z}}\cdot$$

By the monotonicity of $\mathsf{P}(\overline{Y}_n=0)$ and well-known Tauberian theorems one then has

$$\mathsf{P}(\overline{Y}_n=0)\sim\frac{e^r}{\sqrt{\pi n}}\cdot$$

Since $\mathsf{P}(\theta_n=n)=\mathsf{P}^*(\bar{Y}_n=0)$, where P^*, as before, is "induced" by the sequence $-\xi_1, -\xi_2, \ldots$, we obtain

$$\mathsf{P}(\theta_n=n)\sim\frac{e^{-r}}{\sqrt{\pi n}} \quad \text{and} \quad \mathsf{P}(\theta_n=k)\sim\frac{1}{\pi\sqrt{k(n-k)}}\cdot$$

The latter is known as the *local arcsin law*. Recalling the remark of § 15 on the distribution of the time $T_n(+0)$ spent by the trajectory (2) in the domain $x>0$, we get

Corollary 8. The Arcsin Law. *If* $|r|<\infty$ *(for this it is sufficient, for example, that* $\mathsf{M}\xi=0$ *and* $\mathsf{D}\xi<\infty$ *or that the* ξ_k *be symmetric), then*

$$\lim_{n\to\infty}\mathsf{P}\left(\frac{\theta_n}{n}<\alpha\right)=\lim_{n\to\infty}\mathsf{P}\left(\frac{T_n(+0)}{n}<\alpha\right)=\frac{2}{\pi}\arcsin\sqrt{\alpha}\,.$$

If $\mathsf{M}\xi=0$, $\gamma_3=\mathsf{M}|\xi|^3$ exists and $F(x)$ is nonlattice, we can obtain asymptotic relations which are more precise than (12), such as

$$\mathsf{P}(\eta(+0)=n)\sim\frac{e^r}{2\sqrt{\pi}n^{3/2}} \quad \text{and} \quad \mathsf{P}(\eta(0)=n)\sim\frac{e^r\phi(1)}{2\sqrt{\pi}n^{3/2}}\cdot$$

These assertions remain valid for integer-valued random variables with lattice step equal to 1. The proofs can be found in [13].

§ 17. The Second Factorization Identity and Its Consequences

We now consider the transformation of the second distribution at (3).

Theorem 3. *For* $|z|<1$, $\operatorname{Im}\lambda\geqslant 0$ *and* $\operatorname{Im}\mu>0$

$$1-\frac{\lambda-\mu}{\lambda}\int_0^\infty e^{i\lambda x}\,d_x\mathsf{M}(z^{\eta(x)}\,e^{i\mu\chi(x)}\,;\,\eta(x)<\infty)=\frac{\mathfrak{w}_{z+}(\mu)}{\mathfrak{w}_{z+}(\lambda)}\cdot \tag{13}$$

We agree here that the function over which we are integrating is to have a jump at $x=0$ *equal to*

$$\mathsf{M}(z^{\eta(+0)}\,e^{i\mu\chi(+0)}\,;\,\eta(+0)<\infty)\,.$$

For $\lambda=\mu$, *the integral in the identity* (13) *equals* $\lambda\mathfrak{w}'_{z+}(\lambda)/\mathfrak{w}_{z+}(\lambda)$.

Proof. It will be convenient to use the following notation (see also Appendix 2): Let V be the ring of functions of bounded variation with the usual norm and multiplication defined as convolution. Let $\mathfrak{V}$ be the ring of Fourier–Stieltjes transformations of functions from V, which is isometrically isomorphic to V. The sub-rings generated by functions whose variation is zero on $(0,-\infty)$ or $(0,+\infty)$

will be denoted by the subscripts + and −, resp. Thus, for example,

$$\int_{+0}^{\infty} e^{i\lambda x}\, d\mathsf{P}(Y_k<x)\in\mathfrak{B}_+\,,$$

and for $|z|<1$

$$\sum_1^{\infty}\frac{z^k}{k}\int_{+0}^{\infty} e^{i\lambda x}\, d\mathsf{P}(Y_k<x)\in\mathfrak{B}_+\,,\qquad \mathfrak{w}_{z+}^{\pm 1}(\lambda)\in\mathfrak{B}_+\,.$$

The latter implies, in particular, the existence of a function $W_{z+}\in V$ for which

$$\mathfrak{w}_{z+}(\lambda)=\int_0^{\infty} e^{i\lambda t}\, dW_{z+}(t)\,.$$

Now let

$$u_{x,n}^y=\mathsf{P}(\overline{Y}_{n-1}<x,\ Y_n\geqslant x+y),\quad n=1,2,\ldots;\ y\geqslant 0;\ x>0\,.$$

This is a function of bounded variation in x since

$$\begin{aligned}|u_{x+\Delta,n}^y-u_{x,n}^y|&=|\mathsf{P}(\overline{Y}_{n-1}\in(x,x+\Delta),\ Y_n\geqslant x+y+\Delta)\\&\quad-\mathsf{P}(\overline{Y}_{n-1}<x,\ Y_n\in(x+y,x+y+\Delta))|\\&\leqslant\mathsf{P}(\overline{Y}_{n-1}\in(x,x+\Delta))+\mathsf{P}(Y_n\in(x+y,x+y+\Delta))\,.\end{aligned}$$

We have for $x>0$

$$u_{x,n+1}^y=\int_{-\infty}^{x} dF(t)\,u_{x-t,n}^y\,;\quad n\geqslant 1\,,$$

and

$$u_{x,1}^y=1-F(x+y)\,.$$

Moreover, for $|z|<1$ and $x>0$

$$\begin{aligned}&u_x^y(z)\equiv\sum_{n=1}^{\infty} z^n u_{x,n}^y=z\int_{-\infty}^{x} dF(t)\,u_{x-t}^y(z)+z(1-F(x+y))\,,\\&u_0^y(z)=0\,.\end{aligned}$$

Here, $u_x^y(z)\in V$. Hence, for $\operatorname{Im}\lambda=0$ we can write (here $b=b(z,y)$ does not depend on λ)

$$\begin{aligned}u^y(\lambda,z)&\equiv\int_0^{\infty} e^{i\lambda x}\, d_x u_x^y(z)\\&=b+z\int_0^{\infty} e^{i\lambda x}\, d_x\Big(\int_{-\infty}^{x} dF(t)\,u_{x-t}^y(z)\Big)-z\int_0^{\infty} e^{i\lambda x}\, dF(x+y)\\&=zf(\lambda)u^y(\lambda,z)-z\,e^{-i\lambda y}f(\lambda)+b-z\int_{-\infty}^{0} e^{i\lambda x}\, d_x\Big[\int_{-\infty}^{x} dF(t)\,u_{x-t}^y(z)+F(x+y)\Big]\,.\end{aligned}$$

The last integral is obviously a function, h_- say, in $\mathfrak{B}_-$. Further,

$$\begin{aligned}&u^y(\lambda,z)(1-zf(\lambda))=e^{-i\lambda y}(1-zf(\lambda))+h_-+b-e^{-i\lambda y}\,,\\&u^y(\lambda,z)\mathfrak{w}_{z+}(\lambda)=e^{-i\lambda y}\mathfrak{w}_{z+}(\lambda)+(h_-+b-e^{-i\lambda y})\mathfrak{w}_{z-}(\lambda)\phi^{-1}(z)\,.\end{aligned}$$

This implies, as before, that

$$u^y(\lambda,z)\mathfrak{w}_{z+}(\lambda)-e^{-i\lambda y}\int_y^{\infty} e^{i\lambda t}\, dW_{z+}(t)=c(z,y)$$

is an entire bounded function (a function belonging to both $\mathfrak{B}_+$ and $\mathfrak{B}_-$) and necessarily equal to a constant. Since $u_0^y(z) = u_\infty^y(z) = 0$,

$$u^y(0, z) = 0,$$

$$c(z, y) = -\int_y^\infty dW_{z+}(t)$$

and

$$u^y(\lambda, z) = \mathfrak{w}_{z+}^{-1}(\lambda)[\int_0^\infty (e^{i\lambda x} - 1)\, dW_{z+}(x+y)]\,. \tag{14}$$

It remains to find for $\operatorname{Im}\mu > 0$ the transform (all integrals here are absolutely convergent)

$$\begin{aligned} -\int_0^\infty e^{i\mu y}\, d_y u^y(\lambda, z) &= \int_0^\infty e^{i\lambda x}\, d_x(-\int_0^\infty e^{i\mu y}\, d_y u_x^y(z)) \\ &= \mathsf{M}(z^{\eta(+0)}\, e^{i\mu\chi(+0)}\,;\, \eta(+0) < \infty) \\ &\quad + \int_{+0}^\infty e^{i\lambda x}\, d_x \mathsf{M}(z^{\eta(x)}\, e^{i\mu\chi(x)}\,;\, \eta(x) < \infty)\,. \end{aligned}$$

By (14) this is

$$\begin{aligned} &-\frac{1}{\mathfrak{w}_{z+}(\lambda)} \int_0^\infty e^{i\mu y}\, d_y(e^{-i\lambda y} \int_y^\infty e^{i\lambda t}\, dW_{z+}(t) - \int_y^\infty dW_{z+}(t)) \\ &\quad = \frac{i\lambda}{\mathfrak{w}_{z+}(\lambda)} \int_0^\infty e^{i\mu y}\, (e^{-i\lambda y} \int_y^\infty e^{i\lambda t}\, dW_{z+}(t))\, dt = -\frac{\lambda(\mathfrak{w}_{z+}(\lambda) - \mathfrak{w}_{z+}(\mu))}{(\mu-\lambda)\mathfrak{w}_{z+}(\lambda)}\,. \quad \square \end{aligned}$$

Corollary 9. *For* $|z| \leqslant 1$ *and* $\operatorname{Im}\mu \geqslant 0$

$$1 - \mathsf{M}(z^{\eta(+0)}\, e^{i\mu\chi(+0)}\,;\, \eta(+0) < \infty) = \mathfrak{w}_{z+}(\mu)$$

and (15)

$$1 - \mathsf{M}(z^{\eta(0)}\, e^{i\mu\chi(0)}\,;\, \eta(0) < \infty) = \phi(z)\mathfrak{w}_{z+}(\mu)\,.$$

Proof. The first assertion follows immediately from (13) when we set $\lambda = i\infty$. To show the second we use Corollary 5 and note that

$$\begin{aligned} &[1 - \mathsf{M}(z^{\eta(+0)}\, e^{i\lambda\chi(+0)}\,;\, \eta(+0) < \infty)] \\ &\qquad \times [1 - \mathsf{M}(z^{\eta(0)}\,;\, \eta(0) < \infty, \chi(0) = 0)] \\ &\qquad\qquad = 1 - \mathsf{M}(z^{\eta(0)}\, e^{i\lambda\chi(0)}\,;\, \eta(0) < \infty)\,. \end{aligned}$$

In fact, we must show that

$$\begin{aligned} &\mathsf{M}(z^{\eta(+0)}\, e^{i\lambda\chi(+0)});\, \eta(+0) < \infty) \\ &\qquad \times [1 - \mathsf{M}(z^{\eta(0)}\,;\, \eta(0) < \infty, \chi(0) = 0)] \\ &\qquad\qquad = \mathsf{M}(z^{\eta(0)}\, e^{i\lambda\chi(0)}\,;\, \eta(0) < \infty, \chi(0) > 0)\,. \end{aligned}$$

But this follows from the fact that

$$\begin{aligned} &\mathsf{P}(\eta(+0) = k, \chi(+0) \in dx) \\ &\quad = \mathsf{P}(\eta(0) = k, \chi(0) \in dx) + \textstyle\sum_{j=1}^{k-1} \mathsf{P}(\eta(0) = j, \chi(0) = 0) \\ &\qquad\qquad \times \mathsf{P}(\eta(+0) = k - j, \chi(+0) \in dx)\,. \quad \square \end{aligned}$$

We can now find a new relation for the number e^r.

Corollary 10. *If* $\mathsf{M}\xi=0$ *and* $\mathsf{D}\xi=\sigma^2<\infty$, *then* $\mathsf{M}\chi(0)$ *and* $\mathsf{M}\chi(+0)$ *are finite and*

$$\mathsf{M}\chi(+0)=\frac{\sigma e^r}{\sqrt{2}}\geqslant \mathsf{M}\chi(0)=\frac{\sigma e^r\phi(1)}{\sqrt{2}}.$$

Proof. From the identity

$$(1-\mathsf{M}\,e^{i\mu\chi(+0)})(1-\mathsf{M}\,e^{i\mu\chi^*(0)})=1-f(\mu) \tag{16}$$

it follows that

$$\mathsf{P}(\chi(+0)>x)=-\int_0^\infty H^{**}(t)\,dg(x+t)\,, \tag{17}$$

where $H^{**}(t)$ is the renewal function of the random variable $-\chi^*(0)$ and

$$g(t)=\begin{cases}1-F(t)\,, & t\geqslant 0\,,\\ -F(t)\,, & t<0\,.\end{cases}$$

Since any renewal function is majorized by some linear function (Appendix 1), (17) does not exceed $c(g(x)+\int_x^\infty g(t)\,dt)$ for some $c>0$. Consequently, the existence of the variance implies that of $\mathsf{M}\chi(+0)$. In addition, we have from (15)

$$\mathsf{M}(\chi(+0)z^{\eta(+0)})=\mathfrak{w}_{z+}(0)\sum_{k=1}^\infty\frac{z^k}{k}\int_{+0}^\infty x\,d\mathsf{P}(Y_k<x)\,. \tag{18}$$

According to the central limit theorem, which guarantees the convergence of $\mathsf{P}(Y_n/\sigma\sqrt{n}<x)$ along with its first two moments to the normal law, the integral in (18) is asymptotically equivalent to $\sigma\sqrt{k}/\sqrt{2\pi}$ as $k\to\infty$. Hence, as $z\to 1$, the series in (18) is asymptotically equivalent to $\sigma/\sqrt{2(1-z)}$ and

$$\mathsf{M}\chi(+0)z^{\eta(+0)}\sim\sqrt{1-z}\,e^{r(z)}\frac{\sigma}{\sqrt{2(1-z)}}=\frac{\sigma\,e^{r(z)}}{\sqrt{2}}.$$

The convergence of the series r follows from that of $r(z)$ for $z\to 1$ and the simplest Tauberian theorems.

We can find $\mathsf{M}\chi(0)$ analogously. It is simpler to do this with the aid of the equality $\mathsf{M}\chi(+0)\mathsf{M}\chi^*(0)=-\sigma^2/2$ which follows directly from (16). □

We also cite here some relations for the moments of $\chi^*(0)$ and $\chi^*(-0)$ when the series $\sum_1^\infty(1/k)\mathsf{P}(Y_k>0)$ converges. In this case

$$[1-\mathsf{M}(e^{i\lambda\chi(+0)};\eta(+0)<\infty)][1-\mathsf{M}\,e^{i\lambda\chi^*(0)}]$$

$$=[1-\mathsf{M}(e^{i\lambda\chi(0)};\eta(0)<\infty)][1-\mathsf{M}\,e^{i\lambda\chi^*(-0)}]=1-f(\lambda)\,.$$

After dividing by $i\lambda$ and letting $\lambda\to 0$ we get

$$\mathsf{P}(\overline{Y}=0)\mathsf{M}\chi^*(0)=\mathsf{P}(\bar{y}<0)\mathsf{M}\chi^*(-0)=\mathsf{M}\xi\,. \tag{19}$$

Comparing this with Corollary 4. we have

$$\frac{\mathsf{M}\chi^*(0)}{\mathsf{M}\eta^*(0)}=\frac{\mathsf{M}\chi^*(-0)}{\mathsf{M}\eta^*(-0)}=\mathsf{M}\xi\,,$$

which are the well-known Wald identities.

We now obtain another identity, due also to Wald.

Corollary 11. *Assume that* $f_1(\mu)=f(-i\mu)=\mathsf{M}\,e^{\mu\xi}$ *is analytic in some strip* $0<\operatorname{Re}\mu<m, f_1(m)>1$. *Let* $\lambda(z)$ *be the largest root of the equation* $f_1(\mu)=1/z$, $0<z\leqslant 1/\inf_{\operatorname{Im}\mu=0} f_1(\mu)$ *($\lambda(z)$ is defined as the right branch of the double-valued function inverse to $f_1(\mu)$). Then for* $x>0$

$$\mathsf{M}(z^{\eta(x)}\, e^{\lambda(z)\chi(x)};\, \eta(x)<\infty)=e^{-\lambda(z)x}\,. \tag{20}$$

In particular,

$$\mathsf{M}(z^{\eta(+0)}\, e^{\lambda(z)\chi(+0)};\, \eta(+0)<\infty)=1\,.$$

Proof. Since $f_1(\mu)$ is convex, for $1>z>1/f_1(m)$ there exists a unique positive solution $\lambda(z)$ of the equation $f_1(\mu)=1/z$ contained in the interval $(0, m)$. Further, in the factorization identity $\mathfrak{w}_{z+}(\lambda)\phi(z)=\mathfrak{w}_{z-}(\lambda)(1-zf(\lambda))$ the function $f(\lambda)$, as well as $\mathfrak{w}_{z+}(\lambda)$, can be analytically continued into the domain $\operatorname{Im}\lambda>-m$. Since $\inf_{\operatorname{Im}\lambda\leqslant 0}|\mathfrak{w}_{z-}(\lambda)|>0$ in the lower half-plane, we necessarily have

$$\mathfrak{w}_{z+}(-i\lambda(z))=0\,. \tag{21}$$

The left side of the relation (13) can also be analytically continued into the domain $\operatorname{Im}\mu>-m$. At the point $\mu=-i\lambda(z)$ we have by (21)

$$\int_0^\infty e^{i\lambda x}\, d_x\mathsf{M}(z^{\eta(x)}\, e^{\lambda(z)\chi(x)};\, \eta(x)<\infty)=\frac{\lambda}{\lambda+i\lambda(z)}=1-\frac{i\lambda(z)}{\lambda+i\lambda(z)}\,.$$

This proves (20). The validity of (20) for all z for which $\lambda(z)$ is defined is established by analytic continuation w.r.t. z. □

The identity (20) can also be easily obtained directly by means of another useful identity:

$$e^{i\lambda x}\mathsf{M}(e^{i\lambda\chi(x)}z^{\eta(x)};\, \eta(x)<\infty)=1-(1-zf(\lambda))\sum_{n=0}^\infty z^n\mathsf{M}(e^{i\lambda Y_n};\, \overline{Y}_n<x)\,, \tag{22}$$

which follows from the equality

$$f^n(\lambda)=\sum_{k=1}^n \mathsf{M}(e^{i\lambda(x+\chi(x))};\, \eta(x)=k)f^{n-k}(\lambda)+\mathsf{M}(\exp(i\lambda Y_n);\, \overline{Y}_n<x)\,.$$

Setting $z=f(\lambda)^{-1}$ in (22) we find that

$$\mathsf{M}(e^{i\lambda\chi(x)}f(\lambda)^{-\eta(x)};\, \eta(x)<\infty)=e^{-i\lambda x}\,,$$

which is equivalent to (20). □

We mention that relations similar to (20), but differing from it in that they hold only asymptotically in x, can be obtained using Theorems 1 and 2; for example, one can find relations for

$$\sum z^n\mathsf{M}(\rho^{\theta_n};\, \overline{Y}_n>x)\,. \tag{23}$$

The same arguments as in Corollary 11 show that when the right side of the identity in Theorem 1 is analytically continued into the lower half-plane, the function $\mathfrak{w}_{z+}(0)/\mathfrak{w}_{z\rho+}(\lambda)$ will have its first singularity at the point $\lambda=-i\lambda(z\rho)$

which is an isolated simple pole. This allows calculation of the principal part of (23) as

$$-\frac{\mathfrak{w}_{z+}(0)\,e^{-\lambda(z\rho)x}}{\lambda(z\rho)\mathfrak{w}''_{z\rho+}(-i\lambda(z\rho))}.$$

Further inversion of these expressions with the saddle point method leads to limit theorems for Y_n and θ_n (compare with [8]). The determination of the asymptotic behavior of $\mathsf{P}(\bar{Y}>x)$, which is based on this procedure, is presented in Theorems 11 and 12 of Chapter 4.

We turn to the random variables $\chi(x)$ and $\eta(x)$ which occur in the second identity. *There exist only two possibilities when $\chi(x)$ and $\eta(x)$ are independent:*

$$1)\ \mathsf{P}(\xi\geqslant x)=ce^{-\alpha x}(c>0,\alpha>0)\quad \text{for } x>0$$

or

$$2)\ \mathsf{P}(\xi=x)=cp_1^{x-1}(c>0,p_1\geqslant 0)\quad \text{for } x=1,2,\ldots$$

and integer-valued ξ (here $p_1^0=1$ if $p_1=0$). In these cases, $\mathsf{P}(\xi\geqslant x+t\mid\xi\geqslant x)$ does not depend on x (x and t are integers if ξ is) and for the distributions 1) and 2) we have the relations

$$\mathsf{P}(\chi(x)\geqslant t\mid\eta(x)<\infty)=e^{-\alpha t}\quad\text{and}\quad \mathsf{P}(\chi(x)\geqslant t\mid\eta(x)<\infty)=p_1^t,\quad\text{resp.,}$$

and by Wald's identity

$$\mathsf{M}(z^{\eta(x)};\eta(x)<\infty)=\frac{\alpha-\lambda(z)}{\alpha}e^{-\lambda(z)x};$$

$$\mathsf{M}(z^{\eta(x)};\eta(x)<\infty)=\frac{1-p_1e^{\lambda(z)}}{1-p_1}e^{-\lambda(z)x}.$$

With the aid of these equalities we can prove the following exact formulae (compare with Chapter 2).

Corollary 12. *For $x>0$ we have in cases* 1) *and* 2), *resp.*

$$\mathsf{P}(\eta(x)=n)=\frac{x}{n}f(n,x)+\frac{1}{\alpha}\frac{d}{dx}\left(\frac{x}{n}f(n,x)\right),$$

$$\mathsf{P}(\eta(x)=n)=\frac{x}{n}\mathsf{P}(Y_n=x)+\frac{p_1}{1-p_1}\left[\frac{x}{n}\mathsf{P}(Y_n=x)-\frac{(x-1)\mathsf{P}(Y_n=x-1)}{n}\right],\tag{24}$$

where $f(n,x)=(d/dx)\mathsf{P}(Y_n<x)$ is the density of Y_n (for $x>0$, $f(n,x)$ exists and is infinitely differentiable).

For $p_1=0$ ($\xi\leqslant 1$) we obtain the already known formula $\mathsf{P}(\eta(x)=n)=(x/n)\mathsf{P}(Y_n=x)$.

Proof. We obtain the first of the relations at (24). By the total probability formula for $x>0$

$$f(n,x)=\textstyle\sum_{k=1}^{n}\mathsf{P}(\eta(x)=k)m_{n-k},$$

where m_j for $j \geq 0$ is the density at zero of the sum of independent random variables $Y_j + \chi(x)$. Consequently, denoting by $m(z)$ the generating function of the m_j, we get

$$f_x(z) \equiv \sum_{n=1}^{\infty} z^n f(n, x) = \mathsf{M}(z^{\eta(x)}; \eta(x) < \infty) m(z), \quad x > 0.$$

To find $m(z)$ we note that

$$\frac{1}{1 - zf(\lambda)} = 1 + \int_{-\infty}^{0} e^{i\lambda x} \sum_{n=1}^{\infty} z^n \, d\mathsf{P}(Y_n < x) + m(z) \int_0^{\infty} e^{i\lambda x} \mathsf{M}(z^{\eta(x)}; \eta(x) < \infty) \, dx. \tag{25}$$

The last term is equal to

$$-m(z) \frac{\alpha - \lambda(z)}{\alpha} \cdot \frac{1}{i\lambda - \lambda(z)}.$$

Since $\int_{-\infty}^{0} \cdots$ in (25) is an analytic function in the lower half-plane, we obtain by expanding the functions in (25) in a neighborhood of the pole $\lambda = -i\lambda(z)$:

$$\frac{1}{izf'(-i\lambda(z))} = -\frac{m(z)(\alpha - \lambda(z))}{\alpha}.$$

Consequently,

$$f_x(z) = -\frac{e^{-\lambda(z)x}}{izf'(-i\lambda(z))} = e^{-\lambda(z)x} z\lambda'(z) = \frac{z}{x} \frac{d}{dz} (e^{-\lambda(z)x}).$$

Now it is easy to establish that

$$z \frac{d}{dz} \mathsf{M}(z^{\eta(z)}; \eta(x) < \infty) = x f_x(z) + \frac{1}{\alpha} \frac{d}{dx} (x f_x(z)),$$

which is in turn equivalent to (24). The second relation at (24) is proved completely analogously. □

Remark. The distributions possessing properties 1) or 2) obviously belong to the class of distributions for which $\mathfrak{w}_{z\pm}(\lambda)$ can be found explicitly. We will see below that this class is sufficiently "rich" and contains all distributions for which either $\int_0^{\infty} e^{i\lambda x} \, dF(x)$ or $\int_{-\infty}^{0} e^{i\lambda x} \, dF(x)$ is a rational function of λ (or of $e^{i\lambda}$ for integral ξ).

The members of this class possess another noteworthy property. A number of *boundary problems for the random walk* (2) *in a strip* (i.e., with boundaries $x > 0$ and $-y < 0$) *reduce to problems with a single boundary* (compare with [38]). Let us consider as an example the distributions with the property 2), and the problem of determining the probability $p(x, y)$, that the trajectory (2) crosses the upper boundary x before the lower boundary $-y$. This problem is important in sequential analysis and certain areas of queueing theory. If $\mathsf{P}(\overline{Y}^* > -\infty) = 1$, then in view of the properties of the distributions mentioned above we obtain by the total probability formula (x and y are integers)

$$\mathsf{P}(\overline{Y}^* > -y) = p(x, y) \sum_{k=0}^{\infty} p_1^k (1 - p_1) \mathsf{P}(\overline{Y}^* > -x - y - k).$$

In particular, for $p_1=0$ ($\xi\leqslant 1$)

$$p(x,y)=\frac{\mathsf{P}(\bar{Y}^*>-y)}{\mathsf{P}(\bar{Y}^*>-x-y)}.$$

Analogous formulae for $p(x,y)$ obviously hold in Case 1) as well.

Concluding this chapter we remark that Corollaries 6 and 9 imply that

$$\mathsf{M}\rho^{\theta_\infty}e^{i\lambda\bar{Y}}=\frac{1-\mathsf{P}(\eta(+0)<\infty)}{1-\mathsf{M}(\rho^{\eta(+0)}e^{i\lambda\chi(+0)};\eta(+0)<\infty)}. \tag{26}$$

This corresponds to the fact that $\bar{Y}$ and θ_∞ can be written as

$$\bar{Y}=\chi_1+\dots+\chi_\nu \quad\text{and}\quad \theta_\infty=\eta_1+\dots+\eta_\nu,$$

where $(\chi_j,\eta_j), j=1,2,\dots$, are independent and distributed like $(\chi(+0),\eta(+0))$ under the condition that $\eta(+0)<\infty$, and ν is independent of these sequences with

$$\mathsf{P}(\nu=k)=p(1-p)^k,\quad k=0,1,\dots;p=\mathsf{P}(\bar{Y}>0).$$

Factorization identities also allow the discovery of many other properties of distributions of boundary functionals. With their help we can prove limit theorems for the distribution of $(\bar{Y}_n, Y_n)$ as $n\to\infty$, for that of $\eta(x)$ as $x\to\infty$, and several others (see [8]). The distributions of $\chi(+0)$, $\chi(0)$, $\chi(\infty)$ and, partly, $\bar{Y}$, are considered in [15], [5], [9] and [52]. We will continue our investigation of the properties of $\bar{Y}$ below.

Chapter 4

Properties of the Supremum of Sums of Independent Random Variables and Related Problems of Queueing Theory

§ 18. Uniqueness Theorems

In this chapter we retain all of the notation of Chapter 3 and will always assume that $\mathsf{P}(\bar{Y}<\infty)=1$. In this case (see Corollaries 6 and 9, Chapter 3)

$$1-f(\lambda)=\frac{1-p}{\mathsf{M}\, e^{i\lambda\bar{Y}}}[1-\mathsf{M}\, e^{i\lambda\chi^*(0)}]\,,\quad p=\mathsf{P}(\bar{Y}>0)\,. \tag{1}$$

We now assume that we have obtained some representation of the function $1-f(\lambda)$ in the form

$$1-f(\lambda)=\frac{\mathfrak{w}_-(\lambda)}{\mathfrak{w}_+(\lambda)};\quad \operatorname{Im}\lambda=0\,, \tag{2}$$

where $\mathfrak{w}_\pm(\lambda)\in\mathfrak{B}_\pm, \mathfrak{w}_+(i\infty)=1$ (the sub-rings $\mathfrak{B}_\pm$ are defined in § 17 and Appendix 2). In particular, such a representation is effected by Eq. (1). Is it possible to show that

$$\mathsf{M}\, e^{i\lambda\bar{Y}}=(1-p)\mathfrak{w}_+(\lambda)\ ? \tag{3}$$

In other words, we are concerned with the question of *uniqueness of the factorization* (2), which we will call the *V-factorization.*

To avoid misunderstandings we emphasize that in contrast to the factorization (1) in § 15 it is more convenient here to put the positive component in the denominator. This makes no great difference since replacement of λ by $-\lambda$ ($f(-\lambda)$ is the characteristic function of $-\xi$) changes the positive component into the negative and vice versa.

Theorem 1. *The V-factorization of the function* $1-f(\lambda)$ *is unique.*

Proof. Denote by $\mu(B)$ the measure corresponding to $\mathfrak{w}_+(\lambda)$ in the representation (2) and put

$$F(B)=\int_B dF(x)\,.$$

Then the equality $\mathfrak{w}_+(\lambda)(1-f(\lambda))=\mathfrak{w}_-(\lambda)$ implies that for any measurable $B\subset(0,\infty)$

$$\mu(B)=\int_0^\infty \mu(dx)F(B-x)\,.$$

Setting $B=(0, \infty)$ and letting μ_0 be the measure of the positive half of the real line, we get

$$\mu_0-\mu(0)=\mu_0-\int_0^\infty \mu(dx)F((-\infty, -x])$$

or

$$\mu(0)=\int_0^\infty \mu(dx)F((-\infty, -x]). \tag{5}$$

But this means that μ satisfies Eqs. (4) and (5) for the invariant measure of a random walk on the positive axis with jumps ξ_k and a delaying barrier at zero. This walk forms a Markov chain with transition function

$$P(x, B)=F(B-x) \quad \text{if } B\subset(0, \infty),$$
$$P(x, 0)=F((-\infty, -x]).$$

Such a random walk is recurrent since $\mathsf{P}(\overline{Y}<\infty)=1$ implies $\mathsf{P}(\underline{y}^*=\inf(Y_1, Y_2, \ldots)=-\infty)=1$. The time of return to the point 0 has finite expectation (Corollary 4, Chapter 3). From this and well-known theorems there follow immediately the existence and uniqueness of a finite invariant measure, i.e., the existence and uniqueness (up to a multiplicative constant) of a solution of (4) and (5).

This can also be shown directly. Indeed, let $\tau_0, \tau_0+\tau_1, \tau_0+\tau_1+\tau_2, \ldots$ be the times the random walk returns to the point 0 and put

$$\eta_n-1=\max\{k: \tau_0+\cdots+\tau_k\leqslant n\}, \qquad \gamma_n=n-\tau_0-\cdots-\tau_{\eta_n-1}.$$

If ζ_n is the position of the random walk after n steps, then

$$P^{(n)}(x, B)=\mathsf{P}(\zeta_n\in B \mid \zeta_0=x)=\sum_{k=0}\mathsf{P}(\gamma_n=k \mid \zeta_0=x)\mathsf{P}(\zeta_n\in B \mid \gamma_n=k, \zeta_0=x).$$

But here (see Appendix 1) we have as $n\to\infty$

$$\mathsf{P}(\gamma_n=k \mid \zeta_0=x)\to\frac{\mathsf{P}(\tau_1>k)}{\mathsf{M}\tau_1},$$

$$\mathsf{P}(\zeta_n\in B \mid \gamma_n=k, \zeta_0=x)=\mathsf{P}(\zeta_k\in B \mid \zeta_0=0; \zeta_j\neq 0, j=1, 2, \ldots, k).$$

This proves the existence of an invariant probability measure μ_p since

$$P^{(n)}(x, B)\to\mu_p(B)=\sum_{k=0}\frac{\mathsf{P}(\tau_1>k)}{\mathsf{M}\tau_1}\mathsf{P}(\zeta_k\in B \mid \zeta_0=0; \zeta_j\neq 0, j=1, 2, \ldots, k).$$

The uniqueness (up to a multiplicative constant) follows, as usual, from the following relations: if $\mu(B)$ is an invariant measure, then

$$\mu(B)=\int\mu(dx)P^{(1)}(x, B)=\cdots=\int\mu(dx)P^{(n)}(x, B)\to\int\mu(dx)\mu_p(B)=\mu_0\mu_p(B).$$

We now note that (4) and (5) are also satisfied by the distribution of $\overline{Y}$. Hence,

$$\mu(B)=c\mathsf{P}(\overline{Y}\in B).$$

From $\mathfrak{w}_+(i\infty)=1$ it follows that $\mu(0)=1=c\mathsf{P}(\overline{Y}=0)$, and $c=1/(1-p)$. □

The uniqueness of the representation (2) remains unclear in the wider class of ordinary factorizations, i.e., when $\mathfrak{w}_\pm(\lambda)\in\mathfrak{B}_\pm$ is replaced by the weakened requirement of analyticity and boundedness of $\mathfrak{w}_\pm$ in the regions Im $\lambda \lessgtr 0$, resp. and continuity at the finite points of the boundary Im $\lambda = 0$.

One can find additional sufficient conditions for the uniqueness of such a factorization (which obviously coincides in this case with the V-factorization). For example, the requirement of analyticity of $f(\lambda)$ in the strip $0 > \text{Im}\,\lambda > -\varepsilon$ or $\varepsilon > \text{Im}\,\lambda > 0$ for some $\varepsilon > 0$. One can also establish (see Appendix 2) the uniqueness of the factorization of the function $(i\lambda+1)[1-f(\lambda)]/i\lambda$ if $\mathsf{M}\xi$ exists and the distribution of ξ_k is either lattice, or

$$\limsup_{|\lambda|\to\infty} |1-f(\lambda)| > 0 .$$

In the sequel we will consider the functions $\mathfrak{w}_\pm(\lambda)$ *as the components of the V-factorization* (2) *of the function* $1-f(\lambda)$. In this case, by Theorem 1, $\mathfrak{w}_+$ is uniquely determined by the function f: $\mathfrak{w}_+(\lambda) = \mathsf{M}\, e^{i\lambda\bar{Y}}/(1-p)$ or, what is the same, $\mathsf{M}\, e^{i\lambda\bar{Y}} = \mathfrak{w}_+(\lambda)/\mathfrak{w}_+(0)$.

§ 19. Methods of Finding the Distribution of $\bar{Y}$

If $F(x)$ has an absolutely continuous component and $\mathsf{M}|\xi| < \infty$, then (see Appendix 2) for any $\beta > 0$

$$\mathfrak{w}_+(\lambda) = \exp\left\{\frac{1}{2\pi i}\int_{-\infty}^{\infty}\frac{\ln\left(\dfrac{1-f(\mu)}{i\mu}(i\mu+\beta)\right)}{\mu-\lambda}\,d\mu\right\}; \quad \text{Im}\,\lambda > 0 . \tag{6}$$

The integral here is taken in the principal-value sense. Analogous formulae also hold in the lattice case. However, they are far from being universally effective in determining $\mathfrak{w}_+$.

We will define below a class $\mathfrak{R}$ of distribution functions which is *everywhere dense in the sense of weak convergence in the set of all distribution functions and is such that for* $F \in \mathfrak{R}$ *the component* $\mathfrak{w}_+$ *can be found in explicit form*. To this approach one must naturally add theorems on the *continuous dependence* of $\mathfrak{w}_+$ on f.

The class $\mathfrak{R}$ is described as follows: $F \in \mathfrak{R}$ *if at least one of the functions* $f^\pm$ *in the decomposition*

$$f = f^+ + f^- , \qquad f^\pm \in \mathfrak{B}_\pm$$

(we can put $f^+(\lambda) = \int_{+0}^{\infty} e^{i\lambda x}\, dF(x)$) *is rational*. If ξ is integer-valued, then at least one of the functions $f^\pm$ must be a *rational function* of $e^{i\lambda}$.

In order for f^+, say, to be rational it is *necessary and sufficient* that $1-F(x)$ for $x > 0$ be representable as a sum

$$1-F(x) = \textstyle\sum_k P_k(x)\, e^{-\alpha_k x} \quad (\text{Re}\,\alpha_k > 0) , \tag{7}$$

where $P_k(x)$ are polynomials in x. Expressions of this form will be called *exponential polynomials*.

We will show that arbitrary distribution functions can be approximated by functions from $\mathcal{R}$. In fact, let $F(x)$ be a distribution function on $(0, \infty)$. There always exists a sequence of continuous distribution functions $F_n(x)$ such that the Lévy-distance between F and F_n

$$L(F, F_n) \to 0 \tag{8}$$

as $n \to \infty$. Make the monotone substitution $x = -\ln(1-y)$. Then (8) is equivalent to

$$L(F^1, F_n^1) \to 0,$$

where $F^1(y) = F(-\ln(1-y))$ and $F_n^1(y) = F_n(-\ln(1-y))$ are distributions on $[0, 1]$. The problem is thus reduced to one of approximation of distribution functions on $[0, 1]$ by functions of the form

$$\sum_k p_k(-\ln(1-y))(1-y)^{\alpha_k}$$

or by those of the more special form $\sum p_k(1-y)^k$, where the p_k are constants. But for given $\varepsilon > 0$ we can find a distribution F_n^1 which has a continuous and positive density $p_n(x)$ on $[0, 1]$ and which satisfies the inequality $L(F^1, F_n^1) < \varepsilon$. We can then select a polynomial

$$p_{n,m}(y) = \sum_{k=0}^m p_k(1-y)^k, \ \sum_{k=0}^m \frac{p_k}{k+1} = 1$$

which is nonnegative on $[0, 1]$ and for which

$$\sup_{[0,1]} |p_n(x) - p_{n,m}(x)| < \varepsilon.$$

Then

$$F_{n,m}(y) = \int_0^y p_{n,m}(t)\, dt$$

will be a distribution function and a polynomial of the required form, and we have

$$L(F_n^1, F_{n,m}) \leqslant \sqrt{2} \sup_{t \in [0,1]} |F_n^1(t) - F_{n,m}(t)| \leqslant \sqrt{2}\, \varepsilon.$$

Returning to the variable x we see that we have proved the existence of a sequence of functions of the form $\sum_{k=0}^m p_k e^{-kx}$ which is L-convergent to $F(x)$. Distributions on $(-\infty, 0)$ are treated analogously. □

The assertions derived below will clarify the rôle played by the class $\mathcal{R}$ in theorems on the explicit resolvability of a factorization.

Theorem 2. *Assume* $\mathsf{M}\xi$ *exists. Then* $\mathfrak{w}_+(\lambda)$ *is a rational function iff*

$$f^+(\lambda) = \int_{+0}^{\infty} e^{i\lambda x}\, dF(x)$$

is rational.

If $f^+ = P_m/Q_n$ *is the irreducible ratio of polynomials* P_m *and* Q_n *of degrees* m *and* n *and if* $\lambda_1, \ldots, \lambda_N$ *are all the zeros of the function* $1 - f(\lambda)$ *in the region* $\operatorname{Im} \lambda < 0$,

then

$$N=n \quad \text{and} \quad \mathfrak{w}_+(\lambda)=\frac{Q_n(\lambda)}{\prod_{k=1}^n(\lambda-\lambda_k)}.$$

This fraction is irreducible. (*We assume here that the coefficient of* λ^n *in* $Q_n(\lambda)$ *equals* 1.)

The proof of the *sufficiency* will be broken down into several steps.

1. Consider the function

$$\mathfrak{v}(\lambda)=\frac{1-f(\lambda)}{i\lambda}(i\lambda+1), \quad \operatorname{Im}\lambda=0.$$

It is clear that $\mathfrak{v}(\lambda)\neq 0$ for $\operatorname{Im}\lambda=0$ because of the fact that $\mathfrak{v}(0)=-\mathsf{M}\xi>0$ and

$$\limsup_{|\lambda|\to\infty}|f(\lambda)|\leqslant 1-f^+(0)<1. \tag{9}$$

Moreover, $\mathfrak{v}\in\mathfrak{B}$ since

$$\frac{1-f(\lambda)}{i\lambda}=\int_{-\infty}^{\infty} g(t)\,e^{i\lambda t}\,dt, \qquad g(t)=\begin{cases}1-F(t), & t\geqslant 0,\\ -F(t), & t<0.\end{cases}$$

From (9) it also follows that for sufficiently large T

$$\operatorname{ind}_T\mathfrak{v}\equiv\frac{1}{2\pi}\int_{-T}^{T} d(\arg\mathfrak{v}(\lambda))\in\left(-\frac{b}{2},\frac{b}{2}\right); \quad b<1.$$

2. We represent the function $\mathfrak{v}(\lambda)$ as the product

$$\mathfrak{v}(\lambda)=\frac{Q_n-P_m-Qf^-}{i\lambda(i\lambda+1)^{n-1}}\cdot\frac{(i\lambda+1)^n}{Q_n}$$

and denote the factors by $\mathfrak{v}_1$ and $\mathfrak{v}_2$, resp. Then, for large enough T

$$|n+\operatorname{ind}_T\mathfrak{v}_1|<\tfrac{1}{2}. \tag{10}$$

Indeed, since all the zeros of Q_n lie in the domain $\operatorname{Im}\lambda<0$, it follows that $\mathfrak{v}_2\in\mathfrak{B}_+$. Moreover, $\mathfrak{v}_2$ has n zeros in the upper half-plane and $\mathfrak{v}_2(\lambda)\to i^n$ as $|\lambda|\to\infty$. This implies that as $T\to\infty$

$$\operatorname{ind}_T\mathfrak{v}_2\to\operatorname{ind}\mathfrak{v}_2=\frac{1}{2\pi}\int_{-\infty}^{\infty} d\arg\mathfrak{v}_2(\lambda)=n.$$

The assertion (10) then follows from Subsection 1 and the fact that

$$\operatorname{ind}_T\mathfrak{v}=\operatorname{ind}_T\mathfrak{v}_1+\operatorname{ind}_T\mathfrak{v}_2.$$

3. The function $\mathfrak{v}_1$ is in $\mathfrak{B}_-$ and has exactly n zeros in the domain $\operatorname{Im}\lambda<0$. The membership of $\mathfrak{v}_1$ in $\mathfrak{B}_-$ follows from the fact that in the equality

$$f^-(\lambda)=f^-(0)+\frac{i\lambda a(\lambda)}{i\lambda+1}$$

the function a belongs to $\mathfrak{B}_-$ and

$$R_{n-1}(\lambda)=\frac{Q_n(\lambda)-P_m(\lambda)-Q_n(\lambda)f^-(0)}{i\lambda}$$

is a polynomial of degree $n-1$. Thus,

$$\mathfrak{v}_1=\frac{R_{n-1}}{(i\lambda+1)^{n-1}}+\frac{a(\lambda)Q_n(\lambda)}{(i\lambda+1)^n}\in\mathfrak{B}_-.$$

We now consider a positively-oriented contour $\mathscr{T}_T$ formed from the segment $[-T, T]$ and the lower half of the circle $|\lambda|=T$ and will find the value of

$$\frac{1}{2\pi}\int_{\mathscr{T}_T} d\arg\mathfrak{v}_1(\lambda). \tag{11}$$

Since as $|\lambda|\to\infty$ and for $\operatorname{Im}\lambda\leqslant 0$

$$\mathfrak{v}_1(\lambda)\sim i^{-n}(1-f^-(\lambda)),\ |f^-(\lambda)|<1,$$

the part of the integral (11) corresponding to integration over the semi-circular arc will be smaller than $\frac{1}{2}$ for sufficiently large T. Comparing this with (10) we find that the integral (11), being a whole number, is necessarily equal to n. This also means that the function $\mathfrak{v}_1$ has n zeros in the domain $\operatorname{Im}\lambda<0$, which we will denote by $\lambda_1, \ldots, \lambda_n$. Since these cannot be zeros of Q_n (otherwise we would have $P_m(\lambda_j)=0$), the function

$$1-f(\lambda)=\mathfrak{v}_1\frac{(i\lambda-1)^{n-1}i\lambda}{Q_n}$$

has the same zeros as $\mathfrak{v}_1$ in the domain $\operatorname{Im}\lambda<0$.

4. It remains to set

$$\mathfrak{w}_+(\lambda)=\frac{Q_n}{\prod_{k=1}^n(\lambda-\lambda_k)}\quad\text{and}\quad\mathfrak{w}_-(\lambda)=\frac{(Q_n-P_m-Q_nf^-)}{\prod_{k=1}^n(\lambda-\lambda_k)}$$

and to note that $\mathfrak{w}_\pm\in\mathfrak{B}_\pm$.

Necessity. Now assume $\mathfrak{w}_+(\lambda)$ is rational. This means that $\mathfrak{w}_+^{-1}(\lambda)$ is also rational and $\mathfrak{w}_+^{-1}(\lambda)\in\mathfrak{B}_+$. That is, in the representation

$$\mathfrak{w}_+^{-1}(\lambda)=\int_0^\infty e^{i\lambda x}\,dp(x)$$

the function $p(x)$ is for $x>0$ an "exponential polynomial":

$$p(x)=\sum_k p_k(x)\,e^{-\alpha_k x},\quad \operatorname{Re}\alpha_k>0,$$

where the $p_k(x)$ are ordinary polynomials.

From the equalities

$$1-f=\mathfrak{w}_-\mathfrak{w}_+^{-1}\quad\text{and}\quad\mathfrak{w}_-(\lambda)=\int_{-\infty}^0 e^{i\lambda x}\,dW_-(x)$$

it follows that for $x>0$

$$1-F(x)=\int_\infty^0 dW_-(t)p(x-t).$$

Now we need only note that an expression of the form

$$\int_{-\infty}^{0} dW_{-}(t)(x-t)^m e^{-\alpha_k(x-t)} = e^{-\alpha_k x} \sum_{l=0}^{m} (-1)^{m-l}\binom{m}{l} x^l \int_{-\infty}^{0} e^{\alpha_k t} t^{m-l}\, dW_{-}(t)$$

is an exponential polynomial. □

The proof of the discrete analogue of Theorem 2 is somewhat easier. Let ξ be integer-valued:

$$d(z) = \mathsf{M} z^{\xi}, \qquad d^{+}(z) = \mathsf{M}(z^{\xi}; \xi > 0), \qquad d^{-}(z) = \mathsf{M}(z^{\xi}; \xi \leqslant 0).$$

Theorem 2A. *Let $\mathsf{M}|\xi| < \infty$. Then $\mathsf{M}z^{\bar{Y}}$ is a rational function iff $d^{+}(z)$ is a rational function. If $d^{+}(z) = P_m/Q_n$, where P_m and Q_n are polynomials of degrees m and n, and if $z_1, \ldots, z_N$ are all the zeros of the function $1 - d(z)$ in the domain $|z| > 1$, then*

$$\mathsf{M} z^{\bar{Y}} = (1-p) \frac{Q_n(z)}{\prod_{k=1}^{N} (z - z_k)},$$

where $N \leqslant \max(m, n)$ is the maximal power of z contained in the Laurent expansion of the function $Q_n - P_m - d^{-} Q_n$.

The proof of this assertion differs only by simplifications from that of Theorem 2 (here

$$\operatorname{ind} \frac{1-d(z)}{1-z} = \frac{1}{2\pi} \int_{0}^{2\pi} d\left[\arg \frac{1-d(e^{i\varphi})}{1-e^{i\varphi}}\right] \text{ exists}),$$

and we will therefore omit it.

Theorem 3. *Let $\mathsf{M}|\xi| < \infty$. Then $\mathfrak{w}_{+}(\lambda)$ has a representation of the form*

$$\mathfrak{w}_{+}(\lambda) = \frac{R(\lambda)}{1 - f(\lambda)}, \tag{12}$$

where R is a rational function, iff f^{-} is rational. If $f^{-} = P_m/Q_n$, then $1 - f(\lambda)$ has exactly $n-1$ zeros in the domain $\operatorname{Im} \lambda > 0$ (denoted by $\lambda_1, \ldots, \lambda_{n-1}$) and

$$R(\lambda) = \frac{i\lambda \prod_{k=1}^{n-1} (\lambda - \lambda_k)}{Q_n(\lambda)}. \tag{13}$$

For simplicity of notation, the function $\mathfrak{w}_{+}(\lambda)$ is given at (12) with accuracy up to a multiplicative constant.

Proof. Theorem 3 does not differ essentially from Theorem 2, but unfortunately, it is not a direct consequence of it. We give here a short proof of Theorem 3 under the simplifying assumption that $F(x)$ is absolutely continuous. Using the scheme of the proof of Theorem 2, the reader can obtain the general case without difficulty.

Since the path of $\mathfrak{v}(\lambda)$, $-\infty < \lambda < \infty$ does not intersect the radial $\arg \mathfrak{v} = -\pi$, in our case there exists

$$\operatorname{ind} \mathfrak{v} = \lim_{T \to \infty} \operatorname{ind}_T \mathfrak{v} = 0.$$

Again denoting the factors in the product

$$\mathfrak{v}(\lambda)=\frac{Q_n-P_m-Q_nf^+}{i\lambda(i\lambda-1)^{n-1}}\cdot\frac{(i\lambda+1)(i\lambda-1)^{n-1}}{Q_n}$$

by $\mathfrak{v}_1$ and $\mathfrak{v}_2$, we obtain

$$\operatorname{ind}\mathfrak{v}_2=-n+1 \quad\text{and}\quad \operatorname{ind}\mathfrak{v}_1=n-1\,.$$

Since $\mathfrak{v}_1\in\mathfrak{B}_+$, the argument principle says that $\mathfrak{v}_1$, along with $1-f$, has $n-1$ zeros in the domain $\operatorname{Im}\lambda>0$. Setting

$$\mathfrak{w}_+=\frac{i\lambda\prod_{k=1}^{n-1}(\lambda-\lambda_k)}{(1-f(\lambda))Q_n},$$

we obtain the V-factorization.

The proof of the *necessity* follows as before from the factorization identity (2) if we use the fact that $\mathfrak{w}_+^{-1}\in\mathfrak{B}_+$. □

The analogue of Theorem 3 in the discrete case is given by

Theorem 3A. *Let* $\mathsf{M}|\xi|<\infty$. *Then*

$$\mathsf{M}z^{\bar{Y}}=\frac{R(z)}{1-d(z)},$$

where R is a rational function, iff d^- is rational. If

$$d^-(z)=P_m/Q_n,$$

then

$$\mathsf{M}z^{\bar{Y}}=c\frac{(1-z)\prod_{k=1}^{n-1}(z-z_k)}{(1-d(z))Q_n(z)},$$

where $z_1,\ldots,z_{n-1}$ are all the zeros of $1-d(z)$ in the domain $|z|<1$.

We formulate here as a *conjecture* the following assertion:
One can represent $\mathfrak{w}_+^{-1}(\lambda)$ in the form $R_1+f^+R_2$, where R_1 and R_2 are rational functions, iff at least one of the functions $f^\pm$ is rational.

§ 20. Explicit Formulae for the Distribution of $\bar{Y}$ under the Conditions of Queueing Theory

We have seen that the distribution of $\bar{Y}$, when ξ is the difference between two independent, nonnegative random variables (for example, the service time τ^s minus the inter-arrival time τ^e; see Chapter 1) is of interest in queueing theory. For such ξ the function $f(\lambda)$ can be written as

$$f(\lambda)=f_+(\lambda)f_-(\lambda)\,,\quad f_\pm\in\mathfrak{B}_\pm\,,$$

and consequently, it admits a V-factorization. Under what conditions is the factorization resolvable in the sense of Theorems 2 and 3 in terms of the functions $f_\pm$?

If f_+ is a rational function we will call one of its purely imaginary poles λ_Γ *principal* if it has the property that

$$\operatorname{Im} \lambda_j \leqslant \operatorname{Im} \lambda_\Gamma$$

when $\lambda_1, \lambda_2, \ldots$ are the poles of f_+. It is clear that if f_+ is a characteristic function, then a principal pole always exists.

Lemma 1. *If*

$$f = f_+ f_- \quad \textit{and} \quad f_+ = P_m / Q_n$$

is a rational function, then f^+ is also rational:

$$f^+ = \frac{T_M}{Q_N},$$

where $M \leqslant N \leqslant n$, $Q_N = Q_n / \prod_{j=1}^{l} (\lambda - \mu_j)$ and $(\mu_1, \ldots, \mu_l)$ is the intersection of the set of zeros of Q_n and the set of zeros of the function f_- (with multiplicities taken into account). The point λ_Γ does not belong to this intersection.

Proof. We have

$$Q_n f^+ = P_m f_- - Q_n f^- .$$

This means that $Q_n f^+$ is an entire function which grows no faster than $|\lambda|^n$. Such a property can be possessed only by a polynomial. Denoting the latter by T, we get

$$f^+ = T / Q_n .$$

It remains to remark that the poles of f^+ coincide with those of the function f_- / Q_n and that $f_-(\lambda_\Gamma) > 0$. □

Remark. The converse to the statement that the rationality of f^+ implies that of f_+ is not true. This is shown by the following example. Let

$$f_-(\lambda) = \tfrac{1}{4} e^{-2i\lambda} (e^{i\lambda} + 3)$$

and

$$f_+(\lambda) = \frac{\alpha}{1 - i\lambda} + 2^{-1} \beta \left[\sum_{k=1}^{\infty} \frac{2^{-k}}{(\ln 3 - i\lambda - (2k-1)\pi i)} + \sum_{k=1}^{\infty} \frac{2^{-k}}{(\ln 3 - i\lambda + (2k-1)\pi i)} \right],$$

so that the distribution corresponding to f_+ has for suitably chosen α and β a density equal to

$$\alpha e^{-x} + \beta 3^{-x} \sum_{k=1}^{\infty} 2^{-k} \cos (2k-1)\pi x , \quad x > 0 .$$

One can show immediately that for $x > 0$

$$\mathsf{P}(\xi > x) = \frac{\alpha e^{-x}}{4} (e^{-1} + 3 e^{-2}) .$$

Consequently, $f^+(\lambda) = c / (1 - i\lambda)$, although f_+ is not a rational function.

If f_- has a finite number of zeros in the lower half-plane, then the rationality of f^+ implies that of f_+.

As consequences of the lemma and Theorem 2 we can now formulate the required conditions for the resolvability of a factorization.

Theorem 4. *Let* $\mathsf{M}|\xi|<\infty$ *and* f *be representable as* $f=f_+f_-$. *Then the rationality of* f_+ *is sufficient for that of* $\mathfrak{w}_+$.

If $f_+=P_m/Q_n$, *where* P_m *and* Q_n *are polynomials of degrees* m *and* n, *then the function* $Q_n(1-f)$ *has exactly* n *zeros* $\lambda_1, \ldots, \lambda_n$ *in the domain* $\operatorname{Im}\lambda<0$ *and*

$$\mathfrak{w}_+(\lambda)=\frac{Q_n(\lambda)}{\prod_{k=1}^{n}(\lambda-\lambda_k)}.$$

In this fraction we can cancel $l<n$ *factors, where* l *is the number of common zeros (taking account of multiplicities) of the functions* Q_n *and* f_-.

This assertion is a generalization of a number of well-known theorems in queueing theory. The example above shows that the rationality of $\mathsf{M}\,e^{i\lambda\bar{Y}}$ does not, generally speaking, imply that of f_+.

If the random variable S (see (3) § 2) has an *Erlang distribution*, i.e., if $f_+(\lambda)=(1-i\lambda/\alpha)^{-n}$ ($n>0$ an integer), then the function $1-f(\lambda)$ will have n zeros $\lambda_1, \ldots, \lambda_n$ in $\operatorname{Im}\lambda<0$, all different from $-i\alpha$, and

$$\mathfrak{w}_+(\lambda)=\frac{\left(1-\dfrac{i\lambda}{\alpha}\right)^n}{\prod_{k=1}^{n}(\lambda-\lambda_k)}(i\alpha)^n.$$

Theorem 4 also implies that as $x\to\infty$ (see the notation of Corollary 11, Chapter 3)

$$\mathsf{P}(\bar{Y}\geqslant x)=\frac{(1-p)Q_n(-i\lambda(1))}{i\lambda(1)\prod_{k=2}^{n}(-i\lambda(1)-\lambda_k)}e^{-\lambda(1)x}+O(e^{-(\lambda(1)+\varepsilon)x}),$$

where $\lambda(1)>0$ is the unique number for which $f(-i\lambda(1))=1$ ($\lambda_1=-i\lambda(1)$, $\operatorname{Im}\lambda_k<-\lambda(1)$ for $k\neq 1$).

We will see below that the last assertion remains true under more general conditions.

In order that $\bar{Y}$ have a "purely exponential" distribution,

$$\mathsf{P}(\bar{Y}\geqslant x)=\frac{(1-p)Q_1(-i\lambda(1))}{i\lambda(1)}e^{-\lambda(1)x}, \tag{14}$$

it is sufficient, but not necessary, that under the conditions of Theorem 4, n be equal to 1, i.e., that the positive summand of ξ be exponentially distributed. For the systems $\langle G, G, G, 1\rangle$ considered in Chapter 1, the random variable S is this summand (see (3) § 2). As we have already seen, (14) holds iff

$$\mathsf{P}(\xi>x)=c\,e^{-\alpha x} \quad \text{for} \quad \alpha>0 \text{ and } x>0.$$

Theorem 5. *Let* $\mathsf{M}|\xi|<\infty$. *If* f_- *is rational, then* $\mathfrak{w}_+$ *is representable as*

$$\mathfrak{w}_+=\frac{R(\lambda)}{1-f(\lambda)},$$

where R *is rational. If* $f_-=P_m/Q_n$, *then the function* $Q_n(1-f)$ *has* $n-1$ *zeros* $\lambda_1, \ldots, \lambda_{n-1}$ *in the domain* $\operatorname{Im}\lambda>0$ *and* (13) *is satisfied.*

This theorem is also a generalization of a number of formulae from queueing theory.

If τ^e has an Erlang distribution: $f_-(\lambda)=(1+i\lambda/\alpha)^{-n}$, then the function $1-f(\lambda)$ will have $n-1$ zeros $\lambda_1, \ldots, \lambda_{n-1}$ in the domain $\operatorname{Im}\lambda>0$, differing from $i\alpha$, and

$$R(\lambda)=\frac{i\lambda\prod_{k=1}^{n-1}(\lambda-\lambda_k)}{\left(1+\frac{i\lambda}{\alpha}\right)^n}.$$

The discrete analogues of Theorems 4 and 5 look the same except for the essential difference that in this case the *rationality of* $d_+(z)$ $(d(z)=\mathsf{M}z^{\xi_k}=d_+(z)\,d_-(z))$ *is also necessary for the rationality of* $\mathsf{M}z^{\bar{Y}}$ (*the rationality of* $d_-(z)$ *is necessary for that of* $(1-d(z))\mathsf{M}z^{\bar{Y}}$). This is related to the circumstance that the function $d_-(z)$, being analytic in the region $|z|>1$ (including the point at infinity), cannot have an infinite number of zeros outside the circle $|z|\leqslant 1+\varepsilon$, $\varepsilon>0$ (compare with the remark following Lemma 1).

§ 21. Stability Theorems. The Rate of Convergence

1. In the preceding section we singled out a sufficiently wide class of distributions for which factorization is equivalent to determining the zeros of the function $1-f(\lambda)$. Suppose that we have approximated a distribution F by means of functions F_n from this class. Will the components $\mathfrak{w}_+^{(n)}(\lambda)$ of the factorization which correspond to F_n converge to $\mathfrak{w}_+(\lambda)$? From the point of view of interest to us it is enough to establish conditions for the convergence of the ratio $\mathfrak{w}_+^{(n)}(\lambda)/\mathfrak{w}_+^{(n)}(0)$ which will obviously coincide with conditions for the stability of the distribution of $\bar{Y}$.

Assume that we are given a sequence of distributions

$$F_n(x)\Rightarrow F(x) \tag{15}$$

for $n\to\infty$ with $F_n(-\infty)=1-F(\infty)=0$, $F(-0)>0$, $F(+0)<1$.

We consider the double sequence of i.i.d. variables

$$\begin{array}{l}\xi_1^{(1)}, \xi_2^{(1)}, \ldots,\\ \xi_1^{(2)}, \xi_2^{(2)}, \ldots,\\ \quad\ldots\ldots\ldots,\end{array} \tag{16}$$

with the n-th sequence distributed according to $\mathsf{P}(\xi_k^{(n)}<x)=F_n(x)$, and another sequence $\xi_1, \xi_2, \dots$ for which $\mathsf{P}(\xi_k<x)=F(x)$. The boundary functionals considered in Chapter 3 are defined for each of the sequences at (16). We retain the previous notation for these functionals, providing them in each case with the superscript (n) of the corresponding sequence. For example,

$$Y_k^{(n)}=\sum_{j=1}^k \xi_j^{(n)}$$

and

$$\eta^{(n)}(0)=\min\{k\geqslant 1:\ Y_k^{(n)}\geqslant 0\}.$$

Set

$$W_n(x)=\mathsf{P}(\overline{Y}^{(n)}<x), \qquad W(x)=\mathsf{P}(\overline{Y}<x).$$

We will assume everywhere in the sequel that the convergence (15) holds.

Theorem 6. *For the convergence $W_n \Rightarrow W$ it is sufficient that Condition A be satisfied:*

The series $\sum_{k=1}^\infty \mathsf{P}(Y_k^{(n)}>0)/k$ converges uniformly w.r.t. n. $\qquad (A)$

Proof. Since the distribution of $\overline{Y}^{(n)}$ is infinitely divisible, it is sufficient to prove the weak convergence of the spectral functions (see (10), § 16), or that

$$\sum_{k=1}^\infty \frac{\mathsf{P}(0<Y_k^{(n)}<x)}{k} \Rightarrow \sum_{k=1}^\infty \frac{\mathsf{P}(0<Y_k<x)}{k}.$$

But it is clear that this relation is a consequence of Condition A and the convergence $\mathsf{P}(Y_k^{(n)}<x)\Rightarrow \mathsf{P}(Y_k<x)$. □

Now put $\xi^+\equiv\max(0,\xi)$ and consider *Condition B*:

$$\mathsf{M}\xi^{(n)+}\to \mathsf{M}\xi^+ \quad \text{as } n\to\infty \tag{B}$$

(as before we will suppress the subscript k to simplify the notation).

The sufficiency of (B) for the convergence $W_n\Rightarrow W$ was proved under more general assumptions in § 11.

Now let

$$q=\mathsf{P}(\overline{Y}=0)=\mathfrak{w}_+^{-1}(0)=\exp\left\{\sum_{k=1}^\infty \frac{\mathsf{P}(Y_k>0)}{k}\right\}.$$

Finally, we introduce *Condition C*:

For some $\varepsilon>0$, $c<\infty$ and all n greater than some number,

$$\mathsf{M}(\xi^{(n)+})^{1+\varepsilon}<c \tag{C}$$

and *Condition D*:

$$q^{(n)}\to q \quad \text{as } n\to\infty. \tag{D}$$

Theorem 7. *The following relations hold among these conditions:*

$$C \subset B, B \subset A \quad \text{and} \quad D \subset A .$$

Thus, each of the conditions A, B, C and D is sufficient for the convergence $W_n \Rightarrow W$.

It is easy to see that Condition D is also sufficient for the convergence $\mathfrak{w}_+^{(n)}(\lambda) \to \mathfrak{w}_+(\lambda)$ as $n \to \infty$.

In § 11 we remarked that Condition B is not necessary for $W_n \Rightarrow W$ but is in a certain sense close to being so. We can now illustrate this "closeness" of B to necessity by means of the following assertion. Let $\xi^- = \xi - \xi^+ = \min(0, \xi)$.

If $-\infty < \mathsf{M}\xi < 0$ *and in addition to* (15) $\mathsf{M}\xi^{(n)-} \to \mathsf{M}\xi^-$ *as* $n \to \infty$, *then Condition* B *is necessary for* $W_n \Rightarrow W$.

Proof. Since the distribution of $\bar{Y}$ is infinitely divisible, for the convergence of the distributions of $\bar{Y}^{(n)}$ it is necessary and sufficient that the spectral functions converge weakly:

$$\phi^{(n)}(x) = \sum_{k=1}^{\infty} \frac{1}{k} \mathsf{P}(Y_k^{(n)} > x) \Rightarrow \phi(x) = \sum_{k=1}^{\infty} \frac{1}{k} \mathsf{P}(Y_k > x), \quad x > 0 .$$

For this it is in turn necessary and sufficient that the series $\sum (1/k)\mathsf{P}(Y_k^{(n)} > x)$ converge uniformly in n at each point of continuity of $\phi(x)$. Now take $b > 0$ as a number which is a continuity point of the functions $\phi(x)$ and $\mathsf{P}(Y_k > kx)$ and form the sums of the variables $\xi_k^b = \xi_k - b$:

$$Y_k^b = Y_k - bk , \qquad Y_k^{(n)b} = Y_k^{(n)} - bk .$$

Then, putting $\bar{Y}^b = \sup_{k \geq 0} Y_k^b$ and $\bar{Y}^{(n)b} = \sup_{k \geq 0} Y_k^{(n)b}$ we obtain from the convergence $W_n \Rightarrow W$ along with what has been said that

$$\begin{aligned} \mathsf{P}(\bar{Y}^{(n)b} = 0) &= \exp\left\{\sum_{k=1}^{\infty} \frac{1}{k} \mathsf{P}(Y_k^{(n)b} > 0)\right\} \\ &= \exp\left\{\sum_{k=1}^{\infty} \frac{1}{k} \mathsf{P}(Y_k^{(n)} > bk)\right\} \to \exp\left\{\sum_{k=1}^{\infty} \frac{1}{k} \mathsf{P}(Y_k > bk)\right\} \\ &= \mathsf{P}(\bar{Y}^b = 0) . \end{aligned}$$

If, as before, we denote by $\chi^*(0)$ the first nonpositive sum among $Y_1, Y_2, \ldots,$ then for $\mathsf{M}\xi_1 < 0$ we obtain from (19) § 17

$$\mathsf{M}\chi^*(0)\mathsf{P}(\bar{Y} = 0) = \mathsf{M}\xi .$$

Thus, with the obvious agreements as to notation, we have

$$\frac{\mathsf{M}\xi^{(n)b}}{\mathsf{M}\chi^{*(n)b}(0)} \to \frac{\mathsf{M}\xi^b}{\mathsf{M}\chi^{*b}(0)} .$$

Moreover, by assumption and the inequality $\chi^{*(n)b} \leqslant 0$

$$\liminf_{n\to\infty} \mathsf{M}\xi^{(n)b} \geqslant \mathsf{M}\xi^b,$$

$$\limsup_{n\to\infty} \mathsf{M}\chi^{*(n)b}(0) \leqslant \mathsf{M}\chi^{*b}(0)$$

and

$$\limsup_{n\to\infty} \mathsf{M}\xi^{(n)b} \leqslant \frac{\mathsf{M}\xi^b}{\mathsf{M}\chi^{*b}(0)} \limsup_{n\to\infty} \mathsf{M}\chi^{*(n)b}(0) \leqslant \mathsf{M}\xi^b .$$

This implies that

$$\mathsf{M}\xi^{(n)b} \to \mathsf{M}\xi^b, \qquad \mathsf{M}\xi^{(n)} \to \mathsf{M}\xi \quad \text{and} \quad \mathsf{M}\xi^{(n)+} \to \mathsf{M}\xi^+ . \quad \square$$

Proof of Theorem 7. Without loss of generality we can assume that the sequences $\{\xi_k\}$ and $\{\xi_k^{(n)}\}$ are given on the same probability space and $\xi_k^{(n)} \xrightarrow{p} \xi_k$. To show this it suffices to consider a sequence of i.i.d. random variables $\zeta_1, \zeta_2, \ldots$, uniformly distributed on $[0, 1]$ and to set

$$\xi_k^{(n)} = F_n^{-1}(\zeta_k) \quad \text{and} \quad \xi_k = F^{-1}(\zeta_k),$$

defining the inverse functions F_n^{-1} and F^{-1} at points of discontinuity as, say, left continuous. It's easy to verify that then $\mathsf{P}(\xi_k^{(n)} < x) = F_n(x)$, $\mathsf{P}(\xi_k < x) = F(x)$ and for arbitrary $\varepsilon > 0$

$$\mathsf{P}(|\xi_k^{(n)} - \xi_k| > \varepsilon) \to 0 \quad \text{as } n \to \infty .$$

We will prove that $B \subset A$. Introduce the random variables

$$\xi_{k,N}^{(n)} = \max(-N, \xi_k^{(n)}) - a_N^{(n)} \quad \text{and} \quad \xi_{k,N} = \max(-N, \xi_k) - a_N ,$$

where

$$a_N^{(n)} = \mathsf{M}\max(-N, \xi_k^{(n)}) \quad \text{and} \quad a_N = \mathsf{M}\max(-N, \xi_k) .$$

All of the notation for boundary functionals will be retained for the sequences $\{\xi_{k,N}^{(n)}\}$ and $\{\xi_{k,N}\}$ and will be provided with an additional subscript N. Then, obviously,

$$\mathsf{P}(Y_k^{(n)} > 0) \leqslant \mathsf{P}(Y_{k,N}^{(n)} > -ka_N^{(n)}) . \tag{17}$$

Choose N such that $\mathsf{M}\max(-N, \xi_k) = a_N < 0$. Then by Condition B and (15) we have $a_N^{(n)} \to a_N$ as $n \to \infty$. Hence, for some n_0 and $a > 0$ we have $-a_N^{(n)} > a$ for all $n > n_0$.

Thus, by (17)

$$\mathsf{P}(Y_k^{(n)} > 0) \leqslant \mathsf{P}(Y_{k,N}^{(n)} > ak) \leqslant \mathsf{P}\left(Y_{k,N} > \frac{ak}{2}\right) + \mathsf{P}\left(Y_{k,N}^{(n)} - Y_{k,N} > \frac{ak}{2}\right).$$

Since the series formed from the summands $\mathsf{P}(Y_{k,N} - ak/2 > 0)/k$ does not depend on n and is convergent, for the proof of Property A it is sufficient to show that

$$r_n = \sum_{k=1}^{\infty} \frac{\mathsf{P}\left(Y_{k,N}^{(n)} - Y_{k,N} > \dfrac{ak}{2}\right)}{k}$$

converges uniformly w.r.t. n. It is clear that this will be done if we show that $r_n \to 0$ as $n \to \infty$. For this, we note that

$$(Y_{k,N}^{(n)} - Y_{k,N})\frac{3}{a} \leqslant \sum_{j=1}^{k} \varepsilon_j^{(n)},$$

where

$$\varepsilon_j^{(n)} = \frac{3}{a}|\xi_{j,N}^{(n)} - \xi_{j,N}| \underset{p}{\to} 0 \quad \text{and} \quad \mathsf{M}\varepsilon_j^{(n)} \to 0, \quad n \to \infty. \tag{18}$$

Since

$$\varepsilon_j^{(n)} - \tfrac{3}{2} \leqslant \delta_j^{(n)} \equiv [\varepsilon_j^{(n)} - \tfrac{1}{2}]$$

($[x]$ is the integral part of x), one has

$$r_n \leqslant \sum_{k=1}^{\infty} \frac{\mathsf{P}(\sum_{j=1}^{k} \delta_j^{(n)} > 0)}{k}. \tag{19}$$

The random variables $\delta_j^{(n)} \geqslant -1$ are integer-valued and possess by (18) the properties

$$\mathsf{P}(\delta_j^{(n)} = -1) \to 1 \quad \text{and} \quad \mathsf{M}\delta_j^{(n)} \to -1.$$

But the right side of (19) is

$$-\ln \mathsf{P}(\sup_{k \geqslant 0} \sum_{j=0}^{k} \delta_j^{(n)} = 0).$$

Hence, $r_n \to 0$ will follow from the fact that as $n \to \infty$

$$\mathsf{P}(\sup_{k \geqslant 0} \sum_{j=0}^{k} \delta_j^{(n)} = 0) \to 1. \tag{20}$$

But since by Theorem 3A the generating function of $\sup_{k \geqslant 0} \sum_{j=0}^{k} \delta_j^{(n)}$ is $((1-z)/z) \times [1 - \mathsf{M}z^{\delta_1^{(n)}}]^{-1} \mathsf{M}\,\delta_1^{(n)}$, the probability on the left in (20) equals

$$-\frac{\mathsf{M}\,\delta_1^{(n)}}{\mathsf{P}(\delta_1^{(n)} = -1)} \to 1$$

as $n \to \infty$. Hence, (20) and (19) hold, which imply, as already remarked, the validity of Condition A.

Assume now that Condition C holds. Then for $x > 0$ and all n

$$p_n(x) = 1 - F_n(x) < \frac{c}{x^{1+\varepsilon}}.$$

This implies that

$$P(x) = 1 - F(x) \leqslant \frac{c}{x^{1+\varepsilon}}, \quad \mathsf{M}\xi^+ < \infty \quad \text{and} \quad \int_N^\infty p_n(t)\,dt \to 0$$

as $N \to \infty$ uniformly w.r.t. n. Moreover,

$$\int_0^\infty |p_n(x) - p(x)|\,dx \to 0 \tag{21}$$

as $n \to \infty$ since this integral does not exceed

$$\int_0^N |p_n - p|\,dx + \int_N^\infty p_n\,dx + \int_N^\infty p\,dx,$$

where for given $\varepsilon > 0$ we can always choose first $N(\varepsilon)$ such that the latter two integrals are $<\varepsilon/3$ and then $n(\varepsilon)$ such that the first integral admits the same estimate. From (21) it is easily seen that $\mathsf{M}\xi^{(n)+} \to \mathsf{M}\xi^{+}$.

Conversely, it is also not difficult to show that $\mathsf{M}\xi^{(n)+} \to \mathsf{M}\xi^{+}$ implies (21).
We show finally that $D \subset A$, which is the last assertion of Theorem 7.
To this end we note that D implies

$$\mathsf{P}(Y_k^{(n)}>0) \to \mathsf{P}(Y_k>0).$$

Indeed, by the weak convergence (15) we have

$$\liminf_{n\to\infty} \mathsf{P}(Y_k^{(n)}>0) \geqslant \mathsf{P}(Y_k>0).$$

Hence, assuming the contrary, i.e., that for some k_0 and $\delta>0$

$$\limsup_{n\to\infty} \mathsf{P}(Y_{k_0}^{(n)}>0) \geqslant \mathsf{P}(Y_{k_0}>0)+\delta,$$

we arrive at a contradiction to D since by Fatou's lemma

$$\begin{aligned}\limsup_{n\to\infty} \textstyle\sum_{k=1}^{\infty} \frac{1}{k} \mathsf{P}(Y_k^{(n)}>0) &\geqslant \mathsf{P}(Y_{k_0}>0)+\delta+\liminf_{n\to\infty} \textstyle\sum_{k\neq k_0} \frac{1}{k} \mathsf{P}(Y_k^{(n)}>0)\\ &\geqslant \mathsf{P}(Y_{k_0}>0)+\delta+\textstyle\sum_{k\neq k_0} \frac{1}{k} \mathsf{P}(Y_k>0)\\ &=\textstyle\sum_{k=1}^{\infty} \frac{1}{k} \mathsf{P}(Y_k>0)+\delta.\end{aligned}$$

But the convergence $\mathsf{P}(Y_k^{(n)}>0) \to \mathsf{P}(Y_k>0)$ proved above, along with Condition D, obviously implies Condition A. □

2. In §20 we considered a double sequence related to *queueing theory* in which

$$\xi^{(n)} = \xi_+^{(n)} - \xi_-^{(n)}, \tag{22}$$

and $\xi_+^{(n)}$ and $\xi_-^{(n)}$ are independent and positive. We have the

Corollary. *In the case of* (22) *it is sufficient for the convergence* $W_n \Rightarrow W$ *that the distributions of* $\xi_\pm^{(n)}$ *converge weakly to those of some random variables* $\xi_\pm$ *and that*

$$\mathsf{M}\xi_+^{(n)} \to \mathsf{M}\xi_+ < \infty.$$

For the proof of this it is enough to show that Condition B is satisfied. One has

$$\mathsf{M}\xi^{+} \leqslant \mathsf{M}\xi_+ < \infty$$

and

$$\begin{aligned}\mathsf{M}\xi^{(n)+} &= \textstyle\int_0^\infty \int_0^\infty d\mathsf{P}(\xi_-^{(n)}<t)\mathsf{P}(\xi_+^{(n)} \geqslant x+t)\,dx\\ &= \textstyle\int_0^\infty \int_x^\infty d\mathsf{P}(\xi_+<t)\mathsf{P}(\xi_- \leqslant t-x)\,dx\\ &\quad+\textstyle\int_0^\infty d\mathsf{P}(\xi_-^{(n)}<t)\int_0^\infty [\mathsf{P}(\xi_+^{(n)} \geqslant x+t)-\mathsf{P}(\xi_+ \geqslant x+t)]\,dx\\ &\quad+\textstyle\int_0^\infty d\mathsf{P}(\xi_+<t)\int_0^t [\mathsf{P}(\xi_-^{(n)} \leqslant t-x)-\mathsf{P}(\xi_- \leqslant t-x)]\,dx \to \mathsf{M}\xi^{+}\end{aligned}$$

since the last two terms converge to 0—the first because of (21) which, as we have already remarked, follows from the relation $\mathsf{M}\xi_+^{(n)} \to \mathsf{M}\xi_+$, and the other because of the convergence of the distributions of $\xi_-^{(n)}$ to that of ξ_-. □

3. We turn now to conditions for the *uniform convergence* of the distributions of $\bar{Y}^{(n)}$.

Theorem 8. *If* $\sup_x |F_n(x) - F(x)| \to 0$ *and* $W_n \Rightarrow W$ *as* $n \to \infty$, *then*

$$\sup_x |W_n(x) - W(x)| \to 0 .$$

Proof. Necessary and sufficient conditions for the uniform and weak convergence, resp. of W_n to W are convergence of the same type of the spectral functions

$$\phi_n(x) = \sum_{k=1}^{\infty} \frac{\mathsf{P}(Y_k^{(n)} > x)}{k} \quad \text{to} \quad \phi(x) = \sum_{k=1}^{\infty} \frac{\mathsf{P}(Y_k > x)}{k} .$$

It is enough to show that for each $x \geqslant 0$ and $n \to \infty$

$$\phi_n(x) \to \phi(x) . \tag{23}$$

Let $x_0 \leqslant x$ be some point of continuity of ϕ. (It's not hard to see that $\phi(x)$ is a convergent series for negative x as well.) Then (23) holds at x_0 and for all k

$$\mathsf{P}(Y_k^{(n)} > x_0) \to \mathsf{P}(Y_k > x_0) .$$

We have

$$\begin{aligned} \sum_{k \geqslant N} \frac{\mathsf{P}(Y_k^{(n)} > x)}{k} &\leqslant \sum_{k \geqslant N} \frac{\mathsf{P}(Y_k^{(n)} > x_0)}{k} \\ &= [\phi_n(x_0) - \phi(x_0)] + \sum_{k < N} \frac{\mathsf{P}(Y_k > x_0) - \mathsf{P}(Y_k^{(n)} > x_0)}{k} \\ &\quad + \sum_{k \geqslant N} \frac{\mathsf{P}(Y_k > x_0)}{k} . \end{aligned}$$

We can choose $N \geqslant N_0$ and $n \geqslant n_{N_0}$ so that the three terms on the right are smaller than any prescribed constant. This means that by appropriate choice of the bounds for N and n we can make the difference

$$\begin{aligned} \phi_n(x) - \phi(x) &= \sum_{k < N} \frac{\mathsf{P}(Y_k^{(n)} > x) - \mathsf{P}(Y_k > x)}{k} \\ &\quad + \sum_{k \geqslant N} \frac{\mathsf{P}(Y_k^{(n)} > x)}{k} - \sum_{k \geqslant N} \frac{\mathsf{P}(Y_k > x)}{k} \end{aligned}$$

as small as we like, where the convergence to zero of the first sum for fixed N but $n \to \infty$ is guaranteed by the uniform convergence of F_n to F. □

4. We now turn to the determination of the *rate of convergence*. Assume that the following conditions are satisfied:

1. *F_n and F have absolutely continuous components;*
2. $\mathsf{M}|\xi| < \infty$;
3. $\varepsilon_n = \| f^{(n)} - f \| \to 0$, $\delta_n = \|(f^{(n)} - f)/i\lambda\| \to 0$ *as* $n \to \infty$.

Since here

$$\| f^{(n)} - f \| = \int |d(F_n - F)| \quad \text{and} \quad \left\| \frac{f^{(n)} - f}{i\lambda} \right\| = \int |F_n - F| \, dx ,$$

Condition 3 implies the convergence in variation of the functions F_n and their "tails". The convergence in variation of the latter is guaranteed, as we have seen, by the convergence relations

$$F_n \Rightarrow F, \qquad \mathsf{M}\xi^{(n)} \to \mathsf{M}\xi \quad \text{and} \quad \mathsf{M}\xi^{(n)+} \to \mathsf{M}\xi^+ .$$

From Theorem 8 of Appendix 2 it follows that under Conditions 1 and 2

$$\frac{i\lambda}{(1-f)(i\lambda+\beta)} \in \mathfrak{B}$$

for arbitrary $\beta > 0$. Denote by $b(\beta)$ the norm

$$\left\| \frac{i\lambda}{(1-f(\lambda))(i\lambda+\beta)} \right\|$$

of this function in the ring $\mathfrak{B}$.

Theorem 9. *If Conditions* 1–3 *are satisfied, then for arbitrary* $\beta > 0$

$$\sup_x |W_n(x) - W(x)| \leqslant \| W_n - W \| \leqslant \left(1 + \frac{\Delta_n}{1 - \Delta_n} \right)^2 - 1 ,$$

where

$$\Delta_n \leqslant b(\beta)(\varepsilon_n + \beta\delta_n) .$$

Proof. We use Formula (6). Then

$$\frac{\mathfrak{w}_+^{(n)}(\lambda)}{\mathfrak{w}_+(\lambda)} = \exp\left\{ \frac{1}{2\pi i} \int_{-\infty}^{\infty} \frac{\ln \dfrac{\mathfrak{v}^{(n)}(\mu)}{\mathfrak{v}(\mu)}}{\mu - \lambda} \, d\mu \right\}, \qquad \operatorname{Im} \lambda > 0 .$$

Here

$$\mathfrak{v}(\mu) = \frac{1 - f(\mu)}{i\mu} (i\mu + \beta) ,$$

$$\frac{\mathfrak{v}^{(n)}(\mu)}{\mathfrak{v}(\mu)} = \frac{1 - f^{(n)}(\mu)}{1 - f(\mu)} = 1 - \frac{f^{(n)}(\mu) - f(\mu)}{1 - f(\mu)} \equiv 1 - \Delta_n(\mu) ,$$

$$\Delta_n = \|\Delta_n(\lambda)\| \leqslant \left\| \frac{f^{(n)} - f}{i\lambda} (i\lambda + \beta) \right\| \left\| \frac{i\lambda}{(1-f)(i\lambda+\beta)} \right\| \leqslant b(\beta)(\varepsilon_n + \beta\delta_n) .$$

Moreover,

$$l(\mu)=\ln\frac{\mathfrak{v}^{(n)}(\mu)}{\mathfrak{v}(\mu)}\in\mathfrak{B}$$

for $\Delta_n<1$ because

$$\|l\|=\|\ln(1-\Delta_n(\mu))\|\leqslant-\ln(1-\Delta_n).$$

Since the function

$$l_+(\lambda)=\frac{1}{2\pi i}\int_{-\infty}^{\infty}\frac{l(\mu)}{\mu-\lambda}\,d\mu$$

satisfies $\|l_+\|\leqslant\|l\|$, we have

$$\left\|\frac{\mathfrak{w}_+^{(n)}(\lambda)}{\mathfrak{w}_+(\lambda)}-1\right\|\leqslant\exp\|l\|-1\leqslant\frac{1}{1-\Delta_n}-1=\frac{\Delta_n}{1-\Delta_n}.$$

Clearly, such an estimate will also hold for the reciprocal ratio $\mathfrak{w}_+(\lambda)/\mathfrak{w}_+^{(n)}(\lambda)$. Consequently,

$$\left\|\frac{\mathfrak{w}_+(0)}{\mathfrak{w}_+^{(n)}(0)}-1\right\|\leqslant\frac{\Delta_n}{1-\Delta_n}.$$

Now note that

$$\frac{\mathsf{M}\,e^{i\lambda\bar{Y}^{(n)}}}{\mathsf{M}\,e^{i\lambda\bar{Y}}}=\frac{\mathfrak{w}_+^{(n)}(\lambda)\,\mathfrak{w}_+(0)}{\mathfrak{w}_+(\lambda)\,\mathfrak{w}_+^{(n)}(0)}$$

and that

$$\|fg-1\|\leqslant\|(f-1)(g-1)\|+\|f-1\|+\|g-1\|.$$

One can then write

$$\|\mathsf{M}\,e^{i\lambda\bar{Y}^{(n)}}-\mathsf{M}\,e^{i\lambda\bar{Y}}\|\leqslant\left\|\frac{\mathsf{M}\,e^{i\lambda\bar{Y}^{(n)}}}{\mathsf{M}\,e^{i\lambda\bar{Y}}}-1\right\|\leqslant\left(\frac{\Delta_n}{1-\Delta_n}\right)^2+\frac{2\Delta_n}{1-\Delta_n}.\ \square$$

Let us now require that instead of Condition 1 the following one is fulfilled:

1a. *There exist densities $\varphi_n(x)=F_n'(x)$ and $\varphi(x)=F'(x)$ such that the differences $\varphi_n(x)-\varphi(x)$ are functions of bounded variation:*

$$\psi_n=\text{Variation}\,(\varphi_n-\varphi)<\infty\,.$$

Then we can obtain an estimate for $\sup_x|W_n(x)-W(x)|$ which is somewhat different from that in Theorem 9 and which will also be useful when the quantity $\varepsilon_n=\int|d(F_n-F)|$ is not small. Namely, we have

Theorem 9A. *If Conditions* 1a *and* 2 *hold, then the inequality of Theorem 9 is valid with*

$$\Delta_n\leqslant b(\beta)[2\sqrt{(\beta\psi_n+\varepsilon_n)2\delta_n}+\beta\delta_n]$$

for arbitrary $\beta>0$.

As this theorem shows, if ψ_n is uniformly bounded, then $\sup_x |W_n(x)-W(x)| \leqslant c\sqrt{\delta_n}$ where c does not depend on n (obviously, ε_n is always $\leqslant 2$).

We also note that if $F_n(x)-F(x)$ does not change its sign for any x (for example, when $\xi^{(n)}=\xi+\gamma_n$, and $\gamma_n \to 0$ without changing sign), then δ_n coincides with $|\mathsf{M}\xi^{(n)}-\mathsf{M}\xi|$ since then

$$\delta_n=\int |F_n-F|\,dx=|\int (F_n-F)\,dx|=|\int (g_n-g)\,dx|=|\mathsf{M}\xi^{(n)}-\mathsf{M}\xi|,$$

where g_n and g are defined by means of equalities of the form

$$g(x)=\begin{cases}1-F(x), & x\geqslant 0\\ -F(x), & x<0.\end{cases}$$

Proof of Theorem 9A. We appeal to the proof of Theorem 9 and can write

$$\begin{aligned}\Delta_n(\lambda)&=\frac{f^n-f}{i\lambda}(i\lambda+\alpha)\cdot\frac{i\lambda}{(1-f)(i\lambda+\beta)}\cdot\frac{i\lambda+\beta}{i\lambda+\alpha}\\ &=(\varepsilon_n(\lambda)+\alpha\delta_n(\lambda))b(\beta,\lambda)\left[\frac{\beta\alpha(\lambda)}{\alpha}+\frac{i\lambda\alpha(\lambda)}{\alpha}\right],\end{aligned}$$

where

$$\varepsilon_n(\lambda)=f^n-f, \qquad \delta_n(\lambda)=\frac{f^n-f}{i\lambda},$$

$$b(\beta,\lambda)=\frac{i\lambda}{(1-f)(i\lambda+\beta)} \quad \text{and} \quad \alpha(\lambda)=\left(1+\frac{i\lambda}{\alpha}\right)^{-1}.$$

We rewrite the equation for $\Delta_n(\lambda)$ as

$$\Delta_n(\lambda)=b(\beta,\lambda)\left[\frac{\beta\varepsilon_n(\lambda)\alpha(\lambda)}{\alpha}+\beta\alpha(\lambda)\delta_n(\lambda)+\frac{i\lambda\alpha(\lambda)\varepsilon_n(\lambda)}{\alpha}+i\lambda\alpha(\lambda)\delta_n(\lambda)\right].$$

In this relation

$$\|\alpha(\lambda)\|=\|\int_{-\infty}^{0} e^{i\lambda x}\,d(e^{\alpha x})\|=1$$

and

$$\|i\lambda\alpha(\lambda)\|=\|\alpha\int_{-\infty}^{\infty} e^{i\lambda x}\,dA(x)\|\leqslant 2\alpha$$

since

$$A(x)=\begin{cases}-e^{\alpha x}, & x<0\\ 0, & x\geqslant 0.\end{cases}$$

Finally

$$\|i\lambda\varepsilon_n(\lambda)\|=\|\int e^{i\lambda x}\,d(\varphi_n-\varphi)\|\leqslant\psi_n.$$

Hence,

$$\Delta_n=\|\Delta_n(\lambda)\|\leqslant b(\beta)\left[\frac{\beta\varepsilon_n}{\alpha}+\beta\delta_n+\frac{\psi_n}{\alpha}+2\alpha\delta_n\right].$$

The right side of this expression attains a minimum for

$$\alpha = \sqrt{\frac{\varepsilon_n \beta + \psi_n}{2\delta_n}}$$

so that

$$\Delta_n \leqslant b(\beta)[2\sqrt{(\varepsilon_n \beta + \psi_n) 2\delta_n} + \beta\delta_n] . \quad \square$$

We remark that the estimates of the rate of convergence in Theorems 9 and 9A have apparently been obtained under excessively strong conditions. Theorem 7, for example, leads one to imagine that the closeness of W and W_n in terms of Lévy's distance (or that of the characteristic functions of $\bar{Y}$ and $\bar{Y}^{(n)}$) can be estimated merely by the smallness of

$$\int_{-\infty}^{\infty} a(t)\,|F_n(t) - F(t)|\,dt$$

for some positive function $a(t)$ equal to 1 for $t > 0$ and such that

$$\int_{-\infty}^{0} a(t)\,dt < \infty .$$

For uniform closeness of W and W_n (see Theorem 8) the modulus $|F_n - F|$ in the proposed integral can be replaced by

$$\sup_{t \leqslant u \leqslant t+1} |F_n(u) - F(u)| .$$

§ 22. Asymptotic Properties of the Distributions of $\bar{Y}$ and θ

1. Another direction in the investigation of possible approximations for the distributions of the functionals $\bar{Y}$ and θ_∞ (we denote the latter by θ), important in queueing theory, is the study of the behavior of the functions

$$W(x) = \mathsf{P}(\bar{Y} \geqslant x) \quad \text{and} \quad T(n) = \mathsf{P}(\theta = n)$$

as $x \to \infty$ and $n \to \infty$.

The asymptotic behavior of $W(x)$ can be qualitatively quite variable depending on certain numerical characteristics of the distribution. Let

$$f_1(\mu) = f(-i\mu) = \mathsf{M}\, e^{\mu\xi} \quad \text{and} \quad \mu_+ = \sup\{\mu : f_1(\mu) < \infty\} .$$

The value of $f_1(\mu_+)$ turns out to be decisive. If $f_1(\mu_+) > 1$, then the unique root $q > 0$ of the equation

$$f_1(q) = 1$$

is defined (this is $\lambda(1)$ in the notation of Corollary 11, Chapter 3) and we have Cramér's estimate:

$$W(x) = c_1\, e^{-qx}(1 + o(1)) . \tag{24}$$

The case $\mu_+ > 0, f_1(\mu_+) = 1, f_1'(\mu_+) < \infty$ is of the same type.

If the distribution of Y_n has an absolutely continuous component for some n and $f_1(\mu_+)>1$, then $o(1)$ in (24) can be replaced by $o(e^{-\varepsilon x})$ for some $\varepsilon>0$.

However, if $f_1(\mu_+)<1$ or $\mu_+=0$ and $f_1'(0)>-\infty$, then under wide conditions the asymptotic behavior of $W(x)$ is like that of the "double tail" of the distribution of ξ. More precisely, we have

$$W(x)=c_2G(x)(1+o(1)), \qquad G(x)=\int_x^\infty (1-F(t))\,dt. \tag{25}$$

The constants c_1 and c_2 in (24) and (25) are given explicitly in the formulae of Theorems 11 and 12.

The third possibility, $f_1(\mu_+)=1, |f_1'(\mu_+)|=\infty$ is transitional and has been studied less.

The distribution of θ (recall that this distribution coincides with the limiting distribution of the "backward busy period"

$$v_n=\min\{k\geqslant 0: w_{n-k}=0\}$$

of the system in terms of the number of jumps of τ_j^e) is closely related to the distribution of $\eta^*(0)$ (see § 16):

$$\mathsf{M}z^\theta=\frac{1}{\mathsf{M}\eta^*(0)}\exp\left\{\sum\frac{z^k}{k}\mathsf{P}(Y_k>0)\right\}=\frac{\exp\{-\sum(z^k/k)\mathsf{P}(Y_k\leqslant 0)\}}{\mathsf{M}\eta^*(0)(1-z)}=\frac{1-\mathsf{M}z^{\eta^*(0)}}{\mathsf{M}\eta^*(0)(1-z)},$$

or, what is the same,

$$\mathsf{P}(\theta=n)=\frac{\mathsf{P}(\eta^*(0)>n)}{\mathsf{M}\eta^*(0)}.$$

It is not difficult to see that the variable $\eta^*(0)$ can be interpreted as the total busy period of the system in terms of the number of jumps of τ_j^e ($\mathsf{P}(\eta^*(0)=k)$ coincides with the probability that the length of the actual busy period for $w_1=0$ falls in the time interval $(\tau_1^e+\tau_2^e+\dots+\tau_{k-1}^e, \tau_1^e+\dots+\tau_k^e)$).

It turns out that for the distribution of θ we have under wide conditions

$$\mathsf{P}(\theta=n)=c_3\frac{\mathsf{P}(Y_n>0)}{n}(1+o(1)).$$

As is well known, one can distinguish between several estimates for $\mathsf{P}(Y_n>0)$. If, for example, $a=\mathsf{M}\xi>-\infty$ and $1-F(x)$ decreases like some power of x, then

$$\mathsf{P}(Y_n>0)\sim nF(-an).$$

If $\mathsf{M}\,e^{\mu\xi}<\infty$ for some $\mu>0$, then

$$\mathsf{P}(Y_n>0)\sim\frac{1}{\sqrt{2\pi n}}m^n(-a), \qquad m(x)=\inf_\mu e^{-\mu x}\mathsf{M}\,e^{\mu\xi}.$$

To establish these results we must first investigate the properties of the variables $\chi(+0)$ and $\chi^*(0)$.

2. *Some properties of the distributions of* $\chi(+0)$ *and* $\chi^*(0)$. Factorization identities allow one to establish interesting direct connections between the distribution of ξ and those of the random variables $\chi(+0)$, $\chi^*(0)$, $\overline{Y}$, θ, etc. We examine some of these here. As shown by the identity (26) § 17, the distribution of $\overline{Y}$ is closely connected with the distribution

$$U(x)=\mathsf{P}(\chi(+0)\geqslant x|\eta(+0)<\infty). \tag{26}$$

Most of this subsection is devoted to the properties of the distribution U.

We recall some basic notation: $E(t)$ is the identity in the ring V,

$$g(t)=E(t)-F(t)\,, \qquad G(t)=\int_t^\infty g(u)\,du\,,$$

$$p=\mathsf{P}(\overline{Y}>0) \quad \text{and} \quad U^*(t)=\mathsf{P}(\chi^*(0)<t)\,.$$

Thus, for $t>0$

$$g(t)=\mathsf{P}(\xi\geqslant t)\,, \qquad G(-\infty)=a=\mathsf{M}\xi$$

and

$$\int g(t)\,e^{i\lambda t}\,dt=\frac{1-f(\lambda)}{i\lambda}\,.$$

We also set $\xi^+=\max(0,\xi)$, $\xi^-=\min(0,\xi)$ and assume $H^{**}(t)$ is the renewal function of the random variable $\chi^{**}=-\chi^*(0)$:

$$H^{**}(t)=\sum_{k=0}^{\infty} U_k^{**}(t)\,,$$

where U_k^{**} is the k-fold convolution of the distribution $U^{**}(t)$ of the random variable χ^{**}.

Definition. *The function $H(t)$ will be said to behave locally like a power (to be* l.p.) *as $t\to\infty$, if for each b*

$$\lim_{t\to\infty}\frac{H(t+b)}{H(t)}=1\,.$$

For example the functions $e^{\alpha\sqrt{t}}$, $t^\alpha\ln^2 t$, $t^\alpha(2+\sin\sqrt{t})$ are all l.p., but the functions $e^{\alpha t}$, $e^{\alpha t^2}$, $t^\alpha(\sin\sqrt{t}+1)$ are not.

Theorem 10. I. *For $x<0$*

$$F(x)\leqslant U^*(x)\leqslant\frac{F(x)}{1-p}\,.$$

II. *For $U(x)$ we have for $x>0$ the inequalities*

a. $U(x)\geqslant\dfrac{g(x)}{p}$ *and* $U(x)\geqslant\dfrac{p-1}{ap}G(x)$.

b. *let α and β be arbitrary constants with the property that $H^{**}(t)\leqslant\alpha+\beta t$, $t\geqslant 0$ (see Appendix* 1). *Then*

$$U(x)\leqslant\frac{\alpha g(x)}{p}+\frac{\beta G(x)}{p}\,.$$

III. *If* $\mathsf{M}\xi = a$ *exists, then*

$$0 \leqslant U(x) - \frac{p-1}{ap} G(x) \leqslant \frac{\alpha_0 g(x)}{p},$$

where

$$\alpha_0 = \sup_{t \geqslant 0} \left(H^{**}(t) - \frac{t}{\mathsf{M}\chi^{**}} \right) < \infty$$

if $\mathsf{M}(\xi^-)^2 < \infty$.

If $\mathsf{M}\xi = a$ *exists and* $g(x) = o(G(x))$, *then*

$$U(x) \sim \frac{p-1}{ap} G(x). \tag{27}$$

IV. *If the function* $G_1(x) = e^{\mu_+ x} G(x)$ *or the function* $U_{(1)}(x) = e^{\mu_+ x} U(x)$ *are* l.p., *then for* $x \to \infty$

$$U(x) \sim \frac{\mu_+ G(x)}{p(1 - \mathsf{M}\, e^{\mu_+ \chi^*(0)})}.$$

In particular, if $\mu_+ = 0$, *then* (27) *holds. If* $\mu_+ > 0$, *then*

$$U(x) \sim \frac{g(x)}{p(1 - \mathsf{M}\, e^{\mu_+ \chi^*(0)})}.$$

Corollary. *Assume, as above, that*

$$\kappa(\xi) = \sup\{s \colon \mathsf{M}|\xi|^s < \infty\}.$$

Then from the theorem it follows that

$$\kappa(\chi^*(0)) = \kappa(\xi^-) \quad \text{and} \quad \kappa(\xi^+) \geqslant \kappa(\chi) \geqslant \kappa(\xi^+) - 1,$$

where χ *is a random variable with distribution* $U(x)$. *If* $\mathsf{M}\xi$ *exists, then* $\kappa(\chi) = \kappa(\xi^+) - 1$.

We also remark that by the integral renewal theorem and the choice of α, the parameter β can be taken equal to $1/\mathsf{M}\chi^{**} + \varepsilon$ for arbitrarily small $\varepsilon > 0$.

We also have (see Appendix 1)

$$H^{**}(t) < 2 + \frac{2t}{m_0}, \tag{28}$$

where m_0 is the median of the distribution of $-\chi^*(0)$.

Proof of the theorem. We have

$$1 - f(\lambda) = [1 - \mathsf{M}\, e^{i\lambda\chi^*(0)}][1 - p\mathsf{M}\, e^{i\lambda\chi}] \text{ and } 1 - f(\lambda) = \frac{1-p}{\mathsf{M}\, e^{i\lambda\bar{Y}}}[1 - \mathsf{M}\, e^{i\lambda\chi^*(0)}]. \tag{29}$$

From these we find (dividing both sides of (29) by $-i\lambda$ for convenience):

$$\int g(t)\, e^{i\lambda t}\, dt = -\int_{-\infty}^{0} U^*(t)\, e^{i\lambda t}\, dt [1 + p\int_0^\infty e^{i\lambda t}\, dU(t)], \tag{30}$$

$$\int e^{i\lambda t}\, dg(t) \int_{-\infty}^{0} e^{i\lambda t}\, dH^*(t) = [1 + p\int_0^\infty e^{i\lambda t}\, dU(t)] \tag{31}$$

and

$$\int e^{i\lambda t} g(t)\,dt \int_0^\infty e^{i\lambda t}\,dW(t) = -(1-p)\int_{-\infty}^0 e^{i\lambda t} U^*(t)\,dt\,, \tag{32}$$

where $W(x) = \mathsf{P}(\bar{Y} \geqslant x)$, and $H^*(t)$ is the renewal function of $\chi^*(0)$ on the interval $[t, 0]$. From these identities we find:

For $x<0$ (from the first and third)

$$g(x) = -U^*(x) - p\int_0^\infty dU(t)\,U^*(x-t) \tag{33}$$

and

$$(1-p)U^*(x) = -\int_0^\infty dW(t)\,g(x-t)\,; \tag{34}$$

for $x>0$ (from the first and second)

$$g(x) = p\int_{-\infty}^0 U^*(t)\,dU(x-t) \tag{35}$$

and

$$pU(x) = \int_x^\infty g(t)\,dH^{**}(t-x)\,. \tag{36}$$

From (33) and (34) we get

$$g(x) = -F(x) \geqslant -U^*(x) \quad \text{and} \quad (1-p)U^*(x) \leqslant -g(x) = F(x)\,,$$

which proves Part I of the theorem. Next, by (35)

$$g(x) \leqslant pU(x)\,.$$

Integrating (35) from x to $N>x$ and assuming that $G(x)<\infty$, we get

$$G(x) - G(N) = p\int_{-\infty}^0 U^*(t)[U(x-t) - U(N-t)]\,dt\,.$$

By the monotone convergence theorem we obtain as $N \to \infty$

$$G(x) = p\int_{-\infty}^0 U^*(t)\,U(x-t)\,dt\,. \tag{37}$$

Since $\mathsf{M}\chi^*(0)(1-p) = a$, we have

$$G(x) \leqslant pU(x)\int_{-\infty}^0 U^*(t)\,dt = -pU(x)\mathsf{M}\chi^*(0) = -\frac{apU(x)}{(1-p)}\cdot$$

In order to show IIb we use (36). Assuming that $H^{**}(0)=0$ and $H^{**}(+0)=1$, we find that

$$pU(x) = -\int_x^\infty H^{**}(t-x)\,dg(t) \leqslant \alpha g(x) - \int_x^\infty \beta(t-x)\,dg(t) = \alpha g(x) + \beta\int_x^\infty g(t)\,dt\,.$$

The third assertion of the theorem is an almost obvious consequence of the first two. One need merely note that

$$\sup_{t\geqslant 0}\left(H^{**}(t) - \frac{t}{\mathsf{M}\chi^{**}}\right) < \infty\,,$$

if $\mathsf{M}(\chi^{**})^2 < \infty$ (see Appendix 1).

The proof of the fourth part will be broken down into two steps. First let $\mu_+ = 0$ and assume $G(x)$ is l.p. Set

$$b = \frac{1-p}{-ap} \quad \text{and} \quad A_N = -\frac{1}{\mathsf{M}\chi^*(0)}\int_{-N}^0 U^*(t)\,dt\,.$$

Then, if a is finite, we have $A_N \to 1$ as $N \to \infty$ and by (37),

$$bG(x) \leqslant U(x) \leqslant \frac{bG(x-N)}{A_N}.$$

Consequently,

$$1 \leqslant \liminf \frac{U(x)}{bG(x)} \leqslant \limsup \frac{U(x)}{bG(x)} \leqslant \frac{1}{A_N}$$

for arbitrary N, which proves (27). The proof for the case in which $U(x)$ is l.p. goes the same.

Now let $\mu_+ > 0$. We note first that we then have

$$g(x) \sim \mu_+ G(x). \tag{38}$$

Indeed,

$$\frac{G(x)-G(x+N)}{G(x+M)-G(x+M+N)} = \frac{\int_x^{x+N} g(t)\,dt}{\int_{x+M}^{x+M+N} g(t)\,dt} \to e^{\mu_+ M} \quad \text{for } x \to \infty.$$

Consequently,

$$\liminf \frac{g(x)}{g(x+M+N)} \geqslant e^{\mu_+ M}, \qquad \limsup \frac{g(x+N)}{g(x+M)} \leqslant e^{\mu_+ M}$$

and

$$e^{\mu_+(M-N)} \leqslant \liminf \frac{g(x)}{g(x+M)} \leqslant \limsup \frac{g(x)}{g(x+M)} \leqslant e^{\mu_+(M+N)}$$

for arbitrary $N > 0$. This means that $g(x+M) \sim e^{-\mu_+ M} g(x)$ as $x \to \infty$ so that (38) is valid.

Now using (36) we find that

$$\lim p \frac{U(x)}{g(x)} = \lim \int_0^\infty \frac{g(x+t)\, e^{\mu_+ t}}{g(x)} e^{-\mu_+ t}\, dH^{**}(t)$$
$$= \int_0^\infty e^{-\mu_+ t}\, dH^{**}(t) = (1 - \mathsf{M}\, e^{\mu_+ \chi^*(0)})^{-1}.$$

In order to justify taking the limit under the integral sign here we must prove that the integral is uniformly convergent. The function $l(x) = g(x)\, e^{\mu_+ x}$ is l.p. Hence, for arbitrary $\varepsilon > 0$ and all sufficiently large x:

$$\frac{l(x+n)}{l(x)} = \frac{l(x+1)}{l(x)} \cdot \frac{l(x+2)}{l(x+1)} \cdots \frac{l(x+n)}{l(x+n-1)} < e^{\varepsilon n},$$

so that for large enough N

$$\int_N^\infty \frac{l(x+t)}{l(x)} e^{-\mu_+ t}\, dH^{**}(t) < \int_N^\infty e^{\varepsilon t - \mu_+ t}\, dH^{**}(t).$$

Since $H^{**}(t)$ is monotone and $H^{**}(t) < \beta|t| + \alpha$ the last integral converges to zero for $\varepsilon < \mu_+/2$ and $N \to \infty$ and does not depend on x. □

3. We turn now to theorems describing the asymptotic behavior of $W(x)=\mathsf{P}(\bar{Y}\geqslant x)$.

Theorem 11. *Assume $f_1(\mu_+)\geqslant 1$ and $\mu_+>0$. If $f_1(\mu_+)=1$, we assume in addition that $f_1'(\mu_+)<\infty$. Then, as $x\to\infty$*

$$W(x)=\frac{1-p}{q\tilde{a}}e^{-qx}(1+o(1))=-\frac{a\mathsf{M}\,e^{q\chi^*(-\infty)}}{f_1'(q)}e^{-qx}(1+o(1)) \tag{39}$$

(in the lattice case x must be taken as a multiple of the lattice step[1]*), where $q>0$ is the unique root of $f_1(q)=1$,*

$$p=\mathsf{P}(\bar{Y}>0)$$

and

$$\tilde{a}=\int x\,e^{qx}\,d\mathsf{P}(\chi(+0)<x;\eta(+0)<\infty)=f_1'(q)[1-\mathsf{M}\,e^{q\chi^*(0)}]^{-1}.$$

If $f_1(\mu_+)>1$ and the distribution of ξ satisfies the condition

(C_1): Y_n has for some n an absolutely continuous component or is lattice, then $o(1)$ in (39) can be replaced by $o(e^{-\varepsilon x})$ for some $\varepsilon>0$.

Proof. Set

$$\varphi(\mu)=\mathsf{M}\,(e^{\mu\chi(+0)};\eta(+0)<\infty).$$

It follows from (29) that $\varphi(q)=1$ and that $\tilde{\varphi}(\mu)=\varphi(\mu+q)$ is the moment generating function of some random variable, which we will denote by $\tilde{\chi}$:

$$\mathsf{P}(\tilde{\chi}\in dx)=e^{qx}\mathsf{P}(\chi(+0)\in dx;\eta(+0)<\infty).$$

Since

$$\mathsf{M}\,e^{\mu\bar{Y}}=\mathsf{P}(\bar{Y}=0)/(1-\varphi(\mu)),$$

one has

$$\mathsf{M}\,e^{(q+\mu)\bar{Y}}=\frac{1-p}{1-\tilde{\varphi}(\mu)}. \tag{40}$$

But $1/(1-\tilde{\varphi}(\mu))$ is the Laplace transform of the renewal function $\tilde{H}(x)$ of the random variable $\tilde{\chi}$. If $\mathsf{M}\tilde{\chi}=\tilde{a}$, then

$$\frac{1}{1-\tilde{\varphi}(\mu)}=\int_0^\infty e^{\mu x}\,d\tilde{H}(x),\qquad \tilde{H}(x)=\frac{x}{\tilde{a}}+R(x)$$

and

$$R(x+t)-R(x)=o(1)$$

for each t and $x\to\infty$ (if ξ_k is lattice, then x and t must be taken as multiples of the lattice step). By (40)

$$-e^{qx}\,dW(x)=(1-p)\,d\tilde{H}(x)$$

[1] The lattice step is the largest h for which the values of ξ_k are multiples of h.

and

$$\begin{aligned} W(x) &= (1-p)\int_x^\infty e^{-qt}\, d\tilde{H}(t) \\ &= -(1-p)\tilde{H}(x)\, e^{-qx} + q(1-p)\int_x^\infty \tilde{H}(t)\, e^{-qt}\, dt \\ &= \frac{1-p}{q\tilde{a}} e^{-qx} + q(1-p)\int_x^\infty (R(t)-R(x))\, e^{-qt}\, dt\,. \end{aligned} \tag{41}$$

This equality proves the first assertion of the theorem. (In the discrete case the integral must be replaced by a sum.) One need only note that

$$1-f_1(\mu)=(1-\mathsf{M}\, e^{\mu\chi^*(0)})(1-\varphi(\mu))\,,$$

$$\tilde{a}=\int x\, e^{qx}\mathsf{P}(\chi(+0)\in dx;\, \eta(+0)<\infty)=\varphi'(q)=\frac{f_1'(q)}{1-\mathsf{M}\, e^{q\chi^*(0)}}\,. \tag{42}$$

Relation (41) also indicates the possibility of a *refinement of the theorem*, since when a large number of moments of $\tilde{\chi}$ exist, the difference $R(t)-R(x)$ can be estimated to greater accuracy than $o(1)$ as $x\to\infty$ (see Appendix 1).

We will now obtain such a refinement when $f_1(\mu_+)>1$ (the second assertion of the theorem). In this case $\tilde{\varphi}(\mu)=\mathsf{M}\, e^{\mu\tilde{\chi}}$ is an analytic function in a neighborhood of the point $\mu=0$. If Condition (C_1) on the existence of an absolutely continuous component for Y_n holds, then the distribution of $\tilde{\chi}$ will evidently also have an absolutely continuous component. However, if ξ is lattice, then $\tilde{\chi}$ will also be lattice. Thus, the renewal function $\tilde{H}(x)$ of the random variable $\tilde{\chi}$ satisfies the hypotheses of Theorem 3, Appn. 1, by which the difference $R(t)-R(x)$ in Formula (41) can be estimated for some $c>0$ and $\varepsilon>0$ by means of

$$|R(t)-R(x)|<c|e^{-\varepsilon x}+e^{-\varepsilon t}|<2c\, e^{-\varepsilon x}\,. \tag{43}$$

Substitution of the right side of this inequality in (41) completes the proof. □

With respect to the properties of the distribution of ξ, the conditions of Theorem 11 seem to be encountered most frequently in applications. Hence, we now give a numerical example which will characterize to a certain extent how fast $W(x)$ approaches its asymptotic representation under the conditions of Theorem 11. In other words, we are interested in the question of *how soon* (i.e., beginning with what x) *we can replace* $W(x)$ *by* $(1-p)\, e^{-qx}/q\tilde{a}=W_1(x)$ *without much error.*

Our example will be for the systems $\langle G_I, 1, G_I, 1\rangle$ and is similar to the one we treated in § 8. The character of the distributions of τ_j^e and τ_j^s will remain the same except that the exponent α in the distribution

$$\mathsf{P}(\tau^*>x)=e^{-\alpha x}$$

will be taken as 2. Then

$$f_1(\mu)=e^{\mu/2}\,\frac{1+e^{-2\mu}}{2-\mu}\cdot\frac{1-e^{-\mu}}{\mu}$$

and the solution of $f_1(q)=1$ is $q=0.7468$. The values of $1-p$ and $\tilde{a}$ were determined by trials with sufficient accuracy by the formulae of Theorem 11 and this led to the asymptotic representation

$$W_1(x)\approx 0.589\, e^{-qx}\,.$$

To determine the empirical function $W^*(x)$, 2000 trials were carried out. The values of $W^*(x)$ were determined at the points x_k, $k=0, 1, \ldots, 200$, for which $W_1(x_{k-1})-W_1(x_k)=W_1(0)/200$. We present below a plot of the functions $1-W_1(x_k)=1-W_1(0)(1-k/200)$ and $1-W^*(x_k)$ for $k \geqslant 75$ (the values of $1-W^*(x_k)$ are given only for odd k).

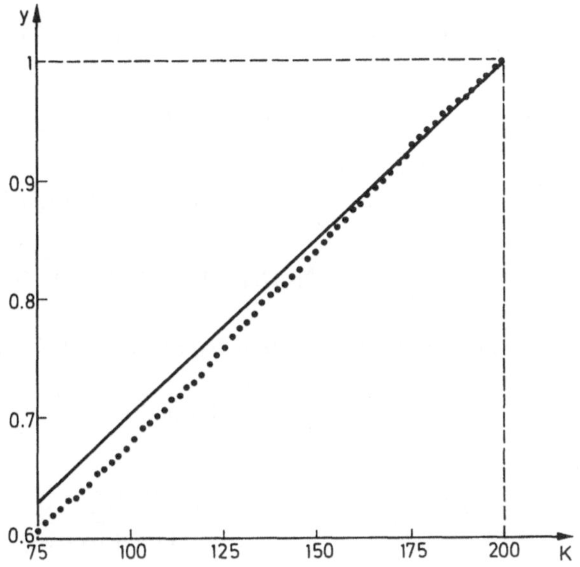

Fig. 12. $1-W_1(x_k)$ and $1-W^*(x_k)$

Table 3. Tabulated values of $1-W^*(x_k)$ in Fig. 12

k	$1-W^*(x_k)$	k	$1-W^*(x_k)$	k	$1-W^*(x_k)$	k	$1-W^*(x_k)$
75	0.6045	107	0.7005	139	0.8080	171	0.9120
77	.6110	109	.7060	141	.8130	173	.9190
79	.6185	111	.7145	143	.8190	175	.9180
81	.6240	113	.7180	145	.8250	177	.9340
83	.6295	115	.7245	147	.8325	179	.9400
85	.6335	117	.7300	149	.8385	181	.9455
87	.6385	119	.7355	151	.8470	183	.9540
89	.6455	121	.7455	153	.8535	185	.9590
91	.6525	123	.7520	155	.8600	187	.9640
93	.6580	125	.7590	157	.8660	189	.9675
95	.6630	127	.7665	159	.8750	191	.9740
97	.6675	129	.7745	161	.8805	193	.9815
99	.6755	131	.7800	163	.8860	195	.9860
101	.6825	133	.7870	165	.8930	197	.9935
103	.6910	135	.7970	167	.8985	199	.9990
105	.6955	137	.8040	169	.9055		

4. We turn now to the case in which

$$f_1(\mu_+)<1 \quad \text{or} \quad \mu_+=0\,. \tag{44}$$

Definition. *The finite-valued function $H(t)$ is said to be an "upper power function" (to be u.p.) if it behaves locally like a power and if for arbitrary $0<\alpha\leqslant 1$*

$$0<\frac{H(\alpha x)}{H(x)}\leqslant c(\alpha)<\infty\,,\quad x>0\,,$$

where $c(\alpha)$ is bounded on any interval $[\alpha_1, 1]$, $\alpha_1>0$.

One sees easily, for example, that an arbitrary l.p. function which satisfies for some $0<c_1\leqslant c_2<\infty$ and m the inequality

$$c_1x^m h(x)\leqslant H(x)\leqslant c_2x^m h(x)\,,$$

where $h(x)$ is a slowly varying function, is u.p.

It is also not difficult to see that a u.p. function always majorizes some power of x, since for $x=2^N$

$$H(x)\geqslant c^{-1}(\tfrac{1}{2})H(x/2)\geqslant c^{-N}(\tfrac{1}{2})H(x/2^N)=H(1)x^{-\log_2 c(1/2)}\,.$$

Theorem 12. *If* (44) *holds and the function $G_1(x)=e^{\mu_+x}G(x)$ is* u.p., *then*

$$W(x)=\frac{G(x)}{A(\mu_+)B(\mu_+)}(1+o(1))\,,$$

where

$$A(\mu)=\frac{1-f_1(\mu)}{\mu}\quad\textit{and}\quad B(\mu)=\frac{1-\mathsf{M}\,(e^{\mu\chi(+0)};\eta(+0)<\infty)}{1-\mathsf{P}(\eta(+0)<\infty)}\,.$$

For $\mu_+=0$, it is obvious that

$$A(0)=-a\quad\textit{and}\quad B(0)=1\,.$$

Theorem 13. *If the function $U(x)$ is majorized by some function $\tilde{U}(x)$ for which $1-\tilde{U}(x)$ is a distribution function and $e^{\mu_+x}\tilde{U}(x)$ is* u.p., *then*

$$\limsup_{x\to\infty}\frac{W(x)}{\tilde{U}(x)}\leqslant\frac{p(1-p)}{(1-\tilde{b}p)^2}\,;\quad \tilde{b}=-\int e^{\mu_+x}\,d\tilde{U}(x)\,.$$

Proof of Theorem 12. We will use again the formula

$$\mathsf{M}\,e^{\mu\bar{\gamma}}=\frac{1-p}{1-p\mathsf{M}\,e^{\mu\chi}}\,,\qquad \mathsf{M}\,e^{\mu\chi}=-\int_0^\infty e^{\mu x}\,dU(x)\,.$$

We obtain

$$\mathsf{M}\,e^{\mu\bar{\gamma}}=(1-p)\sum_{k=0}^\infty p^k(\mathsf{M}\,e^{\mu\chi})^k,\quad \mu\leqslant\mu_+\,,$$

or, what is the same, for $x>0$

$$W(x)=(1-p)\sum_{k=1}^\infty p^k U_k(x)\,, \tag{45}$$

where $1-U_k(x)$ is the k-fold convolution of the distribution

$$1-U(x)\qquad (U_k(x)=U(x)-\int_0^x dU(t)U_{k-1}(x-t))\,.$$

We will need

Lemma 2. *If* $U_{(1)}(x)=e^{\mu+x}U(x)$ *is* u.p., *then*

$$W(x)\sim\frac{p(1-p)}{(1-pb)^2}U(x),\quad b=\mathsf{M}\,e^{\mu+\chi}.$$

Proof. Assume that for each fixed k and $x\to\infty$

$$\frac{U_k(x)}{U(x)}\sim kb^{k-1}. \tag{46}$$

Then, proceeding formally to the limit under the symbol $\sum_1^\infty$, we get

$$\lim_{x\to\infty}\frac{W(x)}{U(x)}=\lim_{x\to\infty}(1-p)\sum_{k=1}^\infty p^k\frac{U_k(x)}{U(x)}=(1-p)p\sum_{k=1}^\infty k(bp)^{k-1}=\frac{p(1-p)}{(1-bp)^2}.$$

Hence, to prove the lemma we must establish (46) and show that the series

$$R_N(x)=\sum_{k=N}^\infty p^k\frac{U_k(x)}{U(x)} \tag{47}$$

converges uniformly w.r.t. x to zero as $N\to\infty$. We first show (46). Assume it holds for $k-1$. Then from the equality (total probability formula)

$$U_k(x)=U\left(\frac{x}{2}\right)U_{k-1}\left(\frac{x}{2}\right)-\int_0^{x/2}dU(t)U_{k-1}(x-t)-\int_0^{x/2}dU_{k-1}(t)U(x-t)$$

we find for arbitrary $N>0$

$$\begin{aligned}\frac{U_k(x)}{U(x)}=&-\int_0^N dU(t)\frac{U_{k-1}(x-t)}{U(x)}-\int_0^N dU_{k-1}(t)\frac{U(x-t)}{U(x)}\\&+O(e^{(\mu+x)/2})U_{k-1}\left(\frac{x}{2}\right)+r_{N,1}(x)+r_{N,2}(x).\end{aligned} \tag{48}$$

For fixed k we get as $x\to\infty$

$$e^{\mu+x}U_{k-1}(x)\leqslant-\int_x^\infty e^{\mu+t}dU_{k-1}(t)=o(1),$$

$$\begin{aligned}r_{N,1}(x)&=-\int_N^{x/2}dU(t)\frac{U_{k-1}(x-t)}{U(x-t)}\frac{U(x-t)}{U(x)}\\&\leqslant-\max_{\alpha\in[1/2,\,1]}c(\alpha)c\int_N^\infty dU(t)(k-1)b^{k-2}e^{\mu+t}=c_k\int_N^\infty e^{\mu+t}dU(t)\end{aligned}$$

and

$$\begin{aligned}r_{N,2}(x)&=-\int_N^{x/2}dU_{k-1}(t)\frac{U(x-t)}{U(x)}\\&\leqslant-\max_{\alpha\in[1/2,\,1]}[c(\alpha)]\int_N^\infty e^{\mu+t}dU_{k-1}(t).\end{aligned}$$

Hence, the remainder in (48) can be made arbitrarily small by suitable choice of N. Since the limiting value as $x \to \infty$ of the main part of (48) is equal to

$$-(k-1)b^{k-2} \int_0^N e^{\mu+t}\, dU(t) - \int_0^N e^{\mu+t}\, dU_{k-1}(t),$$

which can be made arbitrarily close to $(k-1)b^{k-1} + b^{k-1} = kb^{k-1}$ by suitable choice of N, the existence of $\lim_{x\to\infty}(U_k(x)/U(x))$ (which is independent of N) is proved, as well as the validity of (46).

We now show that (47) converges uniformly to 0. Since by Čebyšev's inequality $U_k(x) \leqslant e^{-\mu+x} b^k$, we have

$$R_N(x) \leqslant \sum_N^\infty (pb)^k \frac{1}{U_{(1)}(x)} \quad (pb<1).$$

We divide the domain of values of x and N into two parts:

1) $x \leqslant e^{\varepsilon N}$ for some $\varepsilon > 0$. Then, since $U_{(1)}(x)$ is u.p., we have for some $m < \infty$

$$U_{(1)}^{-1}(x) \leqslant cx^m \leqslant c\, e^{\varepsilon m N}$$

and

$$R_N(x) \leqslant c \sum_{k=N}^\infty (bp)^k e^{\varepsilon m N} < \frac{c}{1-bp} (bp\, e^{\varepsilon m})^N \to 0$$

as $N \to \infty$, provided that ε is chosen so that $bp\, e^{\varepsilon m} < 1$.

2) $x > e^{\varepsilon N}$, where ε is fixed as the value just chosen. We write $R_N(x)$ in the form

$$R_N(x) = \sum_{N \leqslant k < (1/\varepsilon)\ln x} + \sum_{k \geqslant (1/\varepsilon)\ln x}. \tag{49}$$

An estimate of the second sum has already been given. To estimate the first we will need

Lemma 3. *When the conditions of Lemma 2 are satisfied and $\delta > 0$ is arbitrary, then for all sufficiently large k*

$$\sup_{x \geqslant e^{\varepsilon k}} \frac{U_k(x)}{U(x)} \leqslant (b+\delta)^k.$$

Proof. Set $c_k = \sup_{x \geqslant e^{\varepsilon k}} (U_k(x)/U(x))$. One has

$$U_k(x) = U(x(1-e^{-\varepsilon}))U_{k-1}(x\, e^{-\varepsilon}) - \int_0^{x(1-e^{-\varepsilon})} dU(t) U_{k-1}(x-t) - \int_0^{xe^{-\varepsilon}} dU_{k-1}(t) U(x-t).$$

When $x \geqslant e^{\varepsilon k}$, in the first integrand we have $x - t \geqslant x - x(1-e^{-\varepsilon}) = x\, e^{-\varepsilon} \geqslant e^{\varepsilon(k-1)}$, so that $U_{k-1}(x-t) \leqslant c_{k-1} U(x-t)$. Thus,

$$\frac{U_k(x)}{U(x)} \leqslant c(1-e^{-\varepsilon})\, e^{\mu+xe^{-\varepsilon}} U_{k-1}(x\, e^{-\varepsilon}) - c_{k-1} \int_0^{x(1-e^{-\varepsilon})} e^{\mu+t}\, dU(t) \frac{U_{(1)}(x-t)}{U_{(1)}(x)}$$
$$- \int_0^{xe^{-\varepsilon}} e^{\mu+t}\, dU_{k-1}(t) \frac{U_{(1)}(x-t)}{U_{(1)}(x)}.$$

We have

$$e^{\mu+x} U_{k-1}(x) \leqslant -\int_x^\infty e^{\mu+t}\, dU_{k-1}(t) \leqslant b^{k-1}.$$

Next, just as in (48) we can show that

$$-\int_0^{x(1-e^{-\varepsilon})} e^{\mu_+ t}\, dU(t) \frac{U_{(1)}(x-t)}{U_{(1)}(x)} = b + \varepsilon(x),$$

where $\varepsilon(x) = o(1)$ as $x \to \infty$. Hence,

$$\frac{U_k(x)}{U(x)} \leqslant c(1-e^{-\varepsilon})b^{k-1} + c_{k-1}(b+\varepsilon(x)) + \max_{1-e^{-\varepsilon} \leqslant \alpha \leqslant 1} c(\alpha) b^{k-1}. \tag{50}$$

Put $\varepsilon_{k-1} = \sup_{x > e^{\varepsilon k}} \varepsilon(x)$. Then, from (50) we obtain

$$c_k \leqslant c b^{k-1} + c_{k-1}(b + \varepsilon_{k-1}), \qquad c = c(1-e^{-\varepsilon}) + \max_{1-e^{-\varepsilon} \leqslant \alpha \leqslant 1} c(\alpha).$$

Now choose k_0 such that $\varepsilon_{k-1} < \delta/2$ for $k \geqslant k_0$. Then, if we put $B = b + \delta/2$, we have for $k \geqslant k_0$

$$\begin{aligned} c_k &\leqslant B c_{k-1} + c b^{k-1} \leqslant c b^{k-1} + B[c b^{k-2} + B[\cdots + B[c b^{k_0} + c_{k_0} B]]\cdots] \\ &= c b^{k-1} + B c b^{k-2} + B^2 c b^{k-3} + \cdots + B^{k-k_0-1} c b^{k_0} + c_{k_0} B^{k-k_0} \\ &\leqslant c b^{k_0} B^{k-k_0-1}\left(1 + \frac{b}{B} + \frac{b^2}{B^2} + \cdots\right) + c_{k_0} B^{k-k_0} = B^{k-k_0}\left(\frac{2 c b^{k_0}}{\delta} + c_{k_0}\right) \equiv B^k B_0. \end{aligned}$$

If necessary, we can now also choose k_0 so large that $B^k B_0 < (B + \delta/2)^k = (b+\delta)^k$ for $k \geqslant k_0$. Lemma 3 is proved. □

We turn now to the estimation of the first sum in (49). On the basis of Lemma 3 we have for large enough N

$$\sum_{N \leqslant k < (1/\varepsilon)\ln x} p^k \frac{U_k(x)}{U(x)} \leqslant \sum_{k \geqslant N} p^k (b+\delta)^k \leqslant \frac{p^N (b+\delta)^N}{1 - p(b+\delta)},$$

where δ can be chosen so that $p(b+\delta) < 1$. The uniform convergence to zero of the remainder $R_N(x)$ and along with it, Lemma 2, is proved. □

Comparing the assertions of Theorem 10 and Lemma 2, we find that under the conditions of Theorem 12

$$W(x) \sim U(x) p \frac{1-p}{(1-pb)^2} \sim \frac{\mu_+ G(x)}{(1 - \mathsf{M}\, e^{\mu_+ \chi^*(0)})} \frac{(1-p)}{(1-pb)^2}.$$

Using the identity

$$(1 - \mathsf{M}\, e^{\mu_+ \chi^*(0)})(1 - pb) = 1 - f_1(\mu_+),$$

we obtain the assertion of Theorem 12.

The proof of Theorem 13 does not differ essentially from that of Theorem 12. One must prove the inequality

$$\limsup\, (W(x)/\tilde{U}(x)) \leqslant p(1-p)/(1 - p\tilde{b})^2.$$

This can be obtained, for example, as follows: We stipulated that $1-\tilde{U}(x)$ was a distribution function. Then

$$U_k(x) \leqslant \tilde{U}_k(x),$$

where $\tilde{U}_k(x)$ is the k-fold convolution of $\tilde{U}_1 = U$:

$$U_k(x) = \tilde{U}_1(x) - \int_0^x d\tilde{U}_1(t)\tilde{U}_{k-1}(x-t).$$

Thus by (45)

$$\limsup_{x\to\infty} \frac{W(x)p}{\tilde{U}(x)} \leqslant \limsup_{x\to\infty} \frac{(1-p)\sum_1^\infty p^k \tilde{U}_k(x)}{\tilde{U}_1(x)}.$$

It remains to use Lemma 2. □

5. The asymptotic behavior of $W(x)$ remains to be investigated when $f_1(\mu_+)=1$, $|f_1'(\mu_+)|=\infty$. This very special case is "transitional" between the two types of asymptotic behavior discovered in Subsections 3 and 4. The character of this "transition" is rather complicated. We will illustrate this with the following proposition:

Let $\mu_+>0, f_1(\mu_+)=1$ and as $x\to\infty$

$$G(x) \sim e^{-\mu_+ x} x^{-\alpha} \varepsilon(x), \quad 1<\alpha<2,$$

where $\varepsilon(x)$ is some slowly varying function. Then

$$\int_0^x e^{\mu_+ t} W(t)\, dt \sim -\frac{1}{x^{-\alpha+1}\varepsilon(x)} \frac{(1-p)(1-\mathsf{M}\, e^{\mu_+ \chi^*(0)})\sin \pi\alpha}{\pi \mu_+^2}. \tag{51}$$

Thus, if $e^{\mu_+ x} W(x)\downarrow$ beginning with some x, we have

$$W(x) \sim c \frac{e^{-\mu_+ x}}{x^{-\alpha+2}\varepsilon(x)} \sim c \frac{e^{-2\mu_+ x}}{x^2 G(x)}.$$

The constant c is calculated in an obvious way from (51).

Proof. We use (40):

$$\mathsf{M} \exp(\mu+\mu_+)\bar{Y} = \frac{1-p}{1-\tilde{\varphi}(\mu)}, \qquad \tilde{\varphi}(\mu) = -p\int_0^\infty e^{(\mu+\mu_+)t} dU(t); \quad \operatorname{Re}\mu<0.$$

By Theorem 10

$$U(x) \sim \frac{\mu_+ G(x)}{p(1-\mathsf{M}\, e^{\mu_+ \chi^*(0)})} - c_U\, e^{-\mu_+ x} x^{-\alpha}\varepsilon(x).$$

We now employ the following Tauberian theorem:

Suppose $\mathfrak{v}(\mu)=\int_0^\infty e^{-\mu x}\, dV(x)$ and $\rho>0$. Then, if $V(x)\uparrow$, we have $\mathfrak{v}(\mu)\sim \mu^{-\rho}\varepsilon(1/\mu)$ as $\mu\to 0$ iff

$$V(x) \sim \frac{x^\rho \varepsilon(x)}{\Gamma(\rho+1)} \quad \textit{as } x\to\infty. \tag{52}$$

If $dV(x)=v(x)\,dx$ and $v(x)$ is monotone, then (52) is equivalent to $v(x)\sim x^{\rho-1}\varepsilon(x)/\Gamma(\rho)$.

This theorem implies that as $\mu\to 0$, $\mu<0$

$$\frac{1-\tilde{\varphi}(\mu)}{\mu}=p\mu_+\int_0^\infty\left(\int_x^\infty U(t)\,e^{\mu_+t}\,dt\right)e^{\mu x}\,dx-p\int_0^\infty U(x)\,e^{(\mu_++\mu)x}\,dx$$

$$\sim\frac{p\mu_+c_U\Gamma(2-\alpha)}{\alpha-1}(-\mu)^{\alpha-2}\varepsilon(1/-\mu);$$

$$-\int_0^\infty e^{\mu x}\,e^{\mu_+x}\,dW(x)\sim\frac{(1-p)(\alpha-1)}{p\mu_+c_U\Gamma(2-\alpha)}\,\frac{1}{(-\mu)^{\alpha-1}\varepsilon(1/-\mu)}.$$

To obtain (51), we again use the Tauberian theorem:

$$-\int_0^x e^{\mu_+t}\,dW(t)\sim\frac{(1-p)(\alpha-1)}{p\mu_+c_U\Gamma(2-\alpha)}\,\frac{1}{x^{1-\alpha}\varepsilon(x)\Gamma(\alpha)},$$

which, in virtue of $\Gamma(1-\alpha)\Gamma(\alpha)=\pi/\sin\pi\alpha$, is equivalent to (51). □

In this assertion we took $\mu_+>0$. If $\mu_+=0$, then the asymptotic behavior of $W(x)$ is complicated by the fact that it begins to depend essentially on the "negative tail" of the distribution of ξ for $x<0$ (the summarizing character w.r.t. negative tails of limit theorems for $W(x)$ is lost). This can be discerned, for example, from Formula (36) for $U(x)$ (by Lemma 2, $W(x)$ has the same asymptotic behavior as $U(x)$ if the latter is sufficiently smooth). If, for example, $g(t)\sim c_1t^{-\alpha}$, $g(-t)\sim -c_2t^{-\beta}$ as $t\to\infty$ and $1>\alpha>\beta>0$, then

$$H^{**}(t)\sim c_3t^\beta,\qquad \mathsf{P}(\bar{Y}<\infty)=1$$

and

$$pU(x)\sim c_1c_3\beta\int_0^\infty u^{\beta-1}(1+u)^{-\alpha}\,du\,x^{\beta-\alpha}.$$

6. *We now consider the distribution of θ.* We will use the following theorem on the coefficients of expansions of composite functions (Appendix 3). Let $\mathfrak{D}_+$ be the ring of functions representable as absolutely convergent series for $|z|\leqslant 1$:

$$\mathfrak{a}(z)=\sum_{k=0}^\infty a_kz^k,$$

and $\Lambda(z)$ a function analytic in a region $\mathscr{D}$. Then it follows from Assertion (b) of Appendix 3 that if the curve $\mathfrak{a}=\mathfrak{a}(z)$ ($|z|=1$) lies in $\mathscr{D}$, then $\Lambda(\mathfrak{a}(z))\in\mathfrak{D}_+$, and we have the representation

$$\Lambda(\mathfrak{a}(z))=\sum_{k=0}^\infty \lambda_kz^k.$$

Set $\Delta(b_n)=b_n-b_{n+1}$.

Theorem (see Appendix 3). *Assume, in addition to what has been said, that:*

1) $\sum_{n=1}^\infty\frac{1}{n}\sum_{k=n}^\infty|\Delta(ka_k)|<\infty$;

2) *There exists a slowly varying function* $\varepsilon(x)$ *and constants* $\alpha>0$ *and* $c>0$ *such that*

$$c\varepsilon(n)n^{-\alpha}<|a_n|<\varepsilon(n)n^{-\alpha}, \quad n=1,2,\ldots;$$

3) $a_{n+1}\sim a_n$.
Then $\lambda_n\sim a_n\Lambda'(\mathrm{a}(1))$ *as* $n\to\infty$.

As a consequence of this theorem we obtain

Theorem 14. *Set* $p_n=\mathsf{P}(Y_n>0)$. *Assume that*

1. $\sum_{n=1}^{\infty} p_n \ln n<\infty$ *or* $p_n\downarrow$ *for sufficiently large* n;
2. *There exist* $\alpha>0$, $c>0$ *and a slowly varying function* $\varepsilon(x)$ *such that* $c\varepsilon(n)n^{-\alpha}<p_n<\varepsilon(n)n^{-\alpha}$;
3. $p_{n+1}\sim p_n$.

Then $\mathsf{P}(\theta=n)\sim\mathsf{P}(Y_n>0)/n$ *as* $n\to\infty$.

Remark 1. For the validity of Condition 1 it is sufficient that $\mathsf{M}|\xi|^{2+\delta}<\infty$ for some $\delta>0$. This follows easily from a Čebyšev-type inequality with moments of order $2+\delta$ for the sums Y_n-an.

Remark 2. It is well known that if $1-F(x)\sim\varepsilon(x)x^{-\alpha}$, $\alpha>2$, then

$$p_n\sim n(1-F(-an)) \quad (a=\mathsf{M}\xi>-\infty),$$

and all conditions of the theorem will be fulfilled. In this case

$$\mathsf{P}(\theta=n)\sim\mathsf{P}(\xi>-an).$$

Theorem 15. *Assume* $\mu_+>0$ *and that* μ_0 *is a point at which* $\inf_{\operatorname{Im}\mu=0} f_1(\mu)=m<1$ *is attained. If* $\mu_+>\mu_0$ *and either* $\limsup_{|\lambda|\to\infty,\operatorname{Im}\lambda=0}|f(\lambda)|<1$ *or* ξ *is integer-valued, then*

$$\mathsf{P}(\theta=n)\sim\mathsf{M}m^{-\theta}\frac{\mathsf{P}(Y_n>0)}{n}\sim\frac{c}{n^{3/2}}m^n$$

where $c=\mathsf{M}m^{-\theta}/\mu_0\sqrt{2\pi f_1''(\mu_0)/f_1(\mu_0)}$.

Proof of Theorem 14. We have

$$\mathsf{M}z^\theta=\exp\left\{\sum\frac{z^k-1}{k}\mathsf{P}(Y_k>0)\right\}=(1-p)\exp\left\{\sum_{k=1}^{\infty}\frac{z^k\mathsf{P}(Y_k>0)}{k}\right\}.$$

In the present case,

$$\sum\frac{z^k\mathsf{P}(Y_k>0)}{k}\in\mathfrak{D}_+$$

($\mathfrak{D}_+$ is the sub-ring of functions representable as absolutely convergent series for $|z|\leqslant 1$). Since e^x is an entire function we can use the theorem formulated at the beginning of this subsection. We have ($p_n=\mathsf{P}(Y_n>0)$):

$$\Delta(ka_k)=p_k-p_{k+1}$$

and

$$\sum_{n=1}^{\infty}\frac{1}{n}\sum_{k=n}^{\infty}|p_k-p_{k+1}|\leqslant\sum_{n=1}^{\infty}p_n\sum_{k=1}^{n}\left(\frac{1}{k}+\frac{1}{k+1}\right)<\infty,$$

if $\sum_{n=1}^{\infty}p_n\ln n<\infty$. If $p_n\downarrow$, then the series to be estimated is obviously $\sum_{n=1}^{\infty}(1/n)\,p_n<\infty$. Thus,

$$\mathsf{P}(\theta=n)\sim(1-p)\frac{\mathsf{P}(Y_n>0)}{n}\exp\sum_1^{\infty}\frac{\mathsf{P}(Y_k>0)}{k}=\frac{\mathsf{P}(Y_n>0)}{n}. \quad \square$$

Proof of Theorem 15. Under the hypotheses of this theorem the following asymptotic expansion holds (see, for example, [34]):

$$\mathsf{P}(Y_n>0)=\frac{m^n}{\mu_0\sqrt{2\pi n f_1''(\mu_0)/f_1(\mu_0)}}\left(1+\sum_{k=1}^{s}\frac{c_k}{n^k}+O\left(\frac{1}{n^{s+1}}\right)\right). \tag{53}$$

One has

$$\sum_{k=0}^{\infty}z^k m^{-k}\mathsf{P}(\theta=k)=(1-p)\exp\sum_{k=1}^{\infty}\frac{z^k m^{-k}\mathsf{P}(Y_k>0)}{k}.$$

Here again, by (53), the series following the "exp" symbol belongs to $\mathfrak{D}_+$.

Condition 1 of the quoted theorem of Appendix 3 is satisfied since in our case $a_k=[m^{-k}\mathsf{P}(Y_k>0)]/k$ and for some $c<\infty$ (see (53)) $|\Delta(ka_k)|<ck^{-3/2}$.

Conditions 2 and 3 are obviously also satisfied. Hence,

$$m^{-n}\mathsf{P}(\theta=n)\sim(1-p)\frac{m^{-n}\mathsf{P}(Y_n>0)}{n}\exp\sum\frac{m^{-k}\mathsf{P}(Y_k>0)}{k}. \quad \square$$

§ 23. Inequalities for the Distributions of $\bar{Y}_n$ and $\bar{Y}$. The Rate of Approach of the Distributions of w_n and w^1

1. The following inequalities hold for the distribution $W_n(x)=\mathsf{P}(\bar{Y}_n\geqslant x)$. (Recall that $\bar{Y}_n=\max_{0\leqslant k\leqslant n}Y_k$, $Y_k=\sum_{j=1}^{k}\xi_j$, $f_1(\mu)=\mathsf{M}\,e^{\mu\xi_1}$).

Theorem 16. *For arbitrary* $\mu\geqslant 0$, $x\geqslant 0$ *and* $n\geqslant 1$

$$W_n(x)\leqslant e^{-\mu x}f_1^n(\mu) \quad \text{if } f_1(\mu)\geqslant 1;$$
$$W_n(x)\leqslant e^{-\mu x} \quad \text{if } f_1(\mu)\leqslant 1.$$

Corollaries.

1. $W(x)=\mathsf{P}(\bar{Y}\geqslant x)\leqslant e^{-Qx}$, $\quad Q=\sup\{\mu: f_1(\mu)\leqslant 1\}$.

2. *For arbitrary* μ *such that* $f_1(\mu)\leqslant b$, $b\geqslant 1$

$$W_n(x)\leqslant e^{-\mu x}b^n$$

Proof of the theorem. It is sufficient to consider the relations

$$f_1^n(\mu) = \mathsf{M}\, e^{\mu Y_n} \geqslant \sum_{k=1}^n \mathsf{P}(\eta(x)=k)\mathsf{M}(e^{\mu Y_n}|\eta(x)=k) \geqslant \sum_{k=1}^n \mathsf{P}(\eta(x)=k)\, e^{\mu x} f_1^{n-k}(\mu)\,;$$
$$\sum_{k=1}^n \mathsf{P}(\eta(x)=k) = \mathsf{P}(\bar{Y}_n \geqslant x)\,. \quad \square$$

The inequalities of Theorem 16 become vacuous if ξ^+ has only finitely many moments. In this case we can use Čebyšev's inequality, noting that when $\mathsf{M}\xi$ is finite,

$$\kappa(\bar{Y}) = \kappa(\chi) = \kappa(\xi^+) - 1\,, \tag{54}$$

where, as before,

$$\kappa(\xi) = \sup\{\alpha\colon \mathsf{M}|\xi|^\alpha < \infty\}\,.$$

Relation (54) follows directly from the factorization identities (see, for example, Theorem 10 and (45)). Hence, for $\mu_+ = 0$, $W(x) \leqslant \bar{c}_{m-1}/x^{m-1}$, where $\bar{c}_{m-1} = \mathsf{M}\bar{Y}^{m-1} < \infty$ if $c_m = \mathsf{M}(\xi^+)^m < \infty$. However, the calculation of $\bar{c}_{m-1}$ is quite difficult. In this connection, we quote in Appendix 4 precise inequalities for $W(x)$ whose right-hand sides depend only on $a = \mathsf{M}\xi$ and c_m. Asymptotic inequalities for $\bar{Y}$ are given in Theorem 13.

2. In the "exponential" case $\mu_+ > 0$, the convergence of the distributions of $\bar{Y}_n$ and $\bar{Y}$ is described by

Theorem 17. *For any μ for which $f_1(\mu) \leqslant 1$*

$$\mathsf{P}(\bar{Y} \geqslant x) - \mathsf{P}(\bar{Y}_n \geqslant x) \leqslant e^{-\mu x} f_1^n(\mu)\,.$$

Proof. For the values of μ in question we have

$$\mathsf{P}(\bar{Y} \geqslant x) - \mathsf{P}(\bar{Y}_n \geqslant x) = \mathsf{P}(\bar{Y}_n < x,\, \bar{Y} \geqslant x) \leqslant \int_{-\infty}^x d\mathsf{P}(Y_n < t)\mathsf{P}(\bar{Y} \geqslant x - t)$$
$$\leqslant \int_{-\infty}^x e^{-\mu(x-t)}\, d\mathsf{P}(Y_n < t) \leqslant e^{-\mu x} f_1^n(\mu)\,. \quad \square$$

Remark. The rate of convergence and exponential decrease of $W(x)$ for $\mu_+ > 0$ are somewhat invariant w.r.t. the nature of the sequence $\{\xi_k\}$. For example, they are retained when ξ_k is defined on the states of a regular, finite Markov chain.

In the non-exponential case, the estimates are more unwieldy and the proofs more complicated. We will treat all of them in Appendix 4. We quote here merely the following results:

If $\mathsf{D}\xi = 1$, $c_m = \mathsf{M}|\xi^+|^m < \infty$ and $m \geqslant 2$, then $\bar{c}_{m-1} = \mathsf{M}\bar{Y}^{m-1} < \infty$ and for all $n \geqslant n_0$ ($n_0 = n_0(m, c_m)$ will be found in explicit form)

$$\mathsf{P}(\bar{Y} \geqslant x) - \mathsf{P}(\bar{Y}_n \geqslant x) \leqslant \frac{2^{m+1} c_m e}{|a|^m n^{m-1}} + \frac{\bar{c}_{m-1}}{(x - an/2)^{m-1}}\,. \tag{55}$$

Looking over the calculations of Appendix 4, we see easily that this inequality can be sharpened.

3. We have already seen that finding *estimates for the distribution of* w^k *and the rate of convergence of the distributions of* w_n *to that of* w^k *as* $n \to \infty$ are problems closely connected with those of this section. Estimates for the distribution of w^k obviously coincide with those quoted above for $\bar{Y}$. For the convergence of w_n to w^1 we have by Theorem 2, Chapter 1

$$\mathsf{P}(w_{n+1} > x) - \mathsf{P}(w^1 > x) = \mathsf{P}(\bar{Y}_n \leqslant x,\ Y_n > x - w_1) - \mathsf{P}(\bar{Y}_n \leqslant x,\ \bar{Y} > x).$$

Estimates for the second term, which equals $\mathsf{P}(\bar{Y}_n \leqslant x) - \mathsf{P}(\bar{Y} \leqslant x)$, have been given above. For the first summand one has

$$\mathsf{P}(\bar{Y}_n \leqslant x,\ Y_n > x - w_1) \leqslant \mathsf{P}(Y_n \in (x - w_1, x]).$$

In the exponential case, this expression is smaller than $e^{-\mu(x-w_1)} f_1^n(\mu)$ for arbitrary μ. In the non-exponential case there holds an inequality whose proof will be given in Appendix 4. Assume in addition to the conditions for (55) that for $t > t_0$ (the bound t_0 will be given explicitly)

$$F(t+v) - F(t) < \frac{Cv}{t^{m+1}}.$$

Then (see the corollary of Theorem 4 in Appendix 4)

$$\mathsf{P}(Y_n \in (x - w_1, x]) \leqslant \frac{2^{m+1} n}{(-an - w_1 + x)^{m+1}} \left[c w_1 + \frac{2^{m+1} c_m^2 n e^2}{(-an - w_1 + x)^{m-1}} \right].$$

Analogous inequalities can also be obtained for the rate of convergence of the distributions of the sequences $\{w_{k+n}, k \geqslant 0\}$ and $\{w^k, k \geqslant 0\}$ for $n \to \infty$ by using Theorem 4 Chapter 1 and the results given above.

§ 24. Comparison Theorems

Assume that we are faced with the choice between two queueing systems differing from one another through the parameters of the governing sequences. Which of the two should be preferred? If during the waiting time w a loss $L(w)$ is incurred, then the better system is naturally that for which $\mathsf{M}L(w)$ is smaller.

We will consider the value $\mathsf{M}L(w)$ as a numerical characteristic of the systems, where L is some convex (downward) function and w is the stationary waiting time. Thus, under the conditions of this chapter, we will be concerned with finding sufficiently general conditions on the parameters of the two systems, i.e., on the distributions of certain sequences $\{\xi_k\}$ and $\{\xi_k^0\}$ in order that

$$\mathsf{M}L(\bar{Y}^0) \leqslant \mathsf{M}L(\bar{Y})$$
$$(Y^0 = \sup Y_n^0,\ Y_n^0 = \xi_1^0 + \cdots + \xi_n^0).$$

At the end of the section we will discuss several "extremal" problems for service systems in terms of the results obtained.

Definition. *Let $\mathscr{L}$ be the class of continuous, convex, nondecreasing functions. We will say that ξ^0 is the* convex minorant *of* ξ,

$$\xi^0 \prec \xi ,$$

if $\mathsf{M}\xi^+ < \infty$ *and for any* $L \in \mathscr{L}$ *such that* $\mathsf{M}L(\xi)$ *exists*

$$\mathsf{M}L(\xi^0) \leqslant \mathsf{M}L(\xi) . \tag{56}$$

The condition $\mathsf{M}\xi^+ < \infty$ signifies that the class of functions $L \in \mathscr{L}$, $\mathsf{M}L(\xi) < \infty$ contains more than just constants.

The most important results of this section for us will be the fact that *for* $\bar{Y}^0 \prec \bar{Y}$ *it is sufficient that* $\xi^0 \prec \xi$ (see Property 7) and the following

Lemma 4. *For* $\xi^0 \prec \xi$ *it is necessary and sufficient that for all* x

$$G_0(x) \leqslant G(x) . \tag{57}$$

Here, as before

$$G(x) = \int_x^\infty g(t)\, dt , \qquad g(x) = E(x) - F(x) ,$$

$E(x)$ is the distribution concentrated at zero and the index "0" refers to the sequence $\{\xi_k^0\}$.

Proof. By the integral representation of convex functions we can write for arbitrary $L \in \mathscr{L}$:

$$L(x) = c + \int_0^x l(t)\, dt ,$$

where l does not decrease and $l(-\infty) \geqslant 0$. Since $\mathsf{M}L(\xi)$ exists,

$$\begin{aligned} \mathsf{M}L(\xi) &= c - \int_{-\infty}^\infty dg(t) \int_0^t l(u)\, du = c + \int_{-\infty}^\infty g(t) l(t)\, dt \\ &= c + G(-\infty) l(-\infty) + \int_{-\infty}^\infty G(t)\, dl(t) , \end{aligned}$$

where in the case $G(-\infty) = -\infty$, the indeterminacy of $G(-\infty) \times l(-\infty)$ is to be taken as 0. From this representation we see that (57) implies $\xi^0 \prec \xi$.

The necessity is obvious since for an arbitrary distribution function $l(t)$ of a random variable bounded from below and for

$$L(x) = \int_0^x l(t)\, dt ,$$

$\mathsf{M}L(\xi)$ exists (since $\mathsf{M}\xi^+$ exists) and equals $\int_{-\infty}^\infty G(t)\, dl(t)$. Now we need only set $l(t) = E(t - x)$ in the inequality

$$\int_{-\infty}^\infty G_0(t)\, dl(t) \leqslant \int_{-\infty}^\infty G(t)\, dl(t) . \quad \square$$

From the proof of the lemma it is obvious that in the necessity part we need only require that (56) hold for $L \in \mathscr{L}$ such that

$$L(x) \leqslant \alpha + \beta |x| . \tag{58}$$

From this it follows that for $\xi^0 \prec \xi$ it is necessary and sufficient that (56) hold for arbitrary functions $L \in \mathscr{L}$ satisfying (58).

Furthermore, $\xi^0 \prec \xi$ *iff for arbitrary* t

$$\mathsf{M}(\xi^0 - t; \xi^0 \geqslant t) \leqslant \mathsf{M}(\xi - t; \xi \geqslant t).$$

This is obvious by Lemma 4 since

$$\textstyle\int_t^\infty (x-t)\,dF(x) = -t(1-F(t)) + tg(t) + \int_t^\infty g(x)\,dx = -\min(0, t) + G(t).$$

Hence, it is sufficient to establish (56) merely for the family of functions

$$L_t(x) = \begin{cases} x-t, & x \geqslant t, \\ 0, & x < t, \end{cases} \quad -\infty < t < \infty. \tag{59}$$

We now establish several properties connected with the relation $\prec$.

1. $\xi^0 = \mathsf{M}\xi \prec \xi$. This is an immediate consequence of the convexity of L in the definition (56).

2. *If* $L \in \mathscr{L}$, *then* $K(x) = \mathsf{M}L(x+\xi) \in \mathscr{L}$.
Indeed, since $L_1(x) = L(x+t) \in \mathscr{L}$, we have

$$\begin{aligned} K(px_1+qx_2) &= \textstyle\int dF(t) L(px_1+qx_2+t) \\ &\leqslant p \textstyle\int dF(t) L(x_1+t) + q \int dF(t) L(x_2+t) = pK(x_1) + qK(x_2), \\ &\qquad p \geqslant 0,\, q \geqslant 0,\, p+q=1. \end{aligned}$$

The monotonicity and continuity of K are obvious. □

3. *If* ξ_1 *and* ξ_2 *are independent, and* ξ_1^0 *and* ξ_2^0 *are independent and* $\xi_1^0 \prec \xi_1$, $\xi_2^0 \prec \xi_2$, *then* $\xi_1^0 + \xi_2^0 \prec \xi_1 + \xi_2$.

Proof. For $L \in \mathscr{L}$ and with the obvious agreements as to notation, we have

$$\begin{aligned} \mathsf{M}L(\xi_1^0+\xi_2^0) &= \textstyle\int dF_{\xi_1^0}(t)\mathsf{M}L(\xi_2^0+t) \leqslant \int dF_{\xi_1^0}(t)\mathsf{M}L(\xi_2+t) \\ &\leqslant \textstyle\int dF_{\xi_1}(t)\mathsf{M}L(\xi_2+t) = \mathsf{M}L(\xi_1+\xi_2). \quad \square \end{aligned}$$

In particular, it follows from this that we always have

$$\xi_1 \prec \xi_1 + (\xi_2 - \mathsf{M}\xi_2)$$

when ξ_1 and ξ_2 are independent.

4. *Assume that* $\xi_j^0 \prec \xi_j$ *and that the sequences* $\{\xi_j^0\}$ *and* $\{\xi_j\}$ *are formed from non-negative i.i.d. random variables. Assume also that* $\nu^0 \prec \nu$ *are two nonnegative integer-valued random variables independent, resp., of the sequences* $\{\xi_j^0\}$ *and* $\{\xi_j\}$. *Then*

$$Y_{\nu^0}^0 \prec Y_\nu.$$

Remark. If the condition $\xi_j \geqslant 0$ does not hold, then this assertion is generally speaking not true. Suppose, for example, that

$$\mathsf{P}(\xi_j = \pm 1) = \mathsf{P}(\xi_j^0 = \pm 1) = \tfrac{1}{2},$$
$$\mathsf{P}(\nu^0 = 1) = 1 \quad \text{and} \quad \mathsf{P}(\nu = 0) = \mathsf{P}(\nu = 2) = \tfrac{1}{2}.$$

Then the rest of the conditions of 4 remain valid, but for the function $L(x) = \max(0, x) \in \mathscr{L}$ we have

$$\mathsf{M}L(Y_{\nu^0}^0) = \mathsf{M}L(\xi_1^0) = \tfrac{1}{2}$$

and

$$\mathsf{M}L(Y_\nu) = \tfrac{1}{2}\mathsf{M}L(\xi_1 + \xi_2) = \tfrac{1}{2}(0 \cdot \tfrac{3}{4} + 2 \cdot \tfrac{1}{4}) = \tfrac{1}{4} < \mathsf{M}L(Y_{\nu^0}^0).$$

Proof. For $L \in \mathscr{L}$ one has

$$\mathsf{M}L(Y_\nu) = \textstyle\sum_{k=0}^\infty \mathsf{P}(\nu = k)\mathsf{M}L(Y_k) \geqslant \mathsf{M}L^*(\nu),$$

where $L^*(k) = \mathsf{M}L(Y_k)$. Assume that $L^*(k)$ is a convex sequence. Then Property 4 will follow from the inequalities

$$\begin{aligned}\mathsf{M}L^*(\nu) \geqslant \mathsf{M}L^*(\nu^0) &= \textstyle\sum \mathsf{P}(\nu^0 = k)\mathsf{M}L(Y_k) \\ &\geqslant \textstyle\sum \mathsf{P}(\nu^0 = k)\mathsf{M}L(Y_k^0) = \mathsf{M}L(Y_{\nu^0}^0).\end{aligned}$$

Hence, we must show that the sequence $L^*(k)$ is convex, i.e., that

$$L^*(k) \leqslant \tfrac{1}{2} L^*(k-1) + \tfrac{1}{2} L^*(k+1), \quad k = 1, 2, \ldots. \tag{60}$$

But by Property 2

$$L^{**}(x) = \mathsf{M}L(Y_{k-1} + x) \in \mathscr{L},$$

so that for the proof of (60) we must convince ourselves that for any $L \in \mathscr{L}$

$$\mathsf{M}L(\xi_k) \leqslant \tfrac{1}{2}\mathsf{M}L(0) + \tfrac{1}{2}\mathsf{M}L(\xi_k + \xi_{k+1})$$

or that (see (59)) for arbitrary $t \geqslant 0$

$$\mathsf{M}L_t(\xi_k) \leqslant \tfrac{1}{2}\mathsf{M}L_t(\xi_k + \xi_{k+1}). \tag{61}$$

Introduce the truncation $[x]_t = \min(x, t)$. Then $L_t(x) + [x]_t = x$, and in view of the equality $\mathsf{M}\xi_k = \mathsf{M}\xi_{k+1}$, (61) is equivalent to

$$\mathsf{M}[\xi_k]_t \geqslant \tfrac{1}{2}\mathsf{M}[\xi_1 + \xi_{k+1}]_t.$$

But this is obvious since for nonnegative random variables

$$[\xi_k + \xi_{k+1}]_t \leqslant [\xi_k]_t + [\xi_{k+1}]_t. \quad \square$$

5. *If*

$$F_n(x) = \mathsf{P}(\xi_n < x) \Rightarrow F(x) = \mathsf{P}(\xi < x),$$
$$F_n^0(x) = \mathsf{P}(\xi_n^0 < x) \Rightarrow F^0(x) = \mathsf{P}(\xi^0 < x),$$
$$\mathsf{M}\xi_n^+ \to \mathsf{M}\xi^+ \quad \text{and} \quad \mathsf{M}\xi_n^{0+} \to \mathsf{M}\xi^{0+},$$

then $\xi_n^0 \prec \xi_n$, $n = 1, 2, \ldots$, *implies* $\xi^0 \prec \xi$.

Indeed, from the convergence of the expectations it follows that

$$G(x)=\int_x^\infty g(t)\,dt=\lim_{n\to\infty}\int_x^\infty g_n(t)\,dt\geqslant\lim_{n\to\infty}\int_x^\infty g_n^0(t)\,dt=G^0(x)\,. \quad \square$$

It is not difficult to verify that the weak convergence conditions in 5 alone are not sufficient for invariance of the $\prec$-relation under a limit passage.

6. *If* $\xi_1, \xi_2, \ldots, \xi_n$ *and* $\xi_1^0, \xi_2^0, \ldots, \xi_n^0$ *are two sequences of independent random variables and* $\xi_k^0 \prec \xi_k$, $k=1, 2, \ldots n$, *then*

$$\bar{y}_n^0 \prec \bar{y}_n\,. \tag{62}$$

Here, as before, $\bar{y}_n$ denotes $\max(Y_1, \ldots, Y_n)$.

Proof. We use induction.

Let

$$P_1(x)=\mathsf{P}(\xi_1\geqslant x)$$

and

$$Z_{n-1}(x)=\mathsf{P}(\max(\xi_2, \xi_2+\xi_3, \ldots, \xi_2+\cdots+\xi_n)\geqslant x)\,.$$

Then

$$Z_n(x)=\mathsf{P}(\bar{y}_n\geqslant x)=P_1(t)-\int_{-\infty}^x dP_1(t)Z_{n-1}(x-t)$$

and

$$\begin{aligned}
G_n(x)&\equiv\int_x^\infty Z_n(t)\,dt=G_1(x)-\int_x^\infty dt\int_{-\infty}^t dP_1(u)Z_{n-1}(t-u)\\
&=G_1(x)+G_{n-1}(0)P_1(x)-\int_{-\infty}^x G_{n-1}(x-u)\,dP_1(u)\\
&\geqslant G_1(x)+G_{n-1}^0(0)P_1(x)-\int_{-\infty}^x G_{n-1}^0(x-u)\,dP_1(u)\\
&=G_1(x)+\int_{-\infty}^x P_1(u)Z_{n-1}^0(x-u)\,du\\
&=G_1(x)(1-Z_{n-1}^0(0))+\int_{-\infty}^x G_1(u)\,d_uZ_{n-1}^0(x-u)\geqslant\\
&\geqslant G_1^0(x)(1-Z_{n-1}^0(0))+\int_{-\infty}^x G_1^0(u)\,d_uZ_{n-1}^0(x-u)=G_n^0(x)\,. \quad \square
\end{aligned}$$

Under the conditions of Property 6 we also have

$$\bar{Y}_n^0 \prec \bar{Y}_n\,. \tag{63}$$

It is sufficient to set $\xi_1=\xi_1^0=0$.

7. *If the sequences appearing in Property* 6 *are unbounded and each consists of identically distributed random variables, then*

$$\bar{Y}^0 \prec \bar{Y}\,. \tag{64}$$

This relation is a consequence of (63) and Property 5. The conditions of the latter hold since $\mathsf{M}\bar{Y}<\infty$ and $\bar{Y}_n\uparrow\bar{Y}$. $\square$

The assertion at (64) can be strengthened.

8. Set

$$\bar{f}(\mu)=\mathsf{M}\,e^{\mu\bar{Y}} \quad \text{and} \quad \kappa_k(\bar{Y})=\left.\frac{d^k \ln \bar{f}(\mu)}{d\mu^k}\right|_{\mu=0},$$

so that $\kappa_k(\bar{Y})$ are the semi-invariants (cumulants) of the distribution of $\bar{Y}$:

$$\kappa_1(\bar{Y})=\mathsf{M}\bar{Y}, \qquad \kappa_2(\bar{Y})=\mathsf{D}\bar{Y}, \quad \text{etc.}$$

If under the conditions of Properties 6 *and* 7 $\mathsf{M}(\xi^+)^k<\infty$, *then for all* $j\leqslant k-1$

$$\kappa_j(\bar{Y}^0)\leqslant\kappa_j(\bar{Y}).$$

These inequalities can obviously be written in terms of the moments of the spectral functions of the distributions of $\bar{Y}^0$ and $\bar{Y}$.

Proof. We have

$$\ln\bar{f}(\mu)=\sum_{n=1}^{\infty}\frac{1}{n}\int_0^{\infty}(e^{\mu x}-1)\,d\mathsf{P}(Y_n<x)=\sum_{n=1}^{\infty}\frac{1}{n}\mathsf{M}K_\mu(Y_n),$$

where $K_\mu(x)=0$ for $x\leqslant 0$ and $K_\mu(x)=e^{\mu x}-1$ for $x>0$. For $1\leqslant j\leqslant k-1$

$$\left.\frac{d^j \ln\bar{f}(\mu)}{d\mu^j}\right|_{\mu=0}=\kappa_j(\bar{Y})=\sum_{n=1}^{\infty}\frac{1}{n}\mathsf{M}K^{(j)}(Y_n),$$

where

$$K^{(j)}(x)=\left.\frac{\partial^j K_u(x)}{\partial\mu^j}\right|_{\mu=0}=\begin{cases}0, & x<0,\\ x^j, & x>0,\end{cases} \quad j=1,2,\ldots.$$

Since $K^{(j)}(x)\in\mathscr{L}$, it remains to use Property 3 by which $\mathsf{M}K^{(j)}(Y_n^0)\leqslant\mathsf{M}K^{(j)}(Y_n)$. □

Property 8 can easily be extended to the semi-invariants (cumulants)

$$\left.\frac{d^k}{dt^k}\ln\frac{\bar{f}(\mu+t)}{\bar{f}(\mu)}\right|_{t=0}=\frac{d^k\ln\bar{f}(\mu)}{d\mu^k}, \quad \mu>0$$

of Cramér's transformation of the distribution of $\bar{Y}$, provided that $\bar{f}(\mu)<\infty$ for $\mu>0$.

As an application of these results we consider the following problem. Assume that the entrance stream (the sequences $\{\tau_j^e\}$ and $\{v_j^e\}$) in the system $\langle G_I, G_I, G_I, 1\rangle$ is fixed but that we can choose the distribution of τ_j^s arbitrarily in the class of distributions $\mathscr{F}_\alpha$ for which $\mathsf{M}\tau_j^s=\alpha$. Then *for the stationary waiting time* w^k, $\inf_{\mathscr{F}_\alpha}\mathsf{M}w^k$ *is attained by the distribution concentrated at the single point* $\tau_j^s\equiv\alpha$. This follows from Properties 3 and 7 since $a\prec\tau_j^s$ for arbitrary τ_j^s from $\mathscr{F}_\alpha$ and, consequently (Property 3), $\xi_j^0\prec\xi_j$, where the sequence ξ_j^0 is formed by replacing τ_j^s by α in the formulae for ξ_j.

Hence, *deterministic service is extremal in the class* $\mathscr{F}_\alpha$. It follows from Property 8 that $\mathsf{D}w^k$ and a number of other parameters will be minimized simultaneously.

An analogous conclusion can be drawn regarding τ_j^e when the distributions of τ_j^s and v_j^e are fixed. However, the extremal distribution of v_j^e for fixed values of

$\mathsf{M}\nu_j^e = \alpha$ will differ from the deterministic one if α is not an integer. In this case, it is easy to see that $\nu^0 \prec \nu$ for an arbitrary integer-valued random variable ν with $\mathsf{M}\nu = \alpha$ if ν^0 assumes the two values $[\alpha]$ and $[\alpha]+1$; one has

$$\mathsf{P}(\nu^0 = [\alpha]+1) = \alpha - [\alpha].$$

It is also easy to see that the same assertions will hold for the prelimiting waiting time w_n. They also remain valid for the virtual waiting time $w(t)$.

§ 25. Conditions for Heavy Traffic. Transitional Phenomena

We have seen that under general assumptions one can obtain explicit asymptotic formulae for the virtual waiting time $w^c(t)$ under conditions of light or heavy traffic ($p = \mathsf{P}(\overline{Y} > 0)$ is close to 0 or 1). It follows from the results of §§ 8 and 9 that under the assumptions of this chapter the limiting distributions of w^k and $w^c(t)$ will coincide for heavy traffic. Here, however, where the nature of the sequence ξ_k is quite simple, these formulae can be made considerably more precise.

We will see later that the limit laws described in § 8 of Chapter 1 and below are manifestations of much more general and stronger laws by which even the processes $\{w^k, k \geq 0\}$, after suitable normalization, converge for $|a| \to 0$ to rather simple and widely-investigated processes.

In order to pose our problem more precisely, it is necessary, as in § 21, to introduce a double sequence, i.e., a family of sequences of independent random variables

$$\begin{array}{l} \xi_1^{(1)}, \xi_2^{(1)}, \ldots \\ \xi_1^{(2)}, \xi_2^{(2)}, \ldots \\ \cdot\ \cdot\ \cdot\ \cdot\ \cdot\ \cdot\ \cdot\ , \end{array}$$

whose n-th sequence is distributed according to the law $\mathsf{P}(\xi_k^{(n)} < x) = F_n(x)$. We will assume that

$$\mathsf{D}\xi_k^{(n)} = \sigma_n^2 \to \sigma^2 > 0 \quad \text{and} \quad \mathsf{M}\xi_k^{(n)} = a_n \to 0 \tag{65}$$

as $n \to \infty$. Let $\{\omega(t), t \geq 0\}$ be a standard Wiener process: $\mathsf{M}\omega(t) = 0, \mathsf{D}\omega(t) = t$, and set

$$Y_{k,n} = \textstyle\sum_{j=1}^k \xi_j^{(n)} \quad \text{and} \quad \overline{Y}_n = \max_{1 \leq k \leq n} Y_{k,n}.$$

Theorem 18. *Suppose in addition to* (65) *that for any* $\varepsilon > 0$

$$\lim_{n \to \infty} \int_{|x| > \varepsilon\sqrt{n}} x^2 \, dF_n(x) = 0. \tag{66}$$

Then, if $a=a_n\to 0$ as $n\to\infty$ without changing sign in such a way that $na^2\to t$, we have

$$\lim_{n\to\infty} \mathsf{P}\left(\bar{Y}_n<\frac{x}{|a|}\right)=\mathsf{P}\left(\omega(u)<\frac{x-u\,\text{sign}\,a}{\sigma},\,0\leqslant u\leqslant t\right)\equiv P(t,x)$$

$$=\begin{cases} 1-e^{-2x/\sigma^2}\left[1-\dfrac{e^{x/\sigma^2}}{\sqrt{2\pi}}\displaystyle\int_{t\sigma^2/x^2}^{\infty} u^{-3/2}\exp\left\{-\frac{1}{2}\left[\frac{1}{u}+\frac{ux^2}{\sigma^2}\right]\right\}du\right], & a<0,\\ \dfrac{e^{x/\sigma^2}}{\sqrt{2\pi}}\displaystyle\int_{t\sigma^2/x^2}^{\infty} u^{-3/2}\exp\left\{-\frac{1}{2}\left[\frac{1}{u}+\frac{ux^2}{\sigma^4}\right]\right\}du, & a>0. \end{cases} \tag{67}$$

If $na^2\to 0$, then

$$\lim_{n\to\infty} \mathsf{P}(\bar{Y}_n<x\sqrt{n})=P(x)\equiv\mathsf{P}\left(\omega(u)\leqslant\frac{x}{\sigma},\,0\leqslant u\leqslant 1\right)$$
$$=\sqrt{\frac{2}{\pi}}\int_0^{x/\sigma} e^{-u^2/2}\,du\,.$$

If $na^2\to\infty$ and a remains negative, then

$$\lim_{n\to\infty} \mathsf{P}\left(\bar{Y}_n<\frac{x}{|a|}\right)=1-e^{-2x/\sigma^2}. \tag{68}$$

If $na^2\to\infty$ and a remains positive, then

$$\lim_{n\to\infty} \mathsf{P}(\bar{Y}_n<an+x\sqrt{n})=\frac{1}{\sqrt{2\pi}}\int_{-\infty}^{x/\sigma} e^{-u^2/2}\,du\,.$$

Proof. We will use the following theorem (which is a corollary of a theorem due to Prohorov; see, for example, Gihman and Skorohod [29], § 3 Chapter 9):

Let $X_{k,n}=\sum_{j=1}^{k}(\xi_j^n-a_n)$ and denote by $X_n(u)$ the polygon formed by the points $(k/n,\,X_{k,n}/\sigma\sqrt{n})$, $k=0,1,\ldots,n$. Then, under Conditions (65) *and* (66), *the distribution of $\varphi(X_n(u))$ for any functional φ which is defined and continuous* (w.r.t. *the uniform metric*) *in the space $C(0,1)$ of continuous functions on* $[0,1]$, *converges weakly to the distribution of $\varphi(\omega(u))$.*

We prove the first part of the theorem. Assume for simplicity that n and $a=a_n$ vary in such a way that $na^2=t$. Then

$$\mathsf{P}\left(\bar{Y}_n<\frac{x}{|a|}\right)=\mathsf{P}\left(\frac{Y_{k,n}-ka}{\sigma\sqrt{n}}+\frac{ka\sqrt{n}}{n\sigma}<\frac{x}{\sigma|a|\sqrt{n}},\,k=1,2,\ldots,n\right)$$
$$=\mathsf{P}\left(\frac{X_{k,n}}{\sigma\sqrt{n}}+\frac{k}{n}\frac{\sqrt{t}\,\text{sign}\,a}{\sigma}<\frac{x}{\sigma\sqrt{t}},\,k=1,2,\ldots,n\right)$$
$$=\mathsf{P}\left(X_n(u)+\frac{u\sqrt{t}\,\text{sign}\,a}{\sigma}<\frac{x}{\sigma\sqrt{t}},\,0\leqslant u\leqslant 1\right).$$

Since $\varphi(\omega)=\sup_{0\leqslant u\leqslant 1}(\omega(u)+cu)$ is obviously a continuous functional of $\omega(u)$ in the uniform metric, we have

$$\begin{aligned}\mathsf{P}\left(\bar{Y}_n<\frac{x}{|a|}\right)&\Rightarrow\mathsf{P}\left(\sup_{0\leqslant u\leqslant 1}\left(\omega(u)+\frac{u\sqrt{t}\,\operatorname{sign}a}{\sigma}\right)<\frac{x}{\sigma\sqrt{t}}\right)\\&=\mathsf{P}\left(\sqrt{t}\,\omega(u)<\frac{x-ut\operatorname{sign}a}{\sigma},\ 0\leqslant u\leqslant 1\right)\\&=\mathsf{P}\left(\omega(u)<\frac{x-u\operatorname{sign}a}{\sigma},\ 0\leqslant u\leqslant t\right).\end{aligned}\tag{69}$$

Replacing $na^2=t$ by $na^2\to t$ in the final equality at (69) will obviously not change anything.

The process $\{Y(u)=\omega(u)+(u\operatorname{sign}a)/\sigma,\ u\geqslant 0\}$ is continuous from above and below and in the notation of Chapter 2 the probability appearing on the right in (69) can be written as

$$\mathsf{P}\left(\bar{Y}(t)<\frac{x}{\sigma}\right)=1-\mathsf{P}\left(\bar{Y}(t)\geqslant\frac{x}{\sigma}\right).$$

Since $Y(t)$ has a density, it follows from (4) § 12 that

$$\begin{aligned}\mathsf{P}\left(\bar{Y}(t)\geqslant\frac{x}{\sigma}\right)&=\int_0^t\frac{x}{\sigma u}\frac{1}{\sqrt{2\pi u}}\exp\frac{\left(\frac{x}{\sigma}-\frac{u\operatorname{sign}a}{\sigma}\right)^2}{-2u}\,du\\&=\int_0^\infty-\int_t^\infty=\int_0^\infty-\exp\left\{\frac{x\operatorname{sign}a}{\sigma^2}\right\}\int_{t\sigma^2/x^2}^\infty\frac{1}{\sqrt{2\pi}u^{3/2}}\\&\quad\times\exp\left[-\frac{1}{2}\left(\frac{1}{u}+\frac{ux^2}{\sigma^4}\right)\right]du.\end{aligned}\tag{70}$$

Here $\int_0^\infty$ is obviously $\mathsf{P}(\bar{Y}(\infty)\geqslant x/\sigma)$, i.e., the value to which $\mathsf{P}(\bar{Y}_n\geqslant x/|a|)$ converges as $na^2\to\infty$. Using Theorem 1 §12 again, we find that

$$\mathsf{P}\left(\bar{Y}(\infty)\geqslant\frac{x}{\sigma}\right)=e^{-x\mu(0)/\sigma},$$

where $\mu(0)$ is the largest solution of the equation

$$\psi_1(\mu)\equiv\frac{\mu\operatorname{sign}a}{\sigma}+\frac{\mu^2}{2}=0,$$

so that $\mu(0)=2/\sigma$ for $a<0$ and $\mu(0)=0$ for $a>0$. Inserting these values into (70) we get assertions (67) and (68) of the theorem.

If $na^2\to 0$, then in analogy to the previous reasoning, we find for each $\varepsilon>0$

$$\begin{aligned}\mathsf{P}(\bar{Y}_n<x\sqrt{n})&\geqslant\mathsf{P}\left(X_n(u)+u\varepsilon<\frac{x}{\sigma},\ 0\leqslant u\leqslant 1\right)\\&\Rightarrow\mathsf{P}\left(\sup_{0\leqslant u\leqslant 1}(\omega(u)+\varepsilon u)<\frac{x}{\sigma}\right).\end{aligned}$$

Using the reverse inequality and the arbitrariness of ε, we obtain

$$\mathsf{P}(\overline{Y}_n < x\sqrt{n}) \Rightarrow \mathsf{P}\left(\sup_{0\leqslant u\leqslant 1} \omega(u) < \frac{x}{\sigma}\right).$$

It remains to consider the case $na^2 \to \infty$, $a>0$. Here

$$\mathsf{P}(\overline{Y}_n < an + x\sqrt{n}) \leqslant \mathsf{P}(X_{n,n} < x\sqrt{n}) \to \frac{1}{\sqrt{2\pi}} \int_{-\infty}^{x/\sigma} e^{-u^2/2}\, du.$$

On the other hand, for arbitrary $N>0$ and sufficiently large n

$$\begin{aligned}\mathsf{P}(\overline{Y}_n < an + x\sqrt{n}) &= \mathsf{P}\left(X_{k,n} < a\sqrt{n}\left(\frac{n-k}{n}\right)\sqrt{n} + x\sqrt{n},\ k=1,2,\ldots,n\right)\\ &\geqslant \mathsf{P}\left(\frac{X_{k,n}}{\sigma\sqrt{n}} < \frac{N}{\sigma}\left(1-\frac{k}{n}\right) + \frac{x}{\sigma},\ k=1,2,\ldots,n\right)\\ &\Rightarrow \mathsf{P}\left(\omega(u) \leqslant \frac{N(1-u)}{\sigma} + \frac{x}{\sigma},\quad 0\leqslant u\leqslant 1\right).\end{aligned}$$

By choosing N suitably we can obviously make this probability arbitrarily close to

$$\mathsf{P}\left(\omega(1) < \frac{x}{\sigma}\right) = \frac{1}{\sqrt{2\pi}} \int_{-\infty}^{x/\sigma} e^{-u^2/2}\, du.$$

Since the limiting distribution of $\overline{Y}_n$ is independent of N, the theorem is proved. □

If instead of (66) we impose more rigid conditions on $\xi_k^{(n)}$, requiring, for example, that for all n $F_n(x)$ contain an absolutely continuous component and that $\mathsf{M}\exp \mu\xi_k^{(n)}$ be an analytic function for $\mu \leqslant \varepsilon_1$ and some $\varepsilon_1 > 0$, then we can obtain more precise asymptotic expansions in terms of powers of a and n^{-1} for the prelimiting distribution of $\overline{Y}_n$. However, the nature of these expansions will be extremely complicated, if the form of the sequence of distributions F_n is not specified exactly.

On the other hand, with applications in mind, we can make the following observation. An arbitrary concrete distribution with small $a = \mathsf{M}\xi_k^{(n)}$ can always be represented as a member of some well-defined and reasonably chosen sequence of distributions. The choice of the sequence is sure to influence the character of the asymptotic expansions, but is not important from the point of view of the numerical improvements which they yield, provided that the sequence converges and is such that the second and higher-order moments (or even $\mathsf{M}\exp \mu\xi_k^{(n)}$ for $\mu>0$) converge for $n\to\infty$ to the corresponding moments of the limiting distribution of ξ_k. It is natural to choose the convergent sequence so as to simplify the conditions needed. We can assume, for example, that

$$\xi_k^{(n)} = \xi_k + a_n, \quad a_n \to 0 \tag{71}$$

as $n\to\infty$, where $\xi_k(\mathsf{M}\xi_k = 0)$ does not depend on n (or, what is the same, on a, if the independent parameter is taken to be a). Such sequences will certainly satisfy the conditions of Theorem 18 if $\mathsf{D}\xi_k < \infty$.

We consider first the stationary distribution of w^k.

Theorem 19. *Assume that the sequences* $\{\xi_k^{(n)}\}$ *can be written as* (71), *where* $a=a_n<0$, *that the function*

$$f_1(\mu)=\mathsf{M}\exp(\mu\xi_k)$$

is analytic for small enough μ *and that the distribution* $F(x)=\mathsf{P}(\xi_k<x)$ *has an absolutely continuous component. Then*

$$\mathsf{P}\left(\overline{Y}^{(n)}>\frac{x}{|a|}\right)=\exp\left[-\frac{2}{\sigma^2}x+x\Xi_1(a)\right]\left[\frac{1+\Xi_2(a)}{1+\Xi_3(a)}(1+O(e^{-\varepsilon x/|a|}))\right], \qquad (72)$$

where $\varepsilon>0$ *is independent of a and the* $\Xi_j(a)$ *are convergent series in powers of a*:

$$\Xi_1(a)=-a\frac{4c_3}{3\sigma^6}+a^2\frac{2}{\sigma^4}\left[\frac{c_4}{3\sigma^4}-\frac{8}{9}\frac{c_3^2}{\sigma^6}-1\right]+\dots,$$

$$\Xi_2(a)=-a\frac{2\mathsf{M}\chi_0}{\sigma^2}-a^2\left[\frac{4c_3}{3\sigma^6}\mathsf{M}\chi_0+\frac{2}{\sigma^2}\frac{d}{da}\mathsf{M}\chi_a\bigg|_{a=0}+\frac{2}{\sigma^4}\mathsf{M}\chi_0^2\right]+\dots,$$

$$\Xi_3(a)=-a\frac{2}{3}\frac{c_3}{\sigma^4}+a^2\left[\frac{4c_4}{3\sigma^6}+\frac{8}{9}\frac{c_3^2}{\sigma^6}+\frac{2}{\sigma^2}\right]+\dots.$$

In these formulae $c_k=\mathsf{M}\xi_j^k$ *and* χ_a *is the magnitude of the overshoot of a barrier at* $-\infty$ *in a random walk with jumps* $\xi_1+a, \xi_2+a, \dots$. *The coefficients of* a^k *in the expansions* Ξ_1 *and* Ξ_3 *depend only on* $k+2$ *moments of* ξ_j. *The coefficient of* a^k *in* Ξ_2 *is defined by* $k+1$ *moments of* ξ_j *and the derivatives* $(d^l/da^l)\mathsf{M}\chi_a^s|_{a=0}$ *for* $l+s\leqslant k$.

Remark 1. It is clear that for x such that $x/|a|\ln|a|\to\infty$ as $a\to 0$, the factor in the square brackets in (72) can be given as an asymptotic expansion in powers of a.

Remark 2. Concerning the coefficients of the expansion Ξ_2, we note that

$$\frac{d}{da}\mathsf{M}\chi_a|_{a=0}=0.$$

This is shown by comparing (72) with the results of [11].

Remark 3. A comparison of the statement of the theorem with the inequality (see Theorem 16)

$$\mathsf{P}\left(\overline{Y}^{(n)}>\frac{x}{|a|}\right)\leqslant\exp\left[-\frac{2}{\sigma^2}x+x\Xi_1(a)\right]$$

shows how accurate the latter is.

Proof of the theorem. Under the assumptions of the theorem it was shown in Theorem 11 that for the distribution of $\overline{Y}^{(n)}=\sup_{k\geqslant 0}Y_{k,n}$

$$\mathsf{P}(\overline{Y}^{(n)}\geqslant x)=\frac{\mathsf{P}(\overline{Y}^{(n)}=0)}{qf'_{1,a}(q)}[1-\mathsf{M}\exp(q\chi_a^*(0))]\,e^{-qx}(1+O(e^{-\varepsilon x})), \qquad (73)$$

where $q=q(a)$ is the largest (in absolute value) root of the equation $f_{1,a}(\mu)=e^{\mu a}f_1(\mu)=1$ and $\chi_a^*(0)$ is a functional of the form $\chi^*(0)$ defined for the sequence (71).

We note first of all that $q(a)$ is an analytic function in a neighborhood of the point $a=0$ which can be represented as the convergent series

$$q=q_1a+q_2a^2+q_3a^3+\cdots=q_1a+a\Xi_1(a),$$

where the q_k are defined by the identity

$$\left(1+qa+\frac{q^2a^2}{2}+\cdots\right)\left(1+\frac{q^2\sigma^2}{2}+\frac{q^3c_3}{6}+\cdots\right)=1.$$

Thus,

$$q_1=-\frac{2}{\sigma^2},\qquad q_2=-\frac{4c_3}{3\sigma^6},\qquad q_3=\frac{2}{\sigma^4}\left[\frac{c_4}{3\sigma^4}-\frac{8}{9}\frac{c_3^2}{\sigma^6}-1\right],\qquad \text{etc.}$$

We now prove that the function $\mathsf{M}\exp[\mu\chi_a^*(0)]$ can be represented as a function analytic in both arguments μ and a in a neighborhood of the point $\mu=0$, $a=0$. Indeed, the function

$$\mathfrak{v}_-(\lambda)=\frac{1-\mathsf{M}\exp[i\lambda\chi_a^*(0)]}{i\lambda}(i\lambda+1)$$

is the negative component of the V-factorization of the function (Theorem 8, Appendix 2)

$$\mathfrak{v}(\lambda)=\frac{1-e^{i\lambda a}f(\lambda)}{i\lambda}(i\lambda+1),\quad f(\lambda)=\mathsf{M}\,e^{i\lambda\xi_k}.$$

Hence, for $\operatorname{Im}\lambda<0$ we have with accuracy up to a constant

$$\mathfrak{v}_-(\lambda)=\exp\left\{\frac{1}{2\pi i}\int_{-\infty}^{\infty}{}' \frac{\ln\mathfrak{v}(\mu)}{\mu-\lambda}\,d\mu\right\}\in\mathfrak{V}_-.$$

Using the analyticity and boundedness of $\ln\mathfrak{v}(\mu)$ in the region $|\operatorname{Im}\mu|\leqslant\varepsilon$ for sufficiently small $\varepsilon>0$, we can shift the contour $\operatorname{Im}\mu=0$ of this integral upwards a distance ε, replacing it by the contour $\operatorname{Im}\mu=\varepsilon$. The resulting representation of $\mathfrak{v}_-(\lambda)$ will obviously be valid for $\operatorname{Im}\lambda<\varepsilon$. Since for small a the function $\ln[i\lambda/(i\lambda-q(a))]$ is analytic in the region $\operatorname{Im}\lambda\geqslant\varepsilon$, we also have for $\operatorname{Im}\lambda<\varepsilon$

$$\mathfrak{v}_-(\lambda)=\exp\left\{\frac{1}{2\pi i}\int_{-\infty}^{\infty}{}' \frac{\ln\mathfrak{v}(\mu)\dfrac{i\mu-1}{i\mu-q(a)}}{\mu-\lambda}\,d\mu\right\}. \tag{74}$$

But the function

$$\ln\mathfrak{v}(\mu)\frac{i\mu-1}{i\mu-q(a)}=\ln\left[\frac{(1-e^{i\mu a}f(\mu))}{i\mu(i\mu+q(a))}(-\mu^2-1)\right]$$

is analytic in μ and a in $|\operatorname{Im}\mu|<\varepsilon_0$, $|a|<a_0$, for small enough ε_0 and a_0. Moreover, the integral (74) of this function converges for all a in the closed region $|a|\leqslant a_1$ for some a_1, $a_0>a_1>0$ (see Theorem 8 of Appendix 2 and the remark accompanying it). Consequently, it converges uniformly in this region.

By known theorems from the theory of functions of several complex variables (Fuks [27]) it follows that $\mathfrak{v}_-(\lambda)$ is analytic in a and hence analytic in both λ and a in $\operatorname{Im}\lambda \leqslant 0$, $|a| \leqslant a_1$. This means that the function

$$\frac{\mathfrak{v}_-(\lambda)}{\mathfrak{v}_-(0)} = -\frac{1-\mathsf{M}\exp[i\lambda\chi_a^*(0)]}{i\lambda\mathsf{M}\chi_a^*(0)}(i\lambda+1), \quad a<0,$$

can also be expanded as a series in powers of a and λ.

We turn now to Eq. (73). In it we have

$$\frac{\mathsf{P}(\overline{Y}^{(n)}=0)}{q}[1-\mathsf{M}\exp(q\chi_a^*(0))] = \frac{a}{q\mathsf{M}\chi_a^*(0)}[1-\mathsf{M}\exp(q\chi_a^*(0))] = -a\mathsf{M}\,e^{q\chi_a},$$

where χ_a is the quantity defined in the formulation of the theorem (see Appendix 1). By the remarks above we see that $\mathsf{M}\,e^{\mu\chi_a}$ along with the moments $\mathsf{M}\chi_a^k$ are analytic for $|a|<a_0$. The coefficients of the expansions Ξ_j, $j=2, 3, \ldots$, are obviously determined as the coefficients of the expansions of the functions (here $m_k^{(j)} = (d^j/da^j)\mathsf{M}\chi_a^k|_{a=0}$, $m_k^{(0)}=m_k$)

$$\begin{aligned}\mathsf{M}\,e^{q\chi_a} &= 1+\Xi_2(a) = 1+q\mathsf{M}\chi_a+\frac{q^2}{2}\mathsf{M}\chi_a^2+\cdots\\ &= 1+(q_1a+q_2a^2+\cdots)(m_1+am_1^{(1)}+\cdots)\\ &\quad+\tfrac{1}{2}(q_1a+\cdots)^2(m_2+am_2^{(1)}+\cdots) = 1+aq_1m_1\\ &\quad+a^2\left(q_2m_1+q_1m_1^{(1)}+\frac{q_1^2m_2}{2}\right)+\cdots;\end{aligned}$$

$$\begin{aligned}-\frac{f'_{1,a}(q)}{a} &= 1+\Xi_3(a)\\ &= -\frac{1}{a}\left(f'_{1,a}(0)+qf''_{1,a}(0)+\frac{q^2}{2}f'''_{1,a}(0)+\cdots\right)\\ &= -[1+(q_1+q_2a+q_3a^2+\cdots)(\sigma^2+a^2)+\\ &\quad\frac{a}{2}(q_1^2+2q_1q_2a+\cdots)(c_3+3a\sigma^2+\cdots)\\ &\quad+\frac{a^2}{6}(q_1^3+\cdots)\times(c_4+\cdots)+\cdots]\\ &= 1-a\left(q_2\sigma^2+\frac{q_1^2c_3}{2}\right)-a^2\left(q_1+q_3\sigma^2+q_1q_2c_3+\tfrac{3}{2}q_1^2\sigma^2+\frac{q_1^3c_4}{6}\right)+\cdots.\end{aligned}$$

To finish the proof of the theorem, it remains to show that in the representation (73) $\varepsilon>\varepsilon_0>0$, where ε_0 does not depend on a (on n). To do this we obtain (73) again, but by means of transformations of the equality

$$(1-p)^{-1}\mathsf{M}\exp(i\lambda\overline{Y}) = \frac{1-\mathsf{M}\exp(i\lambda\chi^*(0))}{1-f(\lambda)}$$

which are somewhat different from those we carried out in Theorem 11.

Set

$$\mathfrak{w}(\lambda)=\frac{1-\mathsf{M}\exp(i\lambda\chi_a^*(0))}{i\lambda}(i\lambda+1)$$

and

$$\mathfrak{v}_a(\lambda)=-\frac{(1-e^{i\lambda a}f(\lambda))}{i\lambda(i\lambda-q)}(\lambda^2+1), \quad f(\lambda)=\mathsf{M}\, e^{i\lambda\xi_k}.$$

Then we can write

$$\begin{aligned}(1-p)^{-1}\mathsf{M}\exp(i\lambda\overline{Y}^{(n)})&=\mathfrak{w}(\lambda)\mathfrak{v}_a^{-1}(\lambda)\frac{i\lambda-1}{i\lambda-q}\\ &=\mathfrak{w}(\lambda)\mathfrak{v}_a^{-1}(\lambda)+\frac{q-1}{i\lambda-q}\mathfrak{w}(q)\mathfrak{v}_a^{-1}(q)\\ &\quad+\frac{q-1}{i\lambda-q}(\mathfrak{w}(\lambda)\mathfrak{v}_a^{-1}(\lambda)-\mathfrak{w}(q)\mathfrak{v}_a^{-1}(q)).\end{aligned}$$

In order to obtain the required result, we need only—as in Theorem 11—study the behavior of the "tails" of the inverse images of the first and last terms on the right. For this it is in turn sufficient (by Theorem 2 of Appendix 2) to show that the "tail" of the inverse image of the function $\mathfrak{v}_a^{-1}(\lambda)$ is majorized by the function $e^{-\varepsilon_0 t}$ for $t\to\infty$ (in fact, the variation of the inverse image of the function $\mathfrak{v}(\lambda)\in\mathfrak{B}$ is uniformly bounded in $a\leqslant 0$ according to the inequalities of Theorem 10). But in the representation

$$\mathfrak{v}_a(\lambda)=\mathfrak{v}_0(\lambda)+\mathfrak{v}^{(a)}(\lambda)$$

the "tails" of the inverse images of all three functions can obviously be written in explicit form, so that for the inverse image $v^{(a)}(t)$ of the function $\mathfrak{v}^{(a)}(\lambda)$ we obtain under the hypothesis of the theorem for some $\delta>0$

$$\int e^{\delta t}\,|dv^{(a)}(t)|\to 0$$

when $a\to 0$ (see also Theorem 2 of Appendix 2).

Using the fact that $\mathfrak{v}_0^{-1}\in\mathfrak{B}(-\varepsilon,\varepsilon)$ for some $\varepsilon>0$ (see again the remarks concerning Theorem 8 of Appendix 2) and the expansion

$$\mathfrak{v}_a^{-1}(\lambda)=\mathfrak{v}_0^{-1}(\lambda)\left(1+\sum_{k=1}^{\infty}\left(\frac{\mathfrak{v}^{(a)}(\lambda)}{\mathfrak{v}_0(\lambda)}\right)^k\right),$$

we discover that for all sufficiently small $a\leqslant 0$, the inverse image $b_a(t)$ of $\mathfrak{v}_a^{-1}(\lambda)$ satisfies

$$\int e^{\varepsilon t}\,|db_a(t)|<c,$$

where c is independent of a. Now we need only use the estimates of Theorems 1 and 2 in Appendix 2. □

Under the assumption (71) we can also carry out a more complete analysis of transitional phenomena for the *pre-limiting* distribution of

$$\bar{Y}_n = \max_{k \leq n} Y_{k,n}$$

and thereby sharpen Theorem 18. We will give here only the formulation of the corresponding result since its proof (see [8], [11]) requires the introduction of more precise methods of analysis going beyond the scope of this book.

Theorem 20. *Assume the conditions of Theorem 19 are fulfilled except for the inequality $a<0$. If $a\sqrt{n} \leq c < \infty$ as $n \to \infty$, we have (the functions $P(t, x)$ and $P(x)$ are defined in Theorem 18, $t = a^2 n$)*

$$\mathsf{P}\left(\bar{Y}_n > \frac{x}{|a|}\right) = (1 - P(t, x))[1 + R_1 a(x^3 + x^{-1})] + R_2\, e^{-\varepsilon x/|a|}.$$

If $a\sqrt{n} \to 0$ as $n \to \infty$, then

$$\mathsf{P}(\bar{Y}_n > x\sqrt{n}) = (1 - P(x))\left[1 + \frac{R_3}{\sqrt{n}}(x^3 + x^{-1}) + R_4 t\right] e^{-xt/\sigma^2} + R_5\, e^{-\varepsilon x\sqrt{n}}.$$

Here, $R_1, \ldots, R_5$ are uniformly bounded functions and $\varepsilon > \varepsilon_0 > 0$.

In conclusion we remark that finding effective and sufficiently simple estimates for the rate of approach for example of $\mathsf{P}(\bar{Y}^{(n)} > x/|a|)$ to $e^{-xq(a)/|a|}$ (with or without correcting terms of the asymptotic expansions) remains, even under the conditions of Theorem 19, a difficult unsolved problem.

§ 26. The Relation between the Waiting Time and Queue Length Distributions

1. Up to now we have considered basically two characteristics of the systems $\langle G, G, G, 1\rangle$—the waiting time w_n until service of the first customer in the n-th lot and the virtual waiting time $w(t)$ which a customer arriving at time t would have to wait until service. However, as already remarked in § 10, two other characteristics are no less widespread in queueing theory: q_n, *the queue length at the arrival of the n-th lot* (not counting the customers of this lot) and $q(t)$, *the queue length at time t*. The queue length in both cases includes the customer undergoing service so that the events $q(t)=0$ and $w(t)=0$ are equivalent.

The relation between the distributions of $w(t)$ and w_n was considered in § 9 Chapter 1. In particular, we showed that

$$w^c(0) \underset{d}{=} \max(0, S_0 - \tau_0^e + w^0),$$

where $w^c(0)$ and w^0 are the corresponding stationary values, $S_0 - \tau_0^e$ and w^0 are independent and the distribution of $S_0 - \tau_0^e$ is defined by (64) § 9.

2. *The Distribution of q_n.* In Subsection 3 § 10 we found a relation between the distribution of q_n and that of the waiting time w_n. Namely, for the systems $\langle G_I, 1, G_I, 1\rangle$ ($v_j^e \equiv v_j^s \equiv 1$), the event $\{q_n > k\}$ occurs iff the waiting time of the $(n-k)$-th request satisfies

$$w_{n-k} > \tau_{n-k}^e + \cdots + \tau_{n-1}^e .$$

In this connection, w_{n-k} and $\tau_{n-k}^e, \tau_{n-k+1}^e, \ldots$ are obviously independent since w_{n-k} is measurable w.r.t. the σ-algebra generated by $w_1, \xi_1, \xi_2, \ldots, \xi_{n-k-1}$. Hence,

$$\mathsf{P}(q_n > k) = \mathsf{P}(w_{n-k} > \tau_{n-k}^e + \cdots + \tau_{n-k}^e), \quad k = 0, 1, \ldots . \tag{75}$$

From this we immediately obtain

Theorem 21. *The stationary distribution of q_n exists (the corresponding random variable is denoted by q^0) and*

$$\lim_{n\to\infty} \mathsf{P}(q_n > k) = \mathsf{P}(q^0 > k) = \mathsf{P}(w^0 > \tau_1^e + \cdots + \tau_k^e),$$

where w^0 is the stationary waiting time and is independent of $\tau_1^e, \tau_2^e, \ldots$. This equality can also be interpreted as

$$q^0 \underset{d}{=} \eta(w^0),$$

where $\eta(t)$, as before, is the first passage time of the level t, but in a random walk with jumps $\tau_1^e, \tau_2^e, \ldots$.

This assertion is also a direct consequence of Theorem 21 Chapter 1.

For the systems $\langle G_I, G_I, G_I, 1\rangle$ we get in place of (75)

$$\mathsf{P}(q_n > v_{n-k}^e + \cdots + v_{n-1}^e) = \mathsf{P}(w_{n-k} > \tau_{n-k}^e + \cdots + \tau_{n-1}^e),$$

where $v_{n-k}^e, v_{n-k+1}^e, \ldots$ obviously depend on q_n. In this case we can determine only the limits $p(k; j_1, \ldots, j_s)$ of the conditional probabilities

$$\mathsf{P}(q_n > k \mid v_{n-1}^e = j_1, \ldots, v_{n-s}^e = j_s) = \mathsf{P}(w_{n-s} > \tau_{n-s}^e + \cdots + \tau_{n-1}^e)$$

for all sets $j_1, \ldots, j_s$ such that $j_1 + \cdots + j_s = k$:

$$p(k; j_1, \ldots, j_s) = P(w^0 > \tau_1^e + \cdots + \tau_s^e), \tag{76}$$

where w^0 and $\tau_1^e, \tau_2^e, \ldots$ are independent.

These probabilities do not always uniquely determine the distribution of q^0 as is obvious from the example $v_j^e \equiv 2$. In this case we will find only $\mathsf{P}(q^0 > 2k)$; $k = 0, 1, \ldots$. The question of uniqueness of the distribution $\mathsf{P}(q^0 > k)$ corresponding to the conditional probabilities (76) requires special consideration.

3. *The distribution of $q(t)$.* We turn again to the system $\langle G_I, 1, G_I, 1\rangle$. An analogous approach shows that for $k \geqslant 0$

$$\{q(t) > k+1\} = \{w_{\eta(t)-k} > \tau_{\eta(t)-k}^e + \cdots + \tau_{\eta(t)-1}^e + \gamma(t)\}, \tag{77}$$

where $\eta(t)$ has the same meaning as before, and $\gamma(t)$ is the "defect" in this random walk up to the level t:

$$\gamma(t)=t-\tau_1^e-\cdots-\tau_{\eta(t)-1}^e .$$

We now prove

Theorem 22. *If* $\mathsf{M}\xi=a<0$ *and is finite, and* τ_j^e *is non-lattice, then the limiting distribution of* $q(t)$ *as* $t\to\infty$ *exists (the corresponding stationary variable is denoted by* q^c*),*

$$\lim_{t\to\infty}\mathsf{P}(q(t)=0)=1-\frac{\mathsf{M}\tau^s}{\mathsf{M}\tau^e}=-\frac{a}{\mathsf{M}\tau^e},$$

and for $k\geqslant 0$

$$\lim_{t\to\infty}\mathsf{P}(q(t)>k+1)=\mathsf{P}(q^c>k+1)=\mathsf{P}(w^0>\tau_1^e+\cdots+\tau_k^e+\gamma), \tag{78}$$

where all components on the right are independent, w^0 *is the stationary waiting time and* γ *has the density* (see Appendix 1) $\mathsf{P}(\tau_1^e>x)/\mathsf{M}\tau_1^e$.

A similar assertion holds for $q(n)$ *as* $n\to\infty$ *if* τ_k^e *is lattice and the conditions for existence of the stationary distribution of* w^0 *hold.*

Proof. We must show that the probability of the event on the right in (77) converges to the right side of (78). This will follow from convergence to the corresponding limit of the joint distribution

$$\begin{aligned}
&\mathsf{P}(w_{\eta(t)-k}\geqslant l,\ \tau_{\eta(t)-k}^e\in dx_1,\ \dots,\ \tau_{\eta(t)-1}^e\in dx_k,\ \gamma(t)>y)\\
&\quad=\textstyle\sum_{n=1}^{\infty}\int_0^{t-x_1-\cdots-x_k-y}\mathsf{P}(X_{n-1}\in du,\ w_n\geqslant l)\mathsf{P}(\tau_n^e\in dx_1)\cdots\\
&\qquad\times\mathsf{P}(\tau_{n+k-1}^e\in dx_k)\mathsf{P}(\tau_{n+k}^e>t-u-x_1-\cdots-x_k)\\
&\quad=\textstyle\prod_{s=1}^{k}\mathsf{P}(\tau_1^e\in dx_s)\sum_{n=1}^{\infty}\int_0^{t-z-y}d_u\mathsf{P}(X_{n-1}<u,\ w_n\geqslant l)P(t-u-z),
\end{aligned}\tag{79}$$

where

$$X_n=\textstyle\sum_{k=1}^{n}\tau_k^e,\qquad z=x_1+\cdots+x_k\quad\text{and}\quad P(t)=\mathsf{P}(\tau^e>t).$$

Furthermore, it is clear that it suffices to consider merely the series appearing in (79). Assume for simplicity that $w_1=0$. Then w_n coincides with $\max(0,\xi_{n-1},\xi_{n-1}+\xi_{n-2},\dots,Y_{n-1}-Y_1,Y_{n-1})$, where $Y_n=\sum_{k=1}^{n}\xi_k$, and we obtain for the series under consideration the expression

$$\textstyle\sum_{n=0}^{\infty}\int_0^{t-z-y}d_u\mathsf{P}(\overline{Y}_n\geqslant l,\ X_n<u)P(t-u-z), \tag{80}$$

which can be viewed as the probability of the event

$$B=\{\overline{Y}_{\eta(t-z)-1}\geqslant l,\ \gamma(t-z)>y\}.$$

Denote by $\eta_Y(l)$ the first passage time to the level l of the trajectory $0, Y_1, Y_2, \dots$:

$$\eta_Y(l)=\min\{k\colon Y_k\geqslant l\},\quad\text{put}\quad \chi_Y(l)=Y_{\eta_Y(l)}-l,$$

and let

$$A=\{\eta_Y(l)<N,\ \chi_Y(l)<N\}.$$

Obviously, as $N\to\infty$

$$\mathsf{P}(A)\to\mathsf{P}(\eta_Y(l)<\infty)=\mathsf{P}(\bar{Y}\geqslant l)\,. \tag{81}$$

uniformly w.r.t. l.

If $\mathfrak{M}$ is the σ-algebra generated by the random variables $\eta_Y(l)$ and $\chi_Y(l)$, then $A\in\mathfrak{M}$ and ($I(A)$ is the indicator of the event A)

$$\begin{aligned}\mathsf{P}(B)&=\mathsf{P}(B\bar{A})+\mathsf{P}(BA)\\ &=\mathsf{P}(B\bar{A})+\mathsf{M}I(A)\mathsf{P}_{\mathfrak{M}}(\gamma(t-z)>y;\eta(t-z)\geqslant N)\\ &\quad+\mathsf{P}(BA\{\eta(t-z)<N\})\,.\end{aligned} \tag{82}$$

For a given $\varepsilon>0$ we choose N such that the first term on the right is $<\varepsilon$. For each fixed N and everywhere on the set A we have as $t\to\infty$

$$\mathsf{P}_{\mathfrak{M}}(\gamma(t-z)>y;\eta(t-z)\geqslant N)\to\lim_{t\to\infty}\mathsf{P}(\gamma(t)>y)=\int_y^\infty\frac{P(u)\,du}{\mathsf{M}\tau_1^e}=\mathsf{P}(\gamma>y)\,.$$

The last term in (82) obviously converges to zero as $t\to\infty$. Hence,

$$\mathsf{P}(\gamma>y)\mathsf{P}(A)\leqslant\liminf_{t\to\infty}\mathsf{P}(B)\leqslant\limsup_{t\to\infty}\mathsf{P}(B)\leqslant\mathsf{P}(\gamma>y)\mathsf{P}(A)+\varepsilon\,.$$

Since N is arbitrary, we conclude by (81) that for (80) there exists

$$\lim_{t\to\infty}\mathsf{P}(B)=\mathsf{P}(\gamma>y)\mathsf{P}(\bar{Y}>l)=\mathsf{P}(\gamma>y)\mathsf{P}(w^0\geqslant l)\,.$$

Thus, the joint distribution of the random variables $w_{\eta(t)-k},\tau^e_{\eta(t)-k},\ldots,\tau^e_{\eta(t)-1}$ and $\gamma(t)$ converges to the distribution of the independent random variables w^0, $\tau_1^e,\ldots,\tau_k^e$ and γ. Along with (77), this proves the assertion (78) of Theorem 22.

Since this assertion refers to $\mathsf{P}(q(t)\geqslant j)$ for $j\geqslant 2$, to prove the existence of the limiting distribution of $q(t)$ we must still show that $\lim_{t\to\infty}\mathsf{P}(q(t)=0)$ exists. But this is a consequence of the equality $\{q(t)=0\}=\{w(t)=0\}$ and the results of §§ 8 and 9. □

The extension of Theorem 22 to the case in which $w_1>0$ or τ_k^e is lattice is left to the reader.

Corollary. *For the system $\langle E,1,G_I,1\rangle$ the limit distributions of q_n and $q(t)$ coincide as $n\to\infty$.*

Indeed, from (75), (78) and the fact that the distribution of γ coincides in this case with that of τ^e, it follows that

$$\lim_{t\to\infty}\mathsf{P}(q(t)=k)=\lim_{n\to\infty}\mathsf{P}(q_n=k)$$

for $k\geqslant 2$, i.e., the limiting distributions of $\max(1,q_n)$ and $\max(1,q(t))$ coincide. But we also have for the systems $\langle E,1,G_I,1\rangle$ (see § 9)

$$\begin{aligned}\lim_{n\to\infty}\mathsf{P}(q_n=0)&=\lim_{n\to\infty}\mathsf{P}(w_n=0)\\ &=\lim_{t\to\infty}\mathsf{P}(w(t)=0)=\lim_{t\to\infty}\mathsf{P}(q(t)=0)=-\frac{a}{\mathsf{M}\tau^e}\,.\end{aligned}$$

Hence, the limiting distributions of q_n and $q(t)$ also coincide. □

With regard to the systems $\langle G_I, G_I, G_I, 1\rangle$, the remark made at the end of the preceding subsection still holds.

4. Theorems 21 and 22 along with the results of § 20 allow us to find *explicit formulae for the stationary distributions of* q_n *and* $q(t)$ when *at least one of the distributions of* τ^e *or* τ^s *has a rational characteristic function* (we are talking about the systems $\langle G_I, 1, G_I, 1\rangle$).

1. Let us consider two simple examples. Suppose τ^e has an exponential distribution $\mathsf{P}(\tau^e > x) = e^{-\alpha x}$,

$$\mathsf{M}\, e^{i\lambda\tau^e} = \frac{1}{1-\dfrac{i\lambda}{\alpha}}.$$

Then

$$P(z, x) \equiv \sum_{k=1}^{\infty} z^k \mathsf{P}(\tau_1^e + \cdots + \tau_k^e < x) = \frac{z}{(1-z)}(1 - e^{-\alpha(1-z)x}).$$

This is easily obtained, for example, from the fact that

$$\int e^{i\lambda x}\, d_x P(z, x) = \frac{1}{1 - z\mathsf{M}\, e^{i\lambda\tau^e}} - 1 = -\frac{z\alpha}{i\lambda - \alpha(1-z)}.$$

Setting $Q_k = \sum_{j=k+1}^{\infty} p_j = \mathsf{P}(q^0 > k)$ and $Q(z) = \sum_{k=0}^{\infty} z^k Q_k$, we get

$$Q(z) = \frac{1 - \sum_{k=0}^{\infty} p_k z^k}{1-z} = \sum_{k=0}^{\infty} z^k \mathsf{P}(\tau_1^e + \cdots + \tau_k^e < w^0)$$

$$= \mathsf{P}(w^0 > 0) + \mathsf{M} P(z, w^0) = \mathsf{P}(w^0 > 0) + \frac{z}{(1-z)}(1 - \mathsf{M}\, e^{-\alpha(1-z)w^0}).$$

But in this case, by Theorem 5 ($f(\lambda) = \mathsf{M}\, e^{i\lambda(\tau^s - \tau^e)}$)

$$\mathsf{M}\, e^{i\lambda w^0} = -\frac{ai\lambda}{(1-f(\lambda))(1-i\lambda/\alpha)}.$$

Thus, if we put $\psi_s(\mu) = \mathsf{M}\, e^{\mu\tau^s}$, we find that

$$Q(z) = 1 + a\alpha + \frac{z}{1-z}\left(1 - \frac{z\alpha a}{z - \psi_s(-\alpha(1-z))}\right).$$

This is equivalent to the well-known Pollaczek–Hinčin formula.

2. Suppose now that τ^s has an exponential distribution

$$\mathsf{M}\, e^{i\lambda\tau^s} = \frac{1}{1-\dfrac{i\lambda}{\alpha}}.$$

Then by Theorem 4

$$\mathsf{M}\, e^{i\lambda w^0} = \frac{1 - i\lambda/\alpha}{1 - i\lambda/\lambda(1)},$$

where $\lambda(1)>0$ is the only zero of the function $1-f(-i\lambda)$ in the region $\operatorname{Re}\lambda>0$. Consequently, the density of w^0 for $x>0$ equals

$$\lambda(1)\left(1-\frac{\lambda(1)}{\alpha}\right)e^{-\lambda(1)x}$$

and

$$\begin{aligned}Q(z)&=\mathsf{P}(w^0>0)+\textstyle\sum_{k=1}^{\infty} z^k\mathsf{P}(\tau_1^e+\cdots+\tau_k^e<w^0)\\&=\mathsf{P}(w^0>0)+\lambda(1)\left(1-\frac{\lambda(1)}{\alpha}\right)\textstyle\sum_{k=1}^{\infty} z^k\int_0^\infty e^{-\lambda(1)x}\mathsf{P}(\tau_1^e+\cdots+\tau_k^e<x)\,dx\\&=\left(1-\frac{\lambda(1)}{\alpha}\right)+\left(1-\frac{\lambda(1)}{\alpha}\right)\textstyle\sum_{k=1}^{\infty} z^k\int_0^\infty e^{-\lambda(1)x}\,d\mathsf{P}(\tau_1^e+\cdots+\tau_k^e<x)\\&=\left(1-\frac{\lambda(1)}{\alpha}\right)\left(1+\frac{z\psi_e(-\lambda(1))}{1-z\psi_e(-\lambda(1))}\right)=\frac{1-\lambda(1)/\alpha}{1-z\psi_e(-\lambda(1))},\end{aligned}$$

where $\psi_e(\mu)=\mathsf{M}\,e^{\mu\tau^e}$. This implies that

$$Q_k=\mathsf{P}(q^0>k)=\left(1-\frac{\lambda(1)}{\alpha}\right)\psi_e^k(-\lambda(1))=\left(1-\frac{\lambda(1)}{\alpha}\right)^{k+1},\quad k\geq 0.$$

$$\left(f(-i\mu)=\mathsf{M}\,e^{\mu(\tau^s-\tau^e)}=\frac{1-\psi_e(-\mu)}{1-\mu/\alpha}.\right)$$

5. Theorems 21 and 22 allow us to find for the stationary distributions of q^0 and q^c analogues of many other theorems obtained in this chapter for the distribution of w^0. For instance, we can find stability theorems (§ 21), theorems on transitional phenomena (§ 25) and a number of others.

Consider, for example, the distribution of q^0 or q^c when the $\xi_k^{(n)}$ satisfy (65) and (66) and the negative components $-\tau_k^e$ of the random variables $\xi_k^{(n)}$ enjoy the property $\mathsf{M}\tau_k^e\to a^e>0$ as $n\to\infty$. If $a=\mathsf{M}\xi_k^{(n)}<0$, then, as we have seen,

$$\lim_{a\to 0}\mathsf{P}\left(w^0>\frac{x}{|a|}\right)=e^{-2x/\sigma^2}.$$

By the law of large numbers $\eta(t)/t \underset{p}{\to} 1/\mathsf{M}\tau^e$ as $t\to\infty$. Thus,

$$\lim_{a\to 0}\mathsf{P}\left(q^0>\frac{x}{a^e|a|}\right)=\lim_{a\to 0}\mathsf{P}\left(w^0\frac{\eta(w^0)}{w^0}>\frac{x}{a^e|a|}\right)=\lim_{a\to 0}\mathsf{P}\left(w^0>\frac{x}{|a|}\right)=e^{-2x/\sigma^2}.$$

Analogous formulae also hold for q^c.

Chapter 5

Multi-Channel Queueing Systems

§ 27. Classes of Systems Which Can Be Described by Recursion Equations. Existence Theorems for a Stationary Solution in the Systems ⟨G, G, G/m, 1⟩. The Relation between the Waiting Time and the Queue Length

1. We recall the definition of queueing systems with $m \geqslant 1$ channels. We have agreed to denote such systems by the symbol $\langle G/m \rangle$ (see § 1). Assume we are given the governing sequences (1), § 1. Customers arrive in groups of sizes $\nu_1^e, \nu_2^e, \ldots$, with interarrival times $\tau_1^e, \tau_2^e, \ldots$. Service occurs also in groups and can be performed in m channels simultaneously. The service of the n-th group of requests of size ν_n^s (or of smaller size if the number of customers waiting for service turns out to be insufficient) requires τ_n^s time units. Arriving customers are immediately directed (in order of their appearance) into arbitrary free channels if they are not all busy, or wait until they can proceed into some channel which has become free.

In the study of classes of systems $\langle G/m \rangle$ for which there exist natural state characteristics describable by explicit equations, we discover certain analogies to the single-channel case. In fact: these classes comprise systems having the symbol E in one of the two latter positions of their designations. We will devote most of our attention to the treatment of the most interesting class of systems $\langle G, G, G/m, 1 \rangle$. The systems $\langle G, G, E/m, G \rangle$ are examined in Subsection 8 of this section.

2. In order to make the exposition clearer, we will consider first the *systems* $\langle G, 1, G/m, 1 \rangle$. Denote by

$$\mathbf{w}_n = (w_{n,1}, \ldots, w_{n,m})$$

the vector of waiting times of the n-th customer, where $w_{n,i}$ is the time this customer must wait until i channels become free of customers arriving before him.

The sequence $\{\mathbf{w}_n;\ n \geqslant 1\}$ will be completely determined if along with the governing sequences $\{\tau_j^e\}$ and $\{\tau_j^s\}$ we also give on the same probability space the initial value $\mathbf{w}_1$. Of interest to us is the "actual" waiting time of the n-th customer, which is obviously $w_{n,1}$. We will prove a theorem on the existence of the limiting distribution of $\mathbf{w}_n$ as $n \to \infty$.

Let $x^+ = \max(0, x)$ and for a vector $\mathbf{x}$ (in the sequel, bold-faced letters will denote m-dimensional vectors) we set

$$\mathbf{x}^+ = (x_1^+, \ldots, x_m^+).$$

$\mathbf{R}(\mathbf{x})$ will denote the vector obtained from $\mathbf{x}$ by arranging its components in ascending order. Thus, the first coordinate of $\mathbf{R}(\mathbf{x})$ is $\min(x_1, \ldots, x_m)$. Further let

$$\mathbf{e} = (1, 0, \ldots, 0) \quad \text{and} \quad \mathbf{i} = (1, 1, \ldots, 1).$$

Then it's not hard to see that the following recursion relation holds for $\mathbf{w}_n$, being a generalization of its one-dimensional analogue (see § 2 Chapter 1):

$$\mathbf{w}_{n+1} = [\mathbf{R}(\mathbf{w}_n + \tau_n^s \mathbf{e}) - \tau_n^e \mathbf{i}]^+. \tag{1}$$

3. The proof of theorems on convergence of (1) to a stationary solution requires introduction of some additional assumptions. We assume that the vectors (τ_n^e, τ_n^s); $-\infty < n < \infty\}$ form a strictly stationary, metrically transitive sequence. Denote $\tau_n^s - m\tau_n^e$ by ξ_n. Then we have

Theorem 1. *If* $\mathsf{M}\xi_n < 0$ *and* $\mathbf{w}_1 = \mathbf{0}$, *then there exists a proper stationary sequence* $\{\mathbf{w}^n\}$ *satisfying* (1) *and such that the distribution function of* $\mathbf{w}_n$ *converges monotonically to that of* $\mathbf{w}^0$ *as* $n \to \infty$.

Proof. We will need some auxiliary propositions. We will say that a vector $\mathbf{x}$ possesses some property if all of its components possess it. For example, $\mathbf{x} > \mathbf{0}$ if all $x_i > 0$.

Lemma 1. *Let the vectors* $\mathbf{w}_n$, $n \geqslant 1$, *be connected by the relations*

$$\mathbf{w}_{n+1} = \mathbf{f}(\mathbf{w}_n, \tau_n^e, \tau_n^s); \quad \mathbf{w}_1 = \mathbf{0},$$

where $\{\tau_n^e, \tau_n^s\}$ *is stationary and metrically transitive and* $\mathbf{f}(\mathbf{x}, y, z) \geqslant 0$ *is a non-decreasing, left-continuous function of* $\mathbf{x}$ *(i.e., from the direction of smaller values of the coordinates of* $\mathbf{x}$*). Then there exists a stationary sequence* $\{\mathbf{w}^n; -\infty < n < \infty\}$ *(possibly improper) such that*

$$\mathbf{w}^{n+1} = \mathbf{f}(\mathbf{w}^n, \tau_n^e, \tau_n^s)$$

and the distribution function of $\mathbf{w}_n$ *converges monotonically as* $n \to \infty$ *to that of* $\mathbf{w}^0$.

Proof. We construct a sequence $\{\mathbf{v}_n^k; -\infty < n < \infty\}$ for each k as follows:

$$\begin{aligned} \mathbf{v}_n^k &= \mathbf{0} && \text{for } n + k \leqslant 1 \\ \mathbf{v}_{n+1}^k &= \mathbf{f}(\mathbf{v}_n^k, \tau_n^e, \tau_n^s) && \text{for } n + k \geqslant 1. \end{aligned} \tag{2}$$

We note that

1. The sequence $\{\mathbf{u}_n^r = \mathbf{v}_n^{r-n}; -\infty < n < \infty\}$ is stationary for each fixed r. Indeed,

for $\quad r \leqslant 1 \quad \mathbf{u}_n^r = \mathbf{0}, \ -\infty < n < \infty$;

for $\quad r = 2 \quad \mathbf{u}_{n+1}^2 = \mathbf{v}_{n+1}^{1-n} = \mathbf{f}(\mathbf{v}_n^{1-n}, \tau_n^e, \tau_n^s) = \mathbf{f}(\mathbf{u}_n^1, \tau_n^e, \tau_n^s) = \mathbf{f}(\mathbf{0}, \tau_n^e, \tau_n^s)$;

for $\quad r = 3 \quad \mathbf{u}_{n+1}^3 = \mathbf{f}(\mathbf{u}_n^2, \tau_n^e, \tau_n^s) = \mathbf{f}(\mathbf{f}(\mathbf{0}, \tau_{n-1}^e, \tau_{n-1}^s), \tau_n^e, \tau_n^s)$,

etc.

2. For fixed n $\mathbf{v}_n^k \uparrow$ as a function of k. Indeed, for $n+k \leqslant 0$

$$\mathbf{0} = \mathbf{v}_{n+1}^k \leqslant \mathbf{v}_{n+1}^{k+1}.$$

Hence,

$$\mathbf{v}_{n+1}^{k+2} = \mathbf{f}(\mathbf{v}_n^{k+2}, \tau_n^e, \tau_n^s) \geqslant \mathbf{f}(\mathbf{v}_n^{k+1}, \tau_n^e, \tau_n^s) = \mathbf{v}_{n+1}^{k+1}.$$

To continue, one obviously uses induction.

Thus, for each ω the sequence $\mathbf{v}_n^k$ converges monotonically as $k \to \infty$ to some function $\mathbf{w}^n$ which by Remark 1 of the proof is stationary. For the one-dimensional distributions of this sequence this is obvious since $\mathbf{v}_{n+r}^k \underset{d}{=} \mathbf{v}_n^{k+r}$, so that the distributions of $\mathbf{w}^n$ and $\mathbf{w}^{n+k}$ coincide. The stationarity of the multi-dimensional distributions is established in the same way.

Proceeding to the limit in (2) we get

$$\mathbf{w}^{n+1} = \mathbf{f}(\mathbf{w}^n, \tau_n^e, \tau_n^s). \quad \square$$

We return to the proof of the theorem. Since $\mathbf{w}_1 = \mathbf{0}$ and the right side of (1) satisfies the conditions on the function $\mathbf{f}$ of Lemma 1, we have $\mathbf{w}_n = \mathbf{v}_n^0$ in the notation of the proof of that lemma (for $n \leqslant 0$ we set $\mathbf{w}_n$ equal to zero). But $\mathbf{v}_n^0 \underset{d}{=} \mathbf{v}_0^n$. This and Lemma 1 imply the assertion of Theorem 1 except for the part on the properness of the sequence $\{\mathbf{w}^n\}$. To show that $\mathbf{w}^0$ is almost everywhere (a.s.) finite, we need

Lemma 2. *There exists a stationary proper sequence* $\{v_n\}$ *such that* $v_n/n \xrightarrow[\text{a.s.}]{} 0$ *for* $n \to \infty$ *and*

$$D\mathbf{w}_n \leqslant (m-1)\tau_{n-1}^s + v_n,$$

where

$$D\mathbf{x} = \sum_{i=1}^m (x_m - x_i).$$

Proof. If $\tau_n^s \leqslant w_{n,m} - w_{n,1}$, then

$$D\mathbf{R}(\mathbf{w}_n + \tau_n^s \mathbf{e}) = D\mathbf{w}_n - \tau_n^s.$$

If $\tau_n^s > w_{n,m} - w_{n,1}$, then

$$\begin{aligned} D\mathbf{R}(\mathbf{w}_n + \tau_n^s \mathbf{e}) &= \sum_{i=2}^m (w_{n,1} + \tau_n^s - w_{n,i}) \\ &= D\mathbf{w}_n - m(w_{n,m} - w_{n,1}) + (m-1)\tau_n^s \leqslant (m-1)\tau_n^s. \end{aligned}$$

From this and the recursion formula for $\mathbf{w}_{n+1}$ we get

$$D\mathbf{w}_{n+1} \leqslant \max(D\mathbf{w}_n - \tau_n^s, (m-1)\tau_n^s).$$

Subtracting $(m-1)\tau_n^s$ from both sides of this inequality and setting

$$u_n = D\mathbf{w}_n - (m-1)\tau_{n-1}^s, \qquad \zeta_n = -\tau_n^s + (m-1)(\tau_{n-1}^s - \tau_n^s),$$

we get $u_{n+1} \leqslant \max(0, u_n + \zeta_n)$.

Since $\mathsf{M}\zeta_n<0$ (we exclude the trivial case $\tau_n^s\equiv 0$), the suprema

$$v_{n+1}=\sup_{k\geqslant 0}\{0, Y_{k,n}\}, \quad Y_{k,n}=\sum_{j=n-k}^{n}\zeta_j$$

form a proper stationary sequence for which (see Theorem 8 of Chapter 1)

$$v_{n+1}=\max(0, v_n+\zeta_n) \quad \text{and} \quad v_n/n \xrightarrow[\text{a.s.}]{} 0 \quad \text{for } n\to\infty .$$

Noting finally that $v_1\geqslant 0\geqslant u_1$, we get $u_n\leqslant v_n$, $n\geqslant 1$. □

We return to the proof of the fact that $\mathbf{w}^0$ is a.s. finite. Write

$$d_n=\sum_{i=1}^{m}(w_{n,i}-w_{n,1}) \quad \text{and} \quad D_n=D\mathbf{w}_n=\sum_{i=1}^{m}(w_{n,m}-w_{n,i}) .$$

Then, as is easily checked, $d_n\leqslant(m-1)D_n$. Suppose $w_{k,1}=0$ is the last of the variables $0=w_{1,1}, w_{2,1}, \ldots, w_{n,1}$ which vanishes. Then, putting

$$[\mathbf{w}_l]=\sum_{i=1}^{m}w_{l,i} ,$$

we get $d_k=[\mathbf{w}_k]$, and by (1), ($w_{j,1}>0$ for $k<j\leqslant n$)

$$[\mathbf{w}_n]=[\mathbf{w}_k]+\sum_{j=k}^{n-1}\xi_j . \tag{3}$$

This yields

$$\begin{aligned}[\mathbf{w}_n]&\leqslant \sup_{1\leqslant k\leqslant n}\Big(d_k+\sum_{j=k}^{n-1}\xi_j\Big)\leqslant \sup_{1\leqslant k\leqslant n}\Big((m-1)D_k+\sum_{j=k}^{n-1}\xi_j\Big)\\ &\leqslant \sup_{1\leqslant k\leqslant n}\Big((m-1)^2\tau_{k-1}^s+(m-1)v_k+\sum_{j=k}^{n-1}\xi_j\Big) .\end{aligned} \tag{4}$$

The right side here is distributed like

$$\begin{aligned}&\sup_{1\leqslant k\leqslant n}\Big\{(m-1)^2\tau_{k-n-1}^s+(m-1)v_{k-n}+\sum_{j=k-n}^{-1}\xi_j\Big\}\\ &\quad=\sup_{0\leqslant k\leqslant n-1}\Big((m-1)^2\tau_{-k-1}^s+(m-1)v_{-k}+\sum_{j=-k}^{-1}\xi_j\Big)\\ &\quad\leqslant \sup_{k\geqslant 0}\Big((m-1)^2\tau_{-k-1}^s+(m-1)v_{-k}+\sum_{j=-k}^{-1}\xi_j\Big) .\end{aligned}$$

By Lemma 2, $v_k/k \xrightarrow[\text{a.s.}]{} 0$ as $k\to\infty$. This property is also enjoyed by the sequence τ_k^s/k since $\mathsf{M}\tau_k^s<\infty$. These facts and the strong law of large numbers imply that for almost all ω and $k\geqslant k(\omega,\varepsilon)$ the expression after the "sup" sign in (4) does not exceed $\varepsilon k+k\mathsf{M}\xi_j/2$ for arbitrary preassigned $\varepsilon>0$. It will be negative for $\varepsilon<-\mathsf{M}\xi_j/2$. This means the supremum in (4) is finite for almost all ω. □

4. We will demonstrate another method of proving the finiteness of $\mathbf{w}^0$ which yields at the same time a useful inequality for the distribution of the component w_m^0 of the vector $\mathbf{w}^0$.

Along with the process $\mathbf{w}_n$ defined by (1), we consider the process $\mathbf{w}_n^*$, $\mathbf{w}_1^*=0$, differing from $\mathbf{w}_n$ by the fact that in the transition from $\mathbf{w}_n^*$ to $\mathbf{w}_{n+1}^*$ the value τ_n^s is added not to the minimal coordinate of $\mathbf{w}_n$ but to the coordinate with index $m\{n/m\}$, where $\{x\}$ is the fractional part of x. Thus, if T is the shift operator $T(a_1, \ldots, a_m)=(a_m, a_1, \ldots, a_{m-1})$, we have

$$\mathbf{w}_{n+1}^*=(\mathbf{w}_n^*+\tau_n^s T^n\mathbf{e}-\tau_n^e\mathbf{i})^+ . \tag{5}$$

This corresponds to a queueing system in which each arriving customer is assigned in advance to the "next" channel. The system is thereby transformed into a set of m service systems and $w^*_{n,i}$ in (5) is the (virtual) waiting time until service in the i-th channel at the moment of arrival of the n-th customer.

However, it is obvious that among all possible "assignment" algorithms, that of (1) minimizes the channel idle time. Hence,

$$w_{n,m} \leqslant \max_{1 \leqslant i \leqslant m} w^*_{n,i}.$$

We remark that the sequence $\{w^*_{km+1,1}; k=0, 1, \ldots\}$ converges monotonically (as we saw in Chapter 1) to a proper stationary sequence the distribution of whose coordinates coincides with that of the supremum of a known sequence (since $w^*_{(k+1)m+1,1} = \max(0, w^*_{km+1,1} + \tau^s_{km+1} - \tau^e_{km+1} - \cdots - \tau^e_{km+m})$ and $\mathsf{M}(\tau^s_k - m\tau^e_k) < 0$). This property is also enjoyed for each fixed j by the sequence $\{w^*_{km+j,j}; k \geqslant 0\}$.

Similar statements hold for the sequences $\{w^*_{km+j,i}; k \geqslant 0\}$; $1 \leqslant i, j \leqslant m$; moreover,

$$\begin{aligned} w^*_{km+j+i,j} &= (w^*_{km+j,j} + \tau^s_{km+j} - \tau^e_{km+j} - \cdots - \tau^e_{km+j+i-1})^+ \\ &\leqslant w^*_{km+j,j} + \tau^s_{km+j} = \max_{1 \leqslant l \leqslant k} (X_{k,j} - X_{l,j}) + \tau^s_{km+j}, \end{aligned} \tag{6}$$

where

$$X_{k,j} = \sum_{l=1}^{k-1} \zeta_{l,j}, \quad \zeta_{l,j} = \tau^s_{lm+j} - \tau^e_{lm+j} - \cdots - \tau^e_{(l+1)m+j-1}. \tag{7}$$

Using the stationarity, we get

$$w^*_{km+j+i,j} \underset{d}{\leqslant} \sup_{k \geqslant 0} Y_{k,j} + \tau^s_j = \bar{Y}_j + \tau^s_j, \tag{8}$$

where

$$Y_{k,j} = \sum_{l=-k}^{-1} \zeta_{l,j}, \qquad \bar{Y}_j = \sup_{k \geqslant 0} Y_{k,j}. \tag{9}$$

This yields for all n and x

$$\mathsf{P}(w_{n,m} > x) \leqslant \sum_{j=1}^{m} \mathsf{P}(\bar{Y}_j + \tau^s_j > x). \quad \square$$

If we sharpen the inequality (6) somewhat, we can obtain from the proof above the following

Corollary. *In the notation of* (7)–(9), *one has for all n and x*

$$\mathsf{P}(w_{n,m} > x) \leqslant m\mathsf{P}(\bar{Y}_1 + \max(0, \tau^s_1 - \tau^e_1) > x).$$

This can be shown by writing in place of (6)

$$w^*_{km+j+i,j} \leqslant w^*_{km+j,j} + \max(0, \tau^s_{km+j} - \tau^e_{km+j}) \underset{d}{\leqslant} \bar{Y}_j + \max(0, \tau^s_j - \tau^e_j). \quad \square$$

5. We turn now to the systems $\langle G, G, G/m, 1\rangle$ and will prove that Theorem 1 also holds for them, although naturally in a more general formulation.

By $\mathbf{w}_n$ here we will understand a vector of waiting times for the first request in the n-th group ($w_{n,i}$ is the time this request would have to wait until i channels become free of requests arriving in previous groups). It will be convenient to give the governing sequence in a somewhat different form, as $\{\tau_j^e, \nu_j^e, \boldsymbol{\tau}_j^s\}$, where

$$\boldsymbol{\tau}_j^s=(\tau_{j,1}^s, \ldots, \tau_{j,\nu_j^e}^s)$$

is the vector of service times for requests which have arrived in the j-th group (here we abandon our previous agreement by which bold-faced letters designate only m-dimensional vectors). Then the quantities $\mathbf{w}_{n+1}$ and $\mathbf{w}_n$ will be related by the following formulae, in which $\mathbf{w}_n(i)$ are auxiliary vectors

$$\begin{aligned} &\mathbf{w}_n(1)=\mathbf{R}(\mathbf{w}_n+\tau_{n,1}^s\mathbf{e}), \\ &\mathbf{w}_n(2)=\mathbf{R}(\mathbf{w}_n(1)+\tau_{n,2}^s\mathbf{e}), \\ &\cdots\cdots\cdots\cdots\cdots \\ &\mathbf{w}_n(\nu_n^e)=\mathbf{R}(\mathbf{w}_n(\nu_n^e-1)+\tau_{n,\nu_n^e}^s\mathbf{e}), \\ &\mathbf{w}_{n+1}=(\mathbf{w}_n(\nu_n^e)-\tau_n^e\mathbf{i})^+ . \end{aligned} \tag{10}$$

Let $[\mathbf{x}]=\sum x_i$ as before.

Theorem 1A. *If $\{\tau_j^e, \nu_j^e, \boldsymbol{\tau}_j^s\}$ is a stationary, metrically transitive sequence, $\mathbf{w}_1=0$ and $\mathsf{M}[\boldsymbol{\tau}_1^s]-m\mathsf{M}\tau_1^e<0$, then there exists a proper stationary sequence $\{\mathbf{w}^n\}$ satisfying* (10) *and such that the distribution function of $\mathbf{w}_n$ converges monotonically as $n\to\infty$ to that of $\mathbf{w}^0$.*

Proof. It is clear that the transformation (10) is also a function of $\mathbf{w}_n$ satisfying the conditions of Lemma 1 (monotone and continuous). Consequently, the existence of a sequence $\mathbf{w}^n$ as described in the formulation of Theorem 1A (not necessarily finite) is proved.

The proof of the fact that $\mathbf{w}^0$ is a.s. finite goes as before.

Lemma 2A. *There exists a proper stationary sequence $\{v_n\}$ such that $v_n/n \xrightarrow[\text{a.s.}]{} 0$ as $n\to\infty$ and*

$$D\mathbf{w}_n \leqslant (m-1)t_n^s+v_n ,$$

where $t_n^s=\max_j \tau_{n,j}^s$, $\mathsf{M}t_n^s<\infty$.

Proof. We get as before

$$D\mathbf{R}(\mathbf{w}_n(j)+\tau_{n,j+1}^s\mathbf{e})\leqslant \max\{D\mathbf{w}_n(j)-\tau_{n,j+1}^s, (m-1)\tau_{n,j+1}^s\},$$

whence it follows that

$$D\mathbf{w}_{n+1}\leqslant \max(D\mathbf{w}_n-[\boldsymbol{\tau}_n^s], (m-1)\max_j \tau_{n,j}^s) .$$

Subtracting $(m-1)t_n$ from both sides of the inequality and setting $u_n=D\mathbf{w}_n-(m-1)t_{n-1}^s$, we find that

$$u_n\leqslant \max(0, u_n+\zeta_n),$$

where $\zeta_n = -[\tau_n^s] + (m-1)(t_{n-1}^s - t_n^s)$. Since $\mathsf{M}t_n^s \leqslant \mathsf{M}[\tau_n^s] < \infty$, $\mathsf{M}\zeta_n$ exists and is less than zero. The remainder of the proof then goes as in Lemma 2. □

The conclusion of the proof of Theorem 1A does not differ essentially from the arguments in the case $v^e \equiv 1$. One need merely use Lemma 2A and assume that $\xi_n = [\tau_n^s] - m\tau_n^e$ in (3). □

6. We now show that Theorem 1 also implies the existence of a *limiting distribution as $n \to \infty$ of the queue length q_n at the moment of arrival of the n-th request.* In connection with this we will obtain a number of useful relations between the limiting distributions of q_n and $\mathbf{w}_n$.

As in the single-channel case, one can establish directly (see § 26) that the event $\{w_{n-k,1} > \tau_{n-k}^e + \cdots + \tau_{n-1}^e\}$ means that at the time of arrival of the n-th request all channels will be busy with requests arriving before the appearance of the $(n-k)$-th. This is in turn equivalent to the event $\{q_n \geqslant m+k\}$. Thus, for $k \geqslant 0$

$$\{q_n \geqslant m+k\} = \{w_{n-k,1} > \tau_{n-k}^e + \cdots + \tau_{n-1}^e\}. \tag{11}$$

Analogously, for $m-1 \geqslant k \geqslant 0$, we get

$$\{q_n \geqslant m-k\} = \{w_{n,k+1} > 0\}. \tag{12}$$

Performing a shift by $n-k$ time units, we obtain in the notation of (2) ($v_{k,j}^n$ is the j-th coordinate of $\mathbf{v}_k^n$) that as $n \to \infty$ and for $k \geqslant 0$

$$\mathsf{P}(q_n \geqslant m+k) = \mathsf{P}(v_{0,1}^{n-k} > \tau_0^e + \cdots + \tau_{k-1}^e) \downarrow \mathsf{P}(v_{0,1}^{\infty} > \tau_0^e + \cdots + \tau_{k-1}^e). \tag{13}$$

Analogously, for $m-1 \geqslant k \geqslant 0$ we find that

$$\mathsf{P}(q_n \geqslant m-k) = \mathsf{P}(v_{0,k+1}^n > 0) \downarrow \mathsf{P}(v_{0,k+1}^{\infty} > 0). \tag{14}$$

We have thus proved

Theorem 2. *When the conditions of Theorem 1 are satisfied, the distribution of q_n converges monotonically to that of some proper random variable q^0. Formulae* (13) *and* (14) *define a relationship between q^0 and the stationary distribution of* $\mathbf{w}^0$.

Equation (11) implies, in particular, the possibility of the following representation of q^0 *for the systems* $\langle G_I, 1, G/m, 1\rangle$, in which the sequence $\{\tau_j^e\}$ does not depend on $\{\tau_j^s\}$. If $\eta(t) = \min\{k : \sum_{j=1}^k \tau_j^e \geqslant t\}$ is, as before, the first passage time of the level t by a walk with jumps $\tau_1^e, \tau_2^e, \ldots$, then Eqs. (11) and (13) for the limiting distributions can be written as

$$\max(0, q^0 - m + 1) \underset{d}{=} \eta(w_1^{(0)})$$

where $\{\tau_j^e\}$ is independent of w_1^0.

If we return to the systems $\langle G, G, E/m, 1\rangle$ and assume that a limiting distribution for q^0 exists, then for these systems we can write the (in a certain sense inverse) relation

$$w_1^0 \underset{d}{=} \tau_1^* + \cdots + \tau_{\max(0, q^0 - m + 1)}^*,$$

where the τ_j^* are independent and also don't depend on q^0, and $\tau_j^* \underset{d}{=} \min\{\tau_1^s, \ldots, \tau_m^s\}$, so that $\mathsf{P}(\tau_j^* > x) = \mathsf{P}^m(\tau_1^s > x) = e^{-\alpha m x}$. The derived equality follows from its prelimiting analogue

$$w_{n,1} \underset{d}{=} \tau_1^* + \cdots + \tau_{\max(0, q_n - m + 1)}^*$$

where $\{\tau_j^*\}$ does not depend on q_n, which follows from the fact that the discharge stream reducing the system to the state $q < m$ forms, by the properties of the exponential distribution, a homogeneous Poisson process with parameter αm which does not depend on the previous history.

Analogous relations for the systems $\langle G, G, E/m, 1\rangle$ will also hold for the virtual waiting time and the queue length (compare Subsection 3 § 10).

7. We mention the possibility of another direct approach to the proof of the existence of a limiting distribution of q_n which will be quite useful in the following chapters. It employs the following representation ($I(A)$ is the indicator of the event A, $X_n^e = \sum_{j=1}^n \tau_j^e$ and the systems under consideration are $\langle G, 1, G/m, 1\rangle$):

$$q_{n+1} = I(u_1 > X_n^e) + I(u_2 > X_n^e) + \cdots + I(u_n > X_n^e),$$

where $u_1 = \tau_1^s$, $u_2 = \tau_2^s + X_1^e$, $u_3 = \tau_3^s + X_2^e$, ..., $u_m = \tau_m^s + X_{m-1}^e$; u_k is defined recursively for $k > m$: $u_{k+1} = \tau_{k+1}^s + \max(X_k^e, \bar{u}_k)$, where $\bar{u}_k$ is the m-th largest value in the sequence $u_1, \ldots, u_k$. Thus,

$$I(u_k > X_n^e) = 1,$$

if the k-th customer is in the system at the moment of arrival of the $(n+1)$-th, and $I(u_k > X_n^e) = 0$ otherwise.

The monotonicity of the sequence $U^{-n} q_{n+1}$ for increasing n (U is the random variable shift-transformation corresponding to a stationary governing sequence; see § 3) is obvious here. Our basic task is to prove the boundedness of the sequence $u_k - X_{k-1}^e$ when $\mathsf{M}\tau_1^s - m\mathsf{M}\tau_1^e < 0$. If $\mathsf{M}\tau_1^s - m\mathsf{M}\tau_1^e > 0$, then we easily obtain

$$\frac{u_n}{n} \underset{\text{a.s.}}{\longrightarrow} \frac{\mathsf{M}\tau_1^s}{m} \quad \text{and} \quad \frac{q_n}{n} \underset{p}{\longrightarrow} 1 - \frac{m\mathsf{M}\tau_1^e}{\mathsf{M}\tau_1^s} \quad \text{as } n \to \infty .$$

8. *The systems* $\langle G, G, E/m, G\rangle$. Let q_n be the length of the "pure" queue before the arrival time t_n of the n-th group of requests (that is, the queue not counting requests already being served) and k_n the number of idle channels before t_n. Then (q_{n+1}, k_{n+1}) can be expressed explicitly by means of (q_n, k_n) and the terms of the governing sequence. If a_n is the number of groups accepted by the system for service before time t_n, then

$$q_{n+1} = \max(0, q_n + v_n^e - v_{a_n+1}^s - \cdots - v_{a_n+\zeta_n}^s) \qquad \text{for } k_n = 0$$

and

$$q_{n+1} = \max(0, v_n^e - v_{a_n+1}^s - \cdots - v_{a_n+k_n}^s - \cdots - v_{a_n+k_n+\zeta_n}^s) \quad \text{for } k_n > 0 .$$

In these equalities ζ_n is the number of jumps of a Poisson process with parameter $m\alpha$ $(1/\alpha = \mathsf{M}\tau^s)$ on the time interval $[t_n, t_{n+1}]$ of length τ_n^e. The reader can derive similar equalities for the characteristic k_{n+1}, which will be determined by the values of k_n, τ_n^e and the first passage time of the level $q_n + v_n^e$ by a random walk with jumps $v_{a_n+1}^s, v_{a_n+2}^s, \ldots$.

For the systems $\langle G_I, G_I, E/m, G_I\rangle$ these relations transform the pair $\{q_n, k_n\}$ into a homogeneous Markov chain which will be ergodic when $(\mathsf{M}v^e/\mathsf{M}\tau^e) - \alpha m \mathsf{M}v^s < 0$. The search for the explicit form of the stationary distribution of the Markov chain for the systems $\langle G_I, 1, E/m, 1\rangle$ is based on the use of the corresponding equations for it and appears in § 29.

§ 28. The Systems $\langle G_I, G_I, G_I/m, 1\rangle$. Stability Theorems. Connection between the Waiting Time and Queue Length. Estimates of Rates of Convergence

1. We consider the systems $\langle G_I, 1, G_I/m, 1\rangle$. Here $\{\tau_n^e\}$ and $\{\tau_n^s\}$ are independent sequences of i.i.d. random variables. In this case one can write an integral equation for the distribution of the stationary solution $\mathbf{w}^n$ of Eq. (1).

Let

$$\mathbf{f}(\mathbf{x}, u, v) = [\mathbf{R}(\mathbf{x} + v\mathbf{e}) - u\mathbf{i}]^+, \qquad \Phi(\mathbf{x}, u, v) = \{\mathbf{y} : \mathbf{f}(\mathbf{y}, u, v) \leqslant \mathbf{x}\}$$

and

$$\Psi = \{\mathbf{x} : 0 \leqslant x_1 \leqslant \cdots \leqslant x_m\}.$$

Then (1) obviously means that for $\mathbf{x} \in \Psi$

$$\mathsf{P}\{\mathbf{w}_{n+1} \leqslant \mathbf{x}\} = \iint \mathsf{P}(\mathbf{w}_n \in \Phi(\mathbf{x}, u, v)\, d\mathsf{P}(\tau_n^e < u)\, d\mathsf{P}(\tau_n^s < v).$$

Letting $n \to \infty$, we obtain the required equation for the stationary distribution.

Since in the case under consideration the sequence $\{\mathbf{w}_n\}$ forms a Markov chain, the convergence of the distribution of $\mathbf{w}_n$ to a stationary one implies the *uniqueness* of the solution of the equation above in the class of m-dimensional distributions (in fact, this is an equivalent notation for the usual transition equation corresponding to the chain: $\mathsf{P}_{n+1}(A) = \mathsf{P}(\mathbf{w}_{n+1} \in A) \int \mathsf{P}_n(d\mathbf{x}) \mathsf{P}(\mathbf{x}, A)$, where $\mathsf{P}(\mathbf{x}, A)$ is the transition function).

We make the following observation: if $\mathbf{x}^{(k)}$ is the $(m-1)$-dimensional vector obtained from $\mathbf{x}$ by deleting its k-th component and $(z, \mathbf{x}^{(k)})$ is an m-dimensional vector with z as its first component, then for $\mathbf{x} \in \Psi$

$$\Phi(\mathbf{x}, u, v) = \bigcup_{k=1}^{m} \{\mathbf{y} : \mathbf{y} \leqslant u\mathbf{i} + (x_k - v, \mathbf{x}^{(k)})\}.$$

Indeed, if $\mathbf{y} \in \Phi(\mathbf{x}, u, v)$, then there exists a k, $1 \leqslant k \leqslant n$, such that

$$\mathbf{x} \geqslant \mathbf{f}(\mathbf{y}, u, v) = ((y_2 - u)^+, (y_3 - u)^+, \ldots, (y_{k-1} - u)^+, (y_1 + v - u)^+, (y_k - u)^+, \ldots, (y_m - u)^+).$$

This in turn implies that

$$\mathbf{y} \leqslant u\mathbf{i} + (x_k - v, \mathbf{x}^{(k)}).$$

The reverse inclusion

$$\{\mathbf{y}:\mathbf{y}\leqslant u\mathbf{i}+(x_k-v,\mathbf{x}^{(k)})\}\in\Phi(\mathbf{x},u,v)$$

is verified similarly.

Furthermore, if $\Phi=\bigcup_{r=1}^m\{\mathbf{y}:\mathbf{y}\leqslant\mathbf{x}_r\}$, where the coordinates of the vector $\mathbf{x}_r=(x_{r,1},\ldots,x_{r,m})$ satisfy

$$\begin{aligned}&x_{1,1}\leqslant x_{2,1}\leqslant\cdots\leqslant x_{m,1},\\&x_{1,k}\geqslant x_{2,k}\geqslant\cdots\geqslant x_{m,k}\quad\text{for } k=2,3,\ldots,m,\end{aligned}\tag{15}$$

and $F(\mathbf{x})$ is the distribution function of the random vector $\mathbf{w}^0$, then it's not hard to verify that

$$\mathsf{P}(\mathbf{w}^0\in\Phi)=\sum_{r=1}^m F(\mathbf{x}_r)-\sum_{k=2}^m F(x_{k-1,1},x_{k,2},\ldots,x_{k,m}).$$

Since the vectors $\mathbf{x}_r=u\mathbf{i}+(x_r-v,\mathbf{x}^{(r)})$, $\mathbf{x}\in\Psi$, satisfy (15), we obtain as a result that the *equation for the stationary distribution of* $\mathbf{w}^0$ *can be written in the form*

$$\begin{aligned}F(\mathbf{x})=&\sum_{k=1}^m\int_0^\infty\int_0^\infty F(u\mathbf{i}+(x_k-v,\mathbf{x}^{(k)}))\,d\mathsf{P}(\tau_1^e<u)\,d\mathsf{P}(\tau_1^s<v)\\&-\sum_{k=2}^m\int_0^\infty\int_0^\infty F(u\mathbf{i}+(x_{k-1}-v,\mathbf{x}^{(k)})\,d\mathsf{P}(\tau_1^e<u)\,d\mathsf{P}(\tau_1^s<u).\end{aligned}\tag{16}$$

As already remarked, *the solution of this equation is unique.*

2. With the aid of (16) we can now establish conditions guaranteeing the *continuous dependence* of the distribution $F(x)$ of the waiting time on the distributions of τ^e and τ^s.

The important rôle played by theorems of this type ought to be clear in the light of §§ 11 and 21. They allow us to use as approximations formulae obtained under their or other "exact" assumptions w.r.t. the distributions of τ^e and τ^s when these assumptions are in fact only "approximately" fulfilled. For example: if $\mathsf{P}(\tau^s>x)$ is close to the function $e^{-\alpha x}$, $x\geqslant 0$ (see § 29), we are interested in whether the waiting time distribution is close to the expression determined in Theorem 12.

We consider a double sequence, i.e., a sequence of systems $\langle G_I, 1, G_I/m, 1\rangle^{(n)}$, $n=1,2,\ldots$, with governing sequences $\{\tau_j^{(n)e},\tau_j^{(n)s};\,-\infty<j<\infty\}$ depending on a parameter $n=1,2,\ldots$. All notation applying to the system $\langle G_I, 1, G_I/m, 1\rangle^{(n)}$ will be furnished with a superscript (n). Further let

$$U_n(x)=\mathsf{P}(\tau^{(n)e}<x),\qquad V_n(x)=\mathsf{P}(\tau^{(n)s}<x)$$

and

$$F_n(\mathbf{x})=F_n(x_1,\ldots,x_m)=\mathsf{P}(w_1^{(n)0}<x_1,\ldots,w_m^{(n)0}<x_m)$$

($w_j^{(n)0}$ are the components of the stationary vector $\mathbf{w}^{(n)0}$).

Theorem 3. *Let $U(x)$ and $V(x)$ be the non-degenerate distribution functions of some random variables τ^e and τ^s and let*

$$U_n(x)\Rightarrow U(x),\qquad V_n(x)\Rightarrow V(x)\quad\textit{and}\quad \mathsf{M}\tau^{(n)s}\to\mathsf{M}\tau^s$$

as $n\to\infty$.

Then, if $a = \mathsf{M}\tau^s - m\mathsf{M}\tau^e < 0$, *we have*

$$F_n(\mathbf{x}) \Rightarrow F(\mathbf{x}),$$

where $F(\mathbf{x})$ *is the distribution function of the stationary vector* $\mathbf{w}^0$ *constructed* w.r.t. *the sequence* $\{\tau_j^e, \tau_j^s\}$ *and* τ_j^e, τ_j^s *are distributed like* τ^e *and* τ^s *respectively.*

Proof. Since $\mathsf{M}\tau^e \leqslant \liminf_{n\to\infty} \mathsf{M}\tau^{(n)e}$, we can assume without loss of generality that $a_n = \mathsf{M}\tau^{(n)s} - m\mathsf{M}\tau^{(n)e} < 0$ for all n. Then the sequence of (nondegenerate) distribution functions $F_n(\mathbf{x})$ is defined. Let $\{F_{n_j}(\mathbf{x})\}$ be an arbitrary convergent subsequence of $\{F_n(\mathbf{x})\}$:

$$F_{n_j}(\mathbf{x}) \Rightarrow F(\mathbf{x}) \quad \text{as } j \to \infty .$$

We will prove that $F(\mathbf{x})$ satisfies (16).

Let $\mathbf{x}$ be a point of continuity of $F(\mathbf{x})$. Choose $N = N(\varepsilon)$ such that for arbitrary $\varepsilon > 0$

$$1 - U_n(N) < \varepsilon \quad \text{and} \quad 1 - V_n(N) < \varepsilon$$

for all n. Furthermore, choose partitions

$$0 = u_0 < u_1 < \cdots < u_M = u > N \quad \text{and} \quad 0 = v_0 < v_1 < \cdots < v_M = v > N$$

in such a way that $u_j, j \geqslant 1$, are points of continuity of $U(x)$; $v_j, j \geqslant 1$, are points of continuity of $V(x)$ and such that the points (see (16)) $u_j\mathbf{i} + (x_k - v_l, \mathbf{x}^{(k)})$, $u_j\mathbf{i} + (x_{k-1} - v_l, \mathbf{x}^{(k)})$; $j, l = 0, \ldots, M$; $k = 1, \ldots, m$, together with $\mathbf{x}$ are points of continuity of $F(\mathbf{x})$. This is always possible since the set of jumps of U and V is no more than countable as is the number of vertices of the "angles of discontinuity" of $F(\mathbf{x})$. ($F(\mathbf{x})$ has discontinuities provided there exist $\mathbf{x}_0 = (x_{01}, \ldots, x_{0m})$ such that $\mathsf{P}(\mathbf{w}^0 = \mathbf{x}_0) > 0$. If $\mathbf{x}_0$ is such a point, then $F(\mathbf{x})$ has discontinuities at the edges of angles $(x_{01} + t_1, \ldots, x_{0m} + t_m)$, $t_1 \geqslant 0, \ldots, t_m \geqslant 0$; $\min t_j = 0$.)

We now write estimates for the terms on the right in (16):

$$\begin{aligned}
\varphi_{k,n}^M - \varepsilon &\equiv \textstyle\sum_{j=1}^M \sum_{l=1}^M F_n(u_{j-1}\mathbf{i} + (x_k - v_l, \mathbf{x}^{(k)}))\, \Delta U_n(u_j)\, \Delta V_n(v_l) - \varepsilon \\
&\leqslant \textstyle\int_0^\infty \int_0^\infty F_n(u\mathbf{i} + (x_k - v, \mathbf{x}^{(k)}))\, dU_n(u)\, dV_n(v) \\
&\leqslant \textstyle\sum_{j=1}^M \sum_{l=1}^M F_n(u_j\mathbf{i} + (x_k - v_{l-1}, \mathbf{x}^{(k)}))\, \Delta U_n(u_j)\, \Delta V_n(v_l) + \varepsilon \\
&\equiv \Phi_{k,n}^M + \varepsilon ,
\end{aligned}$$

where $\Delta U_n(u_j) = U_n(u_j) - U_n(u_{j-1})$, and $\Delta V(v_l)$ is defined analogously. Similar estimates hold for the negative terms in (16). We will denote them by $\psi_{k,n}^M - \varepsilon$ and $\Psi_{k,n}^M + \varepsilon$, resp. In accordance with (16) we find that

$$\textstyle\sum_{k=1}^m \varphi_{k,n}^M - \sum_{k=2}^m \Psi_{k,n}^M - (2m-1)\varepsilon \leqslant F_n(\mathbf{x}) \leqslant \sum_{k=1}^m \Phi_{k,n}^M - \sum_{k=2}^m \psi_{k,n}^M + (2m-1)\varepsilon .$$

Now putting $n = n_j$ and letting $j \to \infty$, we find that $\lim_{j\to\infty} \varphi_{k,n_j}^M = \varphi_k^M$ exist. Obviously, the limits for the rest of the estimates also exist; we denote them by Φ_k^M, ψ_k^M and Ψ_k^M. Proceeding to the limit in the inequalities obtained for $F_n(\mathbf{x})$ yields

$$\textstyle\sum_{k=1}^m \varphi_k^M - \sum_{k=2}^m \Psi_k^M - (2m-1)\varepsilon \leqslant F(\mathbf{x}) \leqslant \sum_{k=1}^m \Phi_k^M - \sum_{k=2}^m \psi_k^M + (2m-1)\varepsilon .$$

Because of the arbitrariness of the partitions $\{u_j\}$ and $\{v_j\}$ and of the numbers ε, N and M, this means that $F(\mathbf{x})$ satisfies (16).

We now show that $F(\mathbf{x})$ is actually a distribution, i.e., that $F(\infty)=1$.

By the corollary to Theorem 1 we have for all x

$$\mathsf{P}(w_m^{(n)0}>x)\leqslant m\mathsf{P}(\overline{Y}^{(n)}+\max(0,\tau^{(n)s}-\tau^{(n)e})>x),$$

where the terms following the last probability symbol are independent and

$$\overline{Y}^{(n)}=\sup_{k\geqslant 0} Y_k^{(n)} \quad \text{and} \quad Y_k^n=\textstyle\sum_{j=1}^k \xi_j^{(n)}, \ \xi_j^{(n)}, \quad j=1,2,\ldots,$$

are independent and distributed like $\tau_1^{(n)s}-\tau_1^{(n)e}-\cdots-\tau_m^{(n)e}$.

It is clearly sufficient to show that

$$\lim_{n\to\infty}\mathsf{P}(\overline{Y}^{(n)}>x)\leqslant P_1(x),$$

where $P_1(x)$ is some non-degenerate distribution. But in virtue of the corollary to Theorem 7, Chapter 4, the distribution of $\overline{Y}^{(n)}$ converges to the distribution of

$$\overline{Y}=\sup_{k\geqslant 0} Y_k; \qquad Y_k=\textstyle\sum_{j=1}^k \xi_j, \ \xi_j \underset{d}{=} \tau_1^s-\tau_1^e-\cdots-\tau_m^e,$$

the ξ_j independent, provided that the distributions of $\tau_1^{(n)s}$ and $\tau_1^{(n)e}+\cdots+\tau_m^{(n)e}$ converge weakly to those of τ_1^s and $\tau_1^e+\cdots+\tau_m^e$ and $\mathsf{M}\tau^{(n)s}\to\mathsf{M}\tau^s$ as $n\to\infty$. Since these conditions are met (see the formulation of the theorem), we get

$$\mathsf{P}(\overline{Y}^{(n)}>x)\Rightarrow\mathsf{P}(\overline{Y}>x).$$

Thus, $F(\mathbf{x})$ is a distribution function and satisfies (16). Since the solution of (16) in the class of distribution functions is unique, $F(\mathbf{x})=\mathsf{P}(\mathbf{w}^0<\mathbf{x})$.

Therefore, the limit of each weakly convergent sub-sequence $F_{n_j}(\mathbf{x})$ equals $\mathsf{P}(\mathbf{w}^0<\mathbf{x})$, whence it follows that as $n\to\infty$

$$F_n(\mathbf{x})\Rightarrow\mathsf{P}(\mathbf{w}^0<\mathbf{x}). \quad \square$$

3. Since for the systems under consideration in this section $\mathbf{w}_n$ and the sequence $\tau_n^e, \tau_{n+1}^e, \ldots$ are independent, Formulae (11) and (12) imply the following refinement of Theorem 2:

Theorem 4. *Let* $\mathbf{w}^0=(w_1^0,\ldots,w_m^0)$. *For* $k\geqslant m-1$

$$\lim_{n\to\infty}\mathsf{P}(q_n>k)=\mathsf{P}(w_1^0>\tau_1^e+\cdots+\tau_{k-m+1}^e) \tag{17$_1$}$$

exists (the sum $\tau_1^e+\cdots+\tau_j^e$ *is* 0 *for* $j=0$). *For* $m>k\geqslant 0$

$$\begin{aligned}\lim_{n\to\infty}\mathsf{P}(q_n\geqslant m-k)&=\mathsf{P}(w_{k+1}^0>0)\\&=\mathsf{P}(w_{k+2}^0>\tau_1^e;\, w_1^0+\tau_1^s>\tau_1^e)\\&\quad+\mathsf{P}(w_{k+1}^0>\tau_1^e;\, w_1^0+\tau_1^s\leqslant\tau_1^e)=\cdots\end{aligned} \tag{17$_2$}$$

Here, all random variables following the probability symbols are independent.

As already remarked, these equalities imply, in particular, that the "stationary queue length" q^0-m+1 can be interpreted with the help of the first passage time

$\eta(t)$ of the level t in a random walk with jumps $\tau_1^e, \tau_2^e, \ldots$:

$$q^0 - m + 1 \underset{d}{=} \eta(w_1^0),$$

where w_1^0 and the sequence $\{\tau_j^e\}$ are independent.

Remark. The following relation will be useful in the sequel:

$$\begin{aligned}\lim_{n\to\infty} \mathsf{P}(q_n=0) &= \lim_{n\to\infty} \mathsf{P}(w_{n,m}=0) = \mathsf{P}(w_m^0=0)\\ &= \lim_{n\to\infty} \mathsf{P}(\max(w_{n-1,m}, w_{n-1,1}+\tau_{n-1}^s) \leqslant \tau_{n-1}^e)\\ &= \mathsf{P}(\max(w_m^0, w_1^0+\tau_1^s) \leqslant \tau_1^e).\end{aligned}$$

It follows from the coincidence of the events following the probability symbols.

4. We now clarify the conditions for the existence of a limiting distribution for the queue length $q(t)$ at time t as $t \to \infty$.

Theorem 5. *Assume the conditions of Theorem* 1 *hold and again let w_j^0 be the coordinates of the stationary vector $\mathbf{w}^0$. If the distribution of τ^e is nonlattice, then for the systems $\langle G_I, 1, G_I/m, 1\rangle$ the limits*

$$\lim_{t\to\infty} \mathsf{P}(q(t)=j), \quad j=0, 1, \cdots$$

exist. Furthermore, for $k \geqslant 0$

$$\begin{aligned}&\lim_{t\to\infty} \mathsf{P}(q(t)\geqslant m+k+1) = \mathsf{P}(w_1^0 > \tau_1^e + \cdots + \tau_k^e + \gamma^e),\\ &\lim_{t\to\infty} \mathsf{P}(q(t)\geqslant m-k+1) = \mathsf{P}(w_{k+1}^0 > \gamma^e;\, w_1^0+\tau_1^s > \gamma^e)\\ &\qquad + \mathsf{P}(w_k^0 > \gamma^e;\, w_1^0+\tau_1^s \leqslant \gamma^e), \quad 0 \leqslant k \leqslant m-1 \end{aligned} \tag{18}$$

and

$$\lim_{t\to\infty} \mathsf{P}(q(t)=0) = \mathsf{P}(\max(w_m^0, w_1^0+\tau_1^s) \leqslant \gamma^e),$$

where the random variables $\mathbf{w}^0 = (w_1^0, \ldots, w_m^0)$, $\tau_1^s, \tau_1^e, \ldots, \tau_k^e$ and γ^e are independent and γ^e has the density $\mathsf{P}(\tau^e > x)/\mathsf{M}\tau^e$.

An analogous statement is also valid for the sequence $q(n)$ if τ^e is an integer-valued random variable with greatest common divisor of its possible values equal to 1.

The proof of the theorem is analogous in many respects to that of Theorem 22, Chapter 4. Assume, as before, that

$$\begin{aligned}\eta(t)-1 &= \max\{k: \tau_1^e + \cdots + \tau_k^e < t\},\\ \gamma^e(t) &= t - \tau_1^e - \cdots - \tau_{\eta(t)-1}^e\end{aligned}$$

and let $w_{n,j}$ be the coordinates of the vector $\mathbf{w}_n$. Then for $k \geqslant 0$

$$\begin{aligned}\{q(t)\geqslant m+k+1\} &= \{w_{\eta(t)-k,1} > \tau_{\eta(t)-k}^e + \cdots + \tau_{\eta(t)-1}^e + \gamma^e(t)\},\\ \{q(t)\geqslant m-k+1\} &= \{w_{\eta(t),k+1} > \gamma^e(t);\, w_{\eta(t),1} + \tau_{\eta(t)}^s > \gamma^e(t)\}\\ &\quad \cup \{w_{\eta(t),k} > \gamma^e(t);\, w_{\eta(t),1} + \tau_{\eta(t)}^s \leqslant \gamma^e(t)\}.\end{aligned}$$

Consequently, for (18) it suffices to show that the joint distribution of $w_{\eta(t)-k,1}$, $\tau^e_{\eta(t)-k}, \ldots, \tau^e_{\eta(t)-1}$ and $\gamma^e(t)$ converges for $t\to\infty$ to that of the independent random variables $w^0_1, \tau^e_1, \ldots, \tau^e_k, \gamma^e$ and the joint distribution of $\mathbf{w}_{\eta(t)}, \gamma^e(t), \tau^e_{\eta(t)}$ and $\tau^s_{\eta(t)}$ to that of independent $\mathbf{w}^0$ and $\gamma^e, \tau^e_1, \tau^s_1$. We have (compare with (79), § 26):

$$\mathsf{P}(w_{\eta(t)-k,1}>l, \tau^e_{\eta(t)-k}\in dx_1, \ldots, \tau^e_{\eta(t)-1}\in dx_k, \gamma^e(t)>y)$$
$$=\prod_{s=1}^k \mathsf{P}(\tau^e_1\in dx_s)\sum_{n=1}^\infty \int_0^{t-z-y} \mathsf{P}(X^e_{n-1}\in du, w_{n,1}>l)$$
$$\times \mathsf{P}(\tau^e>t-u-z). \quad (19)$$

We turn now to the sequence $\mathbf{v}^k_n$ introduced at (2). As was shown, $\mathbf{w}_n=\mathbf{v}^0_n \underset{d}{=} \mathbf{v}^n_0$ and $\mathbf{v}^0_n$ differs from $\mathbf{v}^n_0$ as a function of the governing sequences through a "shift" by time n. Thus,

$$\mathsf{P}(X^e_{n-1}\in du, w_{n,1}>l)=\mathsf{P}(\tau^e_{-n+1}+\cdots+\tau^e_{-1}\in du, v^n_{0,1}>l).$$

This means that the series on the right in (19) can be viewed as the probability of the event

$$A=\{v^{\tilde\eta(t-z)}_{0,1}>l, \tilde\gamma(t-z)>y\},$$

where $\tilde\eta(t)$ and $\tilde\gamma(t)$ are defined in the same way as $\eta(t)$ and $\gamma^e(t)$, but w.r.t. the sequence $\tau^e_{-1}, \tau^e_{-2}, \ldots$. It is clear that the joint distribution of $\tilde\eta(t)$ and $\tilde\gamma(t)$ coincides with that of $\eta(t)$ and $\gamma^e(t)$.

We use now the fact that $\mathbf{v}^n_0\uparrow$ for increasing n. For arbitrary $N>0$

$$\mathsf{P}(A)=\mathsf{P}(\tilde\eta(t-z)\leqslant N; A)+\mathsf{P}(\tilde\eta(t-z)>N; A).$$

Here $\mathsf{P}(B; A)=\mathsf{P}(BA)$. The first term converges to 0 as $t\to\infty$. For the second

$$\varepsilon_t+\mathsf{P}(v^N_{0,1}>l, \tilde\gamma(t-z)>y)\leqslant \mathsf{P}(\tilde\eta(t-z)>N; A)$$
$$\leqslant \varepsilon_t+\mathsf{P}(v^\infty_{0,1}>l, \tilde\gamma(t-z)>y),$$

where $\varepsilon_t\to 0$ as $t\to\infty$.

But $v^N_{0,1}$ is a function only of the sequences $\{\tau^s_j\}$ and $\tau^e_{-N}, \ldots, \tau^e_{-1}$. Hence, for nonlattice τ^e and $t\to\infty$

$$\mathsf{P}(v^N_{0,1}>l, \tilde\gamma(t-z)>y)\to \mathsf{P}(v^N_{0,1}>l)\mathsf{P}(\gamma^e>y). \quad (20)$$

By appropriate choice of N the probability $\mathsf{P}(v^\infty_{0,1}>l, \tilde\gamma(t-z)>y)$ can be made arbitrarily close to the left side of (20). This means that

$$\varepsilon_N+\mathsf{P}(v^N_{0,1}>l)\mathsf{P}(\gamma^e>y)\geqslant \limsup_{t\to\infty} \mathsf{P}(A)\geqslant \liminf_{t\to\infty} \mathsf{P}(A)$$
$$\geqslant \mathsf{P}(v^N_{0,1}>l)\mathsf{P}(\gamma^e>y),$$

so that there exists

$$\lim_{t\to\infty} \mathsf{P}(A)=\mathsf{P}(v^\infty_{0,1}>l)\mathsf{P}(\gamma^e>y)=\mathsf{P}(w^0_1>l)\mathsf{P}(\gamma^e>y).$$

The required convergence of the joint distribution in (19) is proved.

The asymptotic independence of $\mathbf{w}_{\eta(t)}$ and $\gamma^e(t), \tau^e_{\eta(t)}, \tau^s_{\eta(t)}$ can obviously be proved in exactly the same way. By the same token, the relations at (18) are also proved. But they indicate only the existence of $\lim_{t\to\infty}\mathsf{P}(q(t)=k)$ for $k\geqslant 2$. We will now show that $\lim_{t\to\infty}\mathsf{P}(q(t)=0)$ also exists.

It is clear that considerations analogous to the preceding will also lead to the convergence of the joint distribution

$$\mathsf{P}(\mathbf{w}_{\eta(t)} \geqslant \mathbf{x}, \tau^s_{\eta(t)} > y_1, \gamma^e(t) > y_2) \to \mathsf{P}(\mathbf{w}^0 > \mathbf{x})\mathsf{P}(\tau^s_1 > y_1)\mathsf{P}(\gamma^e > y_2).$$

But the event $\{q(t)=0\}$ is equivalent to

$$\{\max(w_{\eta(t),m}, w_{\eta(t),1} + \tau^s_{\eta(t)}) \leqslant \gamma^e(t)\}.$$

Thus, there exists

$$\lim_{t\to\infty} \mathsf{P}(q(t)=0) = \mathsf{P}(\max(w^0_m, w^0_1 + \tau^s_1) \leqslant \gamma^e). \quad \square$$

The proof of the theorem for discrete τ^e is not essentially different from the preceding one.

Remark. It is possible to prove the convergence of the joint distribution of $\mathbf{w}_{\eta(t)}$ and $\gamma^e(t)$ to a distribution of independent components (essentially, in the preceding theorem we required precisely this) in another way, not using the special properties of $\mathbf{w}_n$ (monotonicity of $\mathbf{v}^n_0$). It consists in proving that

$$\mathsf{P}(\eta(t)=n, \gamma^e(t) > y) \sim \mathsf{P}(\eta(t)=n)\mathsf{P}(\gamma^e > y)$$

as $t \to \infty$ with $n \sim t/\mathsf{M}\tau^e$.

5. For the systems $\langle G_I, G_I, G_I/m, 1\rangle$, it is better to prove the existence of a limiting distribution for the queue length $q(t)$ at t as $t\to\infty$ in another way, which does not employ Theorem 1A on the limiting distribution of $\mathbf{w}_n(t)$. This can be done, for example, by augmenting the process $q(t)$ in such a way that it becomes Markovian, and using general ergodic theorems for Markov processes.

In this subsection we will use more direct probabilistic methods (naturally, under more rigid restrictions on the sequences $\{\tau^e_n\}$, $\{v^e_n\}$ and $\{\tau^s_n\}$) to obtain, along with an existence theorem, an exponential estimate for the rate of convergence of $\mathsf{P}(q(t)=k)$ to a limiting distribution. From these considerations it will be easy to establish an exponential estimate for the rate of convergence of the distributions $\mathsf{P}(q_n=k)$ and $\mathsf{P}(\mathbf{w}_n < \mathbf{x})$.

We will assume here (as before, we will delete the lower index in τ^e_n, τ^s_n and v^e_n where this causes no misunderstanding) that:

1. *The functions $P_e(t)=\mathsf{P}(\tau^e \geqslant t)$ and $P_s(t)=\mathsf{P}(\tau^s \geqslant t)$ satisfy for all x the inequalities*

$$P_i(t+x) \leqslant P_i(x)Q_i(t), \quad i=e,s, \tag{21}$$

for some functions $Q_i(t)$ with $Q_i(t)\to 0$ as $t\to\infty$.

We remark that (21) implies that

$$P_i(t+x) < cP_i(x)\,e^{-\alpha t}, \quad i=e,s \tag{22}$$

for all x and t and some $c<\infty$, $\alpha>0$. Therefore, we can immediately take $ce^{-\alpha t}$ as Q_i.

2. $\mathsf{M}\exp(\mu v^e) < \infty$ *for some* $\mu > 0$.
3. *$P_e(t)$ has an absolutely continuous component.*

4. *Let* $\mathfrak{Q}(t)$ *be the* σ*-algebra generated by the "history of the process"* $q(t)$ *on* $[0, t]$ (a more precise definition of the σ-algebra $\mathfrak{Q}(t)$ is in Remark a) of Part 2 of the proof of Theorem 6). *We assume that there exist* t_0 *and* $p>0$ *such that on the set* $\{q(t)\leqslant m\}$

$$\mathsf{P}_{\mathfrak{Q}(t)}(\inf_{u\leqslant t_0} q(t+u)=0)\geqslant p$$

(i.e., the state $q=0$ can be reached during time t_0 from any of the states $q=1, 2, \ldots, m$).

Theorem 6. *If Conditions 1–4 are met and* $m\mathsf{M}\tau^e-\mathsf{M}[\tau^s]=m\mathsf{M}\tau^e-\mathsf{M}\tau^s\mathsf{M}v^e>0$, *then for an arbitrary fixed initial value* $q(0)$ *the limiting distribution* $\{P_k\}$ *of the queue length* $q(t)$ *exists and*

$$|\mathsf{P}(q(t)=k)-P_k|<ce^{-\varepsilon t} \tag{23}$$

for some $c<\infty$ *and* $\varepsilon>0$. *Moreover, the probabilities* P_k *admit the representation*

$$P_k=\frac{1}{M}\int_0^\infty \mathsf{P}(q(u)=k\,;\, q(v)>0,\, v\in(0, u)\,|\, q(0)=0)\, du\,, \quad k=1, 2, \ldots, \tag{24}$$

where M *is the expectation of the time between exits of the system from the state* $q=0$.

Remark. Minimal conditions guaranteeing the convergence of (23) are, aside from the inequality $m\mathsf{M}\tau^e-\mathsf{M}[\tau^s]>0$, apparently Conditions 2, (22) and $\limsup_{|\lambda|\to\infty}|\mathsf{M}\exp(i\lambda\tau^e)-1|>0$.

Proof. 1. Let $t_1, t_2, \ldots$ be the exit times of $q(t)$ from the state $q=0$. Then the random variables $\tau_k=t_{k+1}-t_k$; $k=1, 2, \ldots$, are clearly i.i.d. Assume there exist $c<\infty$ and $\varepsilon>0$ (the letters c and ε will be used in this proof to designate several different pairs of constants) such that

$$\mathsf{P}(\tau>x)<ce^{-\varepsilon x} \quad \text{and} \quad \mathsf{P}(t_1>x)<ce^{-\varepsilon x}. \tag{25}$$

Then for $l>0$

$$\mathsf{P}(q(t)=l)=\mathsf{P}(t_1>t,\, q(t)=l)+\int_0^t \mathsf{P}(t_1\in du)\mathsf{P}(q(t-u)=l\,|\, q(0)=0)\,.$$

From this and (25) it is clear that we need only prove the theorem for $q(0)=0$. In this case $t_1=0$ and ($X_k=\sum_{j=1}^k \tau_j$; $H(t)$ is the renewal function for $\{X_k\}$)

$$\begin{aligned}\mathsf{P}(q(t)=l)&=\sum_{k=0}^\infty \int_0^t \mathsf{P}(X_k\in du)\\&\quad\times \mathsf{P}(q(t-u)=l;\, q(v)>0,\, v\in(0, t-u)\,|\, q(0=0)\\&=-\int_0^t dH(t-u)\mathsf{P}(q(u)=l;\, q(v)>0,\, v\in(0, u)\,|\, q(0)=0)\,.\end{aligned}$$

Denote the second factor in the last integral by $P_l(u)$. Clearly $P_l(u)<ce^{-\varepsilon u}$. Then by (25) and Condition 3 (see Appendix 1)

$$H(t)=\frac{t}{\mathsf{M}\tau}+c_1+\varepsilon(t)\,, \qquad \underset{(t,\,\infty)}{\text{Variation}}\, \varepsilon(u)<ce^{-\varepsilon t}$$

and

$$\mathsf{P}(q(t)=l)=\frac{1}{\mathsf{M}\tau}\int_0^t P_l(u)\,du-\int_0^t \mathsf{P}_l(u)\,d\varepsilon(t-u)$$

$$=\frac{1}{\mathsf{M}\tau}\int_0^\infty \mathsf{P}_l(u)\,du+r(t)\,,$$

where, obviously, $r(t)<ce^{-\varepsilon t}$. The assertion of the theorem for $l=0$ is proved analogously.

2. Thus, to prove the theorem we must demonstrate the validity of (25). These relations are quite natural and seem to allow a simpler proof than that given in abbreviated form below. As a preliminary we make two remarks:

a) We extend $q(t)$ to a Markov process, adjoining to it as additional coordinates the "defect" variables $\gamma_j(t); j=0, \ldots, m$, where $\gamma_0(t)$ is the "defect" up to the level t in a random walk with jumps $\tau_1^e, \tau_2^e, \ldots$; $\gamma_j(t)$ for $j\geqslant 1$ is the time spent by the j-th channel in serving the current request up to time t. Then the process $Q(t)=(q(t), \gamma_0(t), \ldots, \gamma_m(t))$ is clearly Markov. If $\mathfrak{Q}(t)$ is the σ-algebra generated by its trajectories on $[0, t]$, then it's easy to see that Condition 1 implies that for the corresponding "overshoot" variables $\chi_j(t)$ through the level t for $j=0, \ldots, m$ ($\chi_j(t)$ is the time spent by the j-th channel servicing the current request after time t) we have

$$\mathsf{P}_{\mathfrak{Q}(t)}(\chi_j(t)>x)\leqslant ce^{-\varepsilon x} \tag{26}$$

(the left side is in fact measurable w.r.t. the σ-algebra generated by the $\gamma_j(t)$). The inequality (26) remains valid under our conditions if t is replaced by an arbitrary Markov time θ ($\{\theta\leqslant t\}\in\mathfrak{Q}(t)$).

b) Let u_k, $k=1, 2, \ldots$, be the times at which $q(t)$ leaves the states $q\leqslant m$ and $u_k+\zeta_k$ the first time the system reënters these states after time u_k.

Noting that the u_k are Markov times, we have for the case $v^e\equiv 1$

$$\mathsf{P}_{\mathfrak{Q}(u_k)}(\zeta_k>t)\leqslant \mathsf{P}_{\mathfrak{Q}(u_k)}(1+n^e(t-\chi_0)-n_1^s(t-\chi_1)-\cdots-n_m^s(t-\chi_m)>0)\,, \tag{27}$$

where $\chi_j=\chi_j(u_k)$ has the property (26) and $n^e(t)$ and $n_j^s(t)$, $j=1, 2, \ldots, m$, are independent renewal processes for the sequences $\tau_1^e, \tau_2^e, \ldots$ and $\tau_1^s, \tau_2^s, \ldots$, resp. ($n^e(t)=k$ if $\sum_{j=1}^k \tau_j^e<t\leqslant\sum_{j=1}^{k+1}\tau_j^e$; $n_j^s(t)$ for the j-th service channel is defined analogously). From (27) it is easy to conclude that $\mathsf{P}_{\mathfrak{Q}(u_k)}(\zeta_k>t)<ce^{-\varepsilon t}$. It is also easy to see that this inequality remains valid for any v^e satisfying Condition 2.

We have thus established that the region $q\leqslant m$ is visited "exponentially" often uniformly w.r.t. the previous history. Furthermore, at each such visit there is, by Condition 4, a positive probability that the state $q=0$ is entered during time t_0. This implies the validity of (25). □

Without essential changes one can prove

Theorem 6A. *Assume that the conditions of Theorem* 6 *are fulfilled with Condition* 3 *replaced by the requirement that* τ^e *be integer-valued and have greatest common divisor of its possible values equal to* 1. *Then for an arbitrary initial value* $q(0)$

$$\lim_{n\to\infty}\mathsf{P}(q(n)=k)=P_k \quad \textit{exists and} \quad |\mathsf{P}(q(n)=k)-P_k|<ce^{-\varepsilon n}\,.$$

The probabilities P_k, $k=1, 2, \ldots$, *admit the representation*

$$P_k = \frac{1}{M}\sum_{n=1}^{\infty} \mathsf{P}(q(n)=k;\, q(l)>0,\, l=1,2,\ldots,n-1 \mid q(0)=0)\,,$$

and the constants c, ε *and* M *have the same meaning as in Theorem* 6.

One can prove in an analogous way

Theorem 7. *Assume the distribution of* τ_j^s *satisfies Condition* 1 *and that Conditions* 2 *and* 4 *are also fulfilled but in such a way that the numbers of steps through which* q_n *can return to the state* 0 *have greatest common divisor equal to* 1. *Then, if* $m\mathsf{M}\tau^e - \mathsf{M}[\tau^s] > 0$, *we have for an arbitrary fixed value of* q_1

$$|\mathsf{P}(q_n=k) - \mathsf{P}(q^0=k)| < c e^{-\varepsilon n}$$

and

$$|\mathsf{P}(\mathbf{w}_n < x) - \mathsf{P}(\mathbf{w}^0 < \mathbf{x})| < c e^{-\varepsilon n}$$

for some $c<\infty$ *and* $\varepsilon>0$. *Furthermore, representations analogous to* (24) *hold.*

Here, as in the proof of Theorem 6, it is obviously necessary to consider independent cycles of entry into $q=0$ ($\mathbf{w}=0$). Changes in the proof which thereby appear are of a simplifying character since the time here is discrete. Additional conditions on the distribution of τ^e (like Condition 1) are obviously unnecessary.

If $m\mathsf{M}\tau^e = \mathsf{M}[\tau^s]$, then the recurrence time of q_n to the state $q=0$ has infinite expectation so that $\mathsf{P}(q_n=0)\to 0$ as $n\to\infty$. This implies that $\mathsf{P}(q_n=k)\to 0$ for each fixed k.

Thus, under the general conditions formulated at the beginning of Subsection 5 the rate of convergence of the distributions of queue length and waiting time to stationarity is exponential ($c\,e^{-\varepsilon t}$). However, this is a rather qualitative result since the practical use of the obtained estimates is almost impossible without knowledge of the dependence of the constants c and ε on the number of channels and on the distribution of the terms of the governing sequences. In this connection it would no doubt be of interest to find at least rough, but effective, estimates for c and ε under more special conditions.

§ 29. The Systems $\langle G_I, 1, E/m, 1\rangle$ and $\langle E, G_I, G_I/m, 1\rangle$

1. For the systems $\langle G_I, 1, E/m, 1\rangle$ the sequence q_n of queue lengths before the appearance of the n-th request is Markovian (see Subsection 8 § 27). This fact allows us to find explicit formulae for the stationary distribution. In contrast to the case $m=1$, the systems $\langle E, 1, G_I/m, 1\rangle$ do not possess this property. For them one can find no natural one-dimensional system characteristic which would have a sufficiently simple probabilistic nature as a stochastic process. Regarding the systems $\langle E, G_I, G_I/m, 1\rangle$, we merely remark here that for them, as well as for the systems $\langle E, 1, G_I/m, 1\rangle$, one has

Theorem 8. *For the systems $\langle E, 1, G_I/m, 1\rangle$ and all k*

$$\lim_{n\to\infty} \mathsf{P}(q_n=k) = \lim_{t\to\infty} \mathsf{P}(q(t)=k)\,. \tag{28}$$

Proof. For $k\geqslant 2$ this follows at once from Theorems 4 and 5 since in our case the distributions of γ^e and τ^e coincide. For $k=0$ one compares the statement of Theorem 5 with the remark on Theorem 4. □

2. Let us therefore consider the systems $\langle G_I, 1, E/m, 1\rangle$. That is, we assume that $v^e\equiv v^s\equiv 1$ and

$$\mathsf{P}(\tau^s>x)=e^{-\alpha x}, \quad x\geqslant 0\,.$$

Furthermore, let $m\mathsf{M}\tau^e-\mathsf{M}\tau^s>0$ (or, what is the same, $\alpha m\mathsf{M}\tau^e>1$). Then, as we have already seen (Theorem 7), for an arbitrary initial value q_1

$$\lim_{n\to\infty} \mathsf{P}(q_n=k)=p_k\,,$$

exists and the rate of convergence to the limit is exponential. In this case we can find the explicit form of the distribution $\{p_k\}$.

Theorem 9.

$$p_k=\begin{cases}\sum_{j=k}^{m-1}(-1)^{j-k}\binom{j}{k}U_j, & k=0,1,\dots,m-1\,;\\ A\mu^{k-m}, & k=m, m+1, \dots\,.\end{cases}$$

where μ is the unique root of the equation

$$\mu=\psi((\mu-1)m\alpha)\,, \qquad \psi(\mu)=\mathsf{M}\exp(\mu\tau^e)$$

in the domain $|\mu|<1$,

$$U_j=AC_j\sum_{l=j+1}^{m}\frac{\binom{m}{l}}{C_l(1-\psi_l)}\,\frac{m(1-\psi_l)-l}{m(1-\mu)-l},$$

$$C_j=\prod_{l=1}^{j}\frac{\psi_l}{1-\psi_l}, \quad \psi_j=\psi(-j\alpha) \tag{29}$$

and

$$A=\left[\frac{1}{1-\mu}+\sum_{j=1}^{m}\frac{\binom{m}{j}}{C_j(1-\psi_j)}\,\frac{m(1-\psi_j)-j}{m(1-\mu)-j}\right]^{-1}.$$

Proof. By the properties of the exponential distribution the sequence $\{q_n\}$ forms a Markov chain with transition probabilities

$$p_{ij} = \mathsf{P}(q_{n+1} = j \mid q_n = i) = \int d\mathsf{P}(\tau^e < x)\pi_{ij}(x), \tag{30}$$

where $\pi_{ij}(x)$ is the conditional transition probability given that the corresponding interval between arrivals of requests τ^e equals x:

$$\pi_{ij}(x) = \begin{cases} 0 & \text{for } j > i+1; \\ e^{-m\alpha x}\dfrac{(m\alpha x)^{i-j+1}}{(i-j+1)!} & \text{for } i \geqslant m, j \geqslant m; \\ \dbinom{m}{j} e^{-j\alpha x}\left[\displaystyle\int_0^x \frac{(m\alpha y)^{i-m}}{(i-m)!}(e^{-\alpha y} - e^{-\alpha x})^{m-j}\alpha m\, dy\right] & \\ & \text{for } i \geqslant m, j < m; \\ \dbinom{i+1}{j} e^{-j\alpha x}(1-e^{-\alpha x})^{i-j+1} & \text{for } i < m. \end{cases} \tag{31}$$

These formulae can be obtained with the aid of the equality

$$\pi_{i,i+1}(x) = e^{-(i+1)\alpha x} \quad \text{for } i < m$$

and the recursion formulae

$$\pi_{ij}(x) = \int_0^x e^{-(i+1)\alpha y}(i+1)\alpha\pi_{i-1,j}(x-y)\, dy \quad \text{for } i+1 \leqslant m$$

and

$$\pi_{ij}(x) = \int_0^x e^{-m\alpha y} m\alpha\pi_{i-1,j}(x-y)\, dy \quad \text{for } i+1 \geqslant m .$$

It is clear that the chain under consideration is aperiodic and irreducible. Thus, the $\{p_k\}$ are the unique solution of the system of equations

$$p_j = \sum p_i p_{ij}, \qquad j = 0, 1, \ldots; \quad \sum p_j = 1 . \tag{32}$$

These equations have for $j \geqslant m$ the form

$$p_j = \sum_{k=0}^{\infty} p_{j+k-1} \int_0^\infty e^{-m\alpha x}\frac{(m\alpha x)^k}{k!}\, d\mathsf{P}(\tau^e < x) . \tag{33}$$

For $p_{m-1}, p_m, \ldots$, this system coincides, except for the missing equation for p_{m-1}, with that for the stationary distribution of a random walk on the positive real-axis which is continuous from above, has a delaying barrier at the point $m-1$ and has jump sizes whose generating function is

$$\pi(z) = \sum_{k=-\infty}^{1} z^k \int_0^\infty e^{-m\alpha x}\frac{(m\alpha x)^{-k+1}}{(-k+1)!}\, d\mathsf{P}(\tau^e < x) = z\psi\left(m\alpha\left(\frac{1}{z} - 1\right)\right). \tag{34}$$

It is not difficult to verify that the system of equations for the distribution of the supremum of random variables with distribution (34) will also have the form (33). From the results of § 14 Chapter 2 it follows that the solution of (33) will be of the form

$$p_j = A\mu^{j-m}, \quad j = m, m+1, \ldots .$$

Putting this into (33) we obtain

$$A=p_{m-1}\psi(-m\alpha)+\frac{A}{\mu}\left[\psi(m\alpha(\mu-1))-\psi(-m\alpha)\right],$$

or $A=\mu p_{m-1}$. We now determine the constant A and the probabilities $p_{m-1}, \ldots, p_0$. Let

$$U(z)=\sum_{k=0}^{m-1} p_k z^k .$$

Then $U(1)=1-\sum_{k=m}^{\infty} p_k=1-A/(1-\mu)$ and by (32)

$$\begin{aligned}U(z)=&\int_0^\infty (1-e^{-\alpha x}+z\,e^{-\alpha x})U(1-e^{-\alpha x}+z\,e^{-\alpha x})\,d\mathsf{P}(\tau^e<x)\\&+A\int_0^\infty \left[\int_0^x e^{m\alpha\mu y}(e^{-\alpha y}-e^{-\alpha x}+z\,e^{-\alpha x})^m m\alpha\,dy\right]d\mathsf{P}(\tau^e<x)-Az^m .\end{aligned} \tag{35}$$

Now let

$$U_j=\frac{1}{j!}\left(\frac{d^jU(z)}{dz^j}\right)_{z=1} .$$

Then $U_0=1-A/(1-\mu)$. Differentiating (35) j times and setting $z=1$ we get the equations

$$U_j=U_j\psi_j+U_{j-1}\psi_j-A\binom{m}{j}\frac{m(1-\psi_j)-j}{m(1-\mu)-j},$$

$j=0, 1, \ldots, m-1$, or, what is the same,

$$U_j=\frac{\psi_j}{1-\psi_j}U_{j-1}-\frac{A\binom{m}{j}}{1-\psi_j}\,\frac{m(1-\psi_j)-j}{m(1-\mu)-j}. \tag{36}$$

Putting $C_0=1$, $C_j=\prod_{k=1}^{j}\psi_k/(1-\psi_k)$, we can write (36) as

$$\frac{U_j}{C_j}=\frac{U_{j-1}}{C_{j-1}}-\frac{A\binom{m}{j}}{C_j(1-\psi_j)}\,\frac{m(1-\psi_j)-j}{m(1-\mu)-j}. \tag{37}$$

Using the fact that $U_{m-1}=p_{m-1}=A/\mu$ and summing (37) for $j=r+1, \ldots, m-1$ we get for $r=0, 1, \ldots, m-1$

$$\frac{U_r}{C_r}=A\sum_{j=r+1}^{m}\frac{\binom{m}{j}}{C_j(1-\psi_j)}\cdot\frac{m(1-\psi_j)-j}{m(1-\mu)-j}.$$

Since $U_0/C_0=1-A/(1-\mu)$, this implies (29). The probabilities p_k can now be found by the formula

$$p_k=\frac{1}{k!}\left(\frac{d^kU(z)}{dz^k}\right)_{z=0}=\sum_{r=k}^{m-1}(-1)^{r-k}\binom{r}{k}U_r ,$$

which holds because of the equality

$$U(z)=\sum_{j=0}^{m-1}U_j(z-1)^j . \quad \square$$

3. The theorem just proved also allows us to find the limiting distribution of $q(t)$ when $t \to \infty$.

Theorem 10. *If τ^e is nonlattice and $m\alpha \mathsf{M}\tau^e > 1$, then for an arbitrary initial value $q(0)$*

$$\lim_{t \to \infty} \mathsf{P}(q(t)=k) = P_k$$

exists with

$$P_k = \frac{p_{k-1}}{k\alpha \mathsf{M}\tau^e}, \quad k=1,2,\ldots,m-1,$$

and (38)

$$P_k = \frac{p_{k-1}}{m\alpha \mathsf{M}\tau^e}, \quad k=m,m+1,\ldots,$$

where the p_k are given by the preceding theorem.

If τ^e is integer-valued with g.c.d. of its possible values equal to 1, then the assertion of the theorem will hold for the sequence $q(n)$ when $n \to \infty$.

Corollary. *For the systems $\langle E, 1, E/m, 1\rangle$*

$$p_k = \frac{p_0}{k!(\alpha \mathsf{M}\tau^e)^k} \qquad \text{for } k=1,2,\ldots,m,$$

and (39)

$$p_k = \frac{p_0}{m!(\alpha \mathsf{M}\tau^e)^k m^{k-m}} \quad \text{for } k=m,m+1,\ldots.$$

In fact, on the basis of Theorem 8, $p_k = P_k$ for such systems and the relations (38) reduce to simple recursion equations whose solution is (39).

Proof of the theorem. By the total probability formula ($X_k^e = \sum_{j=1}^k \tau_j^e$)

$$P_j(t) = \mathsf{P}(q(t)=j)$$
$$= \sum_{k=0}^{\infty} \int_0^t \sum_{r=j-1}^{\infty} d\mathsf{P}(X_k^e < u, q_k = r)\mathsf{P}(\tau^e > t-u)\pi_{rj}(t-u),$$

where the $\pi_{rj}(x)$ are defined in (31). Here,

$$H_r(u) = \sum_{k=0}^{\infty} \mathsf{P}(X_k^e < u, q_k = r)$$

is the expected number of times the sequence q_n is in the state $E_r = \{q=r\}$ during time u or (what is the same w.p.1), of the number of transitions during this time of the process $q(t)$ from E_r into E_{r+1}. But the integer-valued intervals $\nu_1, \nu_2, \ldots$ between entries of q_n into E_r are i.i.d. and have expectation $\mathsf{M}\nu_j = 1/p_r$. The time intervals between transitions $E_r \to E_{r+1}$ of the process $q(t)$ are clearly distributed like the sums $\zeta = \tau_1^e + \cdots + \tau_{\nu_1}^e$ of a random number of random variables, where ν_1 "does not depend on the future" (the event $\{\nu_1 \leqslant n\}$ and the random variables $\tau_{n+1}^e, \tau_{n+2}^e, \ldots$ are independent). Thus, by Wald's identity $\mathsf{M}\zeta = \mathsf{M}\tau^e/p_r$, and by the renewal theorem it follows that as $u \to \infty$ and for arbitrary $h > 0$

$$\frac{H_r(u)}{u} \to \frac{p_r}{\mathsf{M}\tau^e}, \qquad \frac{H_r(u+h) - H_r(u)}{h} \to \frac{p_r}{\mathsf{M}\tau^e}.$$

Consequently (see Appendix 1),

$$P_j(t)=\sum_{r=j-1}^{\infty}\int_0^t \mathsf{P}(\tau^e>t-u)\pi_{rj}(t-u)\,dH_r(u)$$
$$\to\frac{1}{\mathsf{M}\tau^e}\sum_{r=j-1}^{\infty}\int_0^{\infty} p_r\mathsf{P}(\tau^e>u)\pi_{rj}(u)\,du\,. \tag{40}$$

We will now find relations between P_k and p_{k-1}. Let $N_k(t)$ be the expectation of the number of transitions $E_k\to E_{k-1}$ of the process $q(u)$ on the interval $[0,t]$. Then, clearly

$$N_k(t)=\begin{cases} k\alpha\int_0^t P_k(u)\,du\,, & k\leqslant m\,,\\ m\alpha\int_0^t P_k(u)\,du\,, & k\geqslant m\,.\end{cases}$$

By the foregoing, the limits

$$\lim_{t\to\infty}\frac{N_k(t)}{t}=\begin{cases} k\alpha P_k\,, & k\leqslant m\,,\\ m\alpha P_k\,, & k\geqslant m\end{cases}$$

exist. But $N_k(t)$ differs from $H_{k-1}(t)$ by no more than 1. Consequently,

$$\lim_{t\to\infty}\frac{N_k(t)}{t}=\lim_{t\to\infty}\frac{H_{k-1}(t)}{t}\,,$$

or, what is the same,

$$\frac{p_{k-1}}{\mathsf{M}\tau^e}=\begin{cases} k\alpha P_k\,, & k\leqslant m\,,\\ m\alpha P_k\,, & k\geqslant m\,.\end{cases}$$

The proof of the theorem in the case where τ^e is integer-valued goes quite analogously. □

4. *The Waiting Time.* Assume, in accordance with the notation of § 27, that $w_{n,1}$ is the time the n-th request waits until the start of service. Then, as we have seen, for an arbitrary initial condition

$$W(x)=\lim_{n\to\infty}\mathsf{P}(w_{n,1}>x)$$

exists and the rate of convergence is exponential.

Theorem 11. *If* $\alpha m\mathsf{M}\tau^e>1$, *then in the notation of Theorem 9*

$$W(x)=\frac{A\,e^{-m\alpha(1-\mu)x}}{1-\mu}\,.$$

Proof. By the total probability formula

$$\mathsf{P}(w_{n,1}>x)=\sum_{j=m}^{\infty}\mathsf{P}(q_n=j)\sum_{r=0}^{j-m}e^{-\alpha mx}\frac{(\alpha mx)^r}{r!}\,.$$

It remains to pass to the limit and substitute the values of p_k, $k\geqslant m$, found in Theorem 9. □

5. We turn now to the *discrete* analogue of the systems $\langle G_I, 1, E/m, 1\rangle$, where τ^e is integer-valued and $\mathsf{P}(\tau^s=k)=(1-p)p^{k-1}$, $k=1, 2, \ldots; 1>p>0$.

Let q_n be the queue length before the arrival of the n-th customer and after termination of service if the moment of this termination coincides with the arrival of the n-th customer. The sequence $\{q_n\}$ for the systems under consideration also forms a Markov chain. The transition probabilities p_{ij} of this chain are defined by the relations $p_{ij}=\sum_{k=1}^{\infty} \mathsf{P}(\tau^e=k)\pi_{ij}(k)$, where $\pi_{ij}(k)$ is the probability that the queue q_n goes from state i to j, given that the interval between the arrivals of the n-th and $(n+1)$-st customers equals k. We find that

$$\pi_{ij}(k)=\begin{cases} 0 & \text{for } j>i+1 \\ \mathsf{P}(\nu_1+\cdots+\nu_{km}=i-j+1) & \text{for } i\geqslant m, j\geqslant m\,, \end{cases}$$

where the ν_j are independent random variables and $\mathsf{P}(\nu_j=0)=1-\mathsf{P}(\nu_j=1)=p$. Thus, for the indicated values of the indices i and j

$$\pi_{ij}(k)=\binom{km}{i-j+1}(1-p)^{i-j+1}p^{km-(i-j+1)}\,.$$

If $i<m$, then

$$\pi_{ij}(k)=\binom{i+1}{j}\mathsf{P}^j(\nu_1+\cdots+\nu_k=0)\mathsf{P}^{i-j+1}(\nu_1+\cdots+\nu_k\geqslant 1)$$

$$=\binom{i+1}{j}p^{kj}(1-p^k)^{i-j+1}\,.$$

Finally, for $i\geqslant m$, $j<m$ we use the total probability formula w.r.t. the first entry time l of the queue into the region $q<m$ and get

$$\pi_{ij}(k)=\textstyle\sum_{l=1}^{k}\sum_{r=0}^{m-1} \mathsf{P}(N_1+\cdots+N_{l-1}\leqslant i+1-m,$$
$$N_1+\cdots+N_l=i+1-r)\pi_{r-1,j}(k-l)\,,$$

where $N_{j+1}=\nu_{mj+1}+\nu_{mj+2}+\cdots+\nu_{m(j+1)}$. We can also write

$$\pi_{ij}(k)=\textstyle\sum_{l=1}^{k-1}\sum_{r=m}^{2m-1}\sum_{s>r-m}\pi_{ir}(l)\mathsf{P}(N_1=s)\pi_{r-s-1,j}(k-l-1)\,.$$

Since the random walk $\{q_n\}$ is continuous from above (see Chapter 2), we find, as in Theorem 9, that for $j\geqslant m$

$$p_j=A\mu^{j-m}\,,$$

where μ is the unique root in the domain $|\mu|<1$ of the equation

$$\mu=\psi([\mu(1-p)+p]^m)\,,\qquad \psi(z)=\mathsf{M}z^{\tau^e}$$

(the generating function for the jumps of the walk q_n in the region $q\geqslant m$ equals

$$\textstyle\sum_{n=-\infty}^{1} z^n\sum_{k=1}^{\infty}\mathsf{P}(\tau^e=k)\mathsf{P}(\nu_1+\cdots+\nu_{km}=-n+1)$$
$$=z\textstyle\sum_{k=1}^{\infty}\mathsf{P}(\tau^e=k)[p+z^{-1}(1-p)]^{mk}=z\psi([p+z^{-1}(1-p)]^m))\,.$$

The quantities $p_0, \ldots, p_{m-1}$ can be found as in Subsection 2 or directly, by solving the finite system of equations $p_j=\sum_{i=0}^{m} p_i p_{ij}$, $j=0, 1, \ldots, m-1$, where $p_m=A$.

Chapter 6

The Systems $\langle G, G, G/\infty, 1\rangle$ with an Infinite Number of Service Channels

§ 30. Theorems on Convergence to Stationary Processes

1. Here the waiting time is clearly always equal to 0, and we will characterize the state of the system by the number of busy channels. Suppose, as before, that q_n is the number of busy channels at the moment of arrival of the n-th group of requests. For simplicity we will assume that $q_1 = 0$. The reader can carry over almost all of the results of this chapter without great difficulty to the case of arbitrary initial conditions.

We denote by $I(A)$ the indicator of the event A and consider first the case $\nu^e \equiv 1$.

Theorem 1. *Assume the sequence* $\{\tau_j^e, \tau_j^s; -\infty < j < \infty\}$ *is strictly stationary and that the sequence* $\{\tau_j^e\}$ *is also metrically transitive.*

Then, if $\mathsf{M}\tau^s < \infty$, *the distribution of the sequence* $\{q_{n+k}; k \geqslant 0\}$ *converges monotonically for* $n \to \infty$ *to that of the proper stationary sequence*

$$q^k = I(\tau_k^s > \tau_k^e) + I(\tau_{k-1}^s > \tau_{k-1}^e + \tau_k^e) + I(\tau_{k-2}^s > \tau_{k-2}^e + \tau_{k-1}^e + \tau_k^e) + \cdots. \quad (1)$$

Proof. We have

$$q_n = I(\tau_1^s > \tau_1^e + \cdots + \tau_{n-1}^e) + I(\tau_2^s > \tau_2^e + \cdots + \tau_{n-1}^e) + \cdots + I(\tau_{n-1}^s > \tau_{n-1}^e). \quad (2)$$

Using the stationarity, we find that

$$q_{n+k} \underset{d}{=} q_{n,k} \equiv I(\tau_k^s > \tau_k^e) + I(\tau_{k-1}^s > \tau_{k-1}^e + \tau_k^e) + \cdots + I(\tau_{-n+2}^s > \tau_{-n+2}^e + \cdots + \tau_k^e).$$

Hence, $q_{n,k} \uparrow q^k$ as $n \to \infty$. It remains to show that q^0 is a proper random variable. Indeed, for some $b > 0$

$$\begin{aligned} \mathsf{P}(q^0 > N) &\leqslant \mathsf{P}\left(\textstyle\bigcup_{j=N}^{\infty} \{I(\tau_{-j}^s > \tau_{-j}^e + \cdots + \tau_0^e) = 1\}\right) \\ &\leqslant \mathsf{P}\left(\textstyle\bigcup_{j=N}^{\infty} \{\tau_0^e + \cdots + \tau_{-j}^e < b(j+1)\}\right) + \mathsf{P}\left(\textstyle\bigcup_{j=N}^{\infty} \{\tau_{-j}^s > b(j+1)\}\right). \end{aligned} \quad (3)$$

The first term here converges for suitably chosen $b > 0$ to 0 as $N \to \infty$ by the strong law of large numbers. The second is majorized by the sum

$$\sum_{j=N}^{\infty} \mathsf{P}(\tau_0^s > b(j+1))$$

and also converges to 0 since $\mathsf{M}\tau_0^s < \infty$. □

Theorem 2. *Assume in addition to the conditions of Theorem 1 that the sequence* $\{\tau_j^s\}$ *consists of independent random variables not depending on* $\{\tau_j^e\}$, *and that* $\mathsf{M}\tau^e < \infty$. *Then the condition* $\mathsf{M}\tau^s < \infty$ *is necessary and sufficient for the finiteness of* q^k.

Proof. For large enough L and preassigned $\varepsilon > 0$ we have by the strong law of large numbers

$$\mathsf{P}\left(\bigcap_{j=0}^{\infty} \{\tau_{-j}^e + \cdots + \tau_0^e < 2\mathsf{M}\tau^e(j+1) + L\}\right) \geqslant 1 - \varepsilon .$$

Thus, using the 0-1-criterion of Borel we can write, by (1), that $\mathsf{P}(q^0 = \infty) \geqslant 1 - \varepsilon$ if

$$\sum_{j=0}^{\infty} \mathsf{P}(\tau_{-j}^s > 2(j+1)\mathsf{M}\tau^e + L) = \infty .$$

This series diverges if $\mathsf{M}\tau^e = \infty$. This proves the necessity. □

If $\mathsf{M}\tau^e = \infty$, then the assertion of Theorem 2 obviously does not hold. In this case more precise conditions on the distributions of τ^e and τ^s guaranteeing the finiteness of $\{q^k\}$ are of interest.

Under the conditions of Theorem 1, it is not difficult to estimate the *rate of convergence* of the distributions of the sequences $\{q_{n+k}\}$ and $\{q^k\}$. For example, for the one-dimensional distributions ($q_n \underset{d}{=} q_{n,0}$)

$$\mathsf{P}(q_{n,0} \neq q^0) = \mathsf{P}\left(\bigcup_{j=n-1}^{\infty} \{\tau_{-j}^s > \tau_{-j}^e + \cdots + \tau_0^e\}\right). \tag{4}$$

To estimate this expression we can use Relations (3) or the majorant

$$\sum_{j=n-1}^{\infty} \mathsf{P}(\tau_{-j}^s > \tau_{-j}^e + \cdots + \tau_0^e). \tag{5}$$

If the τ_j^e are independent and also don't depend on $\{\tau_j^s\}$, then (5) does not exceed

$$\sum_{j=n-1}^{\infty} \mathsf{M}\, e^{\lambda\tau^s}(\mathsf{M}\, e^{-\lambda\tau^e})^{j+1} = \frac{\mathsf{M}\, e^{\lambda\tau^s}(\mathsf{M}\, e^{-\lambda\tau^e})^n}{1 - \mathsf{M}\, e^{-\lambda\tau^e}} \tag{6}$$

for arbitrary $\lambda > 0$.

From this one can easily obtain a precise exponential estimate for the rate of convergence provided that the distribution of τ^s satisfies Cramér's condition ($\mathsf{M}\, e^{\lambda\tau^s} < \infty$ for some $\lambda > 0$).

2. All of what was said in Subsection 1 will, as we shall see, remain valid for more precise characteristics of the state of the system and can be generalized without difficulty to systems with arbitrarily distributed ν_j^e. In the latter case, as in Subsection 5 § 27 of the last chapter, it will be convenient to give the governing sequence as $\{\tau_n^e, \nu_n^e, \tau_n^s\}$, where τ_n^s is the vector of service times of the n-th group.

The sum of the components $\tau^s_{n,i}$ of this vector will be denoted as before by

$$[\tau^s_n]=\sum_{i=1}^{\nu^e_n}\tau^s_{n,i}\,.$$

Let us introduce the processes

$$q_n(x)=\sum_{j=1}^{\nu^e_1}I(\tau^s_{1,j}>\tau^e_1+\cdots+\tau^e_{n-1}+x)+\cdots$$
$$+\sum_{j=1}^{\nu^e_{n-1}}I(\tau^s_{n-1,j}>\tau^e_{n-1}+x)\,, \tag{7}$$

so that $q_n(0)=q_n$. The process $\{q_n(x);\,x\geq 0\}$ is obviously nonincreasing and has the form of jumps. It can be described uniquely by the initial value $q_n(0)=q_n$ and the positions $x_1,\ldots,x_{q_n}$ of the points of isolated unit jumps. The vector $(x_1,\ldots,x_{q_n})$ is the vector of times during which requests arriving in the system before the arrival of the n-th group are still being served. The variable $q_n(x)$ itself *indicates the number of requests, out of those appearing before the arrival time t_n of the n-th group, which remain in the system longer than time x after t_n.*

The corresponding stationary sequence of processes $q^k(x)$ will be defined as

$$q^k(x)=\sum_{j=1}^{\nu^e_k}I(\tau^s_{k,j}>\tau^e_k+x)+\sum_{j=1}^{\nu^e_{k-1}}I(\tau^s_{k-1,j}>\tau^e_{k-1}+\tau^e_k+x)+\cdots.$$

As before, one has

$$\mathsf{P}(q^0(x)>\sum_{j=-N+1}^{0}\nu^e_j)\leq\mathsf{P}(\textstyle\bigcup_{j=N}^{\infty}\{\tau^e_0+\cdots+\tau^e_{-j}<b(j+1)\})$$
$$+\mathsf{P}(\textstyle\bigcup_{j=N}^{\infty}\{\max_k\tau^s_{-j,k}>b(j+1)\})\,.$$

Since $\max_k\tau^s_{0,k}\leq[\tau^s_0]$ we obtain

Theorem 3. *Assume the sequence $\{\tau^e_n,\nu^e_n,\boldsymbol{\tau}^s_n;\,-\infty<n<\infty\}$ is strictly stationary and that $\{\tau^e_j\}$ is also metrically transitive.*

Then, if $\mathsf{M}[\boldsymbol{\tau}^s]<\infty$, the distribution of the sequence of processes $\{q_{n+k}(x);\,k\geq 0,\,x\geq 0\}$ converges monotonically as $n\to\infty$ to the distribution of the proper stationary (w.r.t. k) process $\{q^k(x);\,k\geq 0,\,x\geq 0\}$.

(More precisely: there exist processes $\{\tilde q^k(x)=\tilde q^k_n(x);\,k\geq 0,\,x\geq 0\}$, $n=1,2,\ldots$, distributed for all n like $\{q^k(x);\,k\geq 0,\,x\geq 0\}$ and such that $q_{n+k}(x)\leq\tilde q^k_n(x)$, $U^{-n-k}q_{n+k}\uparrow U^{-n-k}\tilde q^k_n$ as $n\to\infty$ (U is defined in § 3), *and*

$$\mathsf{P}(\textstyle\bigcup_{k=0}^{\infty}\bigcup_{x\geq 0}\{q_{n+k}(x)\neq\tilde q^k_n(x)\})\downarrow 0\,.)$$

In an analogous way one can establish

Theorem 4. *If for the systems $\langle G_I,G_I,G_I/\infty,1\rangle$ under the conditions of Theorem 3 $\mathsf{M}\tau^e<\infty$, then the condition $\mathsf{M}\max_j\tilde\tau^s_{k,j}<\infty$ is necessary and sufficient for the finiteness of $q^k=q^k(0)$.*

To prove this one uses the computations of Theorem 2. □

The estimate obtained above of the rate of convergence remains valid if τ^s_k is replaced by $\max_j\tau^s_{k,j}$ or $[\boldsymbol{\tau}^s]$ in (5) and (6).

3. We remark that the representation (2) allows us to obtain theorems on the convergence of the sequence $\{q_{n+k}\}$ as $n\to\infty$ to stationarity under *minimal* assumptions when the sequences

$$\{\tau^e_{n+j}, \tau^s_{n+j}; j\geq 0\}. \tag{8}$$

converge to a stationary sequence as $n\to\infty$.

Indeed, with the help of relations of the form (3) we can show directly, for example, the following: If the joint finite-dimensional distributions of the sequence (8) converge weakly to the corresponding distributions of the proper stationary metrically transitive sequence $\{\tau^e_{(j)}, \tau^s_{(j)}\}$, so that $\mathsf{M}\tau^s_{(j)}<\infty$ and

$$\mathsf{P}(\tau^s_n > N) < p(N), \quad \text{where} \quad \int_0^\infty p(N)\, dN < \infty, \tag{9}$$

then as $n\to\infty$

$$\mathsf{P}(q_n = k) \to \mathsf{P}(q^0 = k),$$

where q^0 is constructed w.r.t. the sequence $\{\tau^e_{(j)}, \tau^s_{(j)}\}$. The same conditions are clearly sufficient for the convergence of arbitrary finite-dimensional distributions of $\{q_{n+k}\}$.

§ 31. Stability Theorems

1. We consider the systems $\langle G, G, G/\infty, 1\rangle$ governed by the stationary sequence $\{\tau^e_n, \nu^e_n, \boldsymbol{\tau}^s_n\}$, where $\boldsymbol{\tau}^s_n$ is the vector of service times for requests in the n-th lot. Let the vector $(q_1; \rho_1, \ldots, \rho_{q_1})$ denote the initial conditions, where $\rho_1, \ldots, \rho_{q_1}$ are the service times of requests present in the system at $t=0$. The vector $(q_1; \rho_1, \ldots, \rho_{q_1})$ is assumed given on the basic probability space.

We will be interested in conditions for the stability (or continuous dependence) of the stationary distribution of $\{q^k\}$ under a change of the distribution of the governing sequence. This question was considered for the systems $\langle G, G, G, 1\rangle$ in § 11.

To illustrate the problem to be examined under the conditions of this chapter we treat the following example. As we will show in § 33, the distribution of q^k for the systems $\langle E, 1, G_I/\infty, 1\rangle$ is Poisson. Will the distribution of q^k for such systems be "close" to Poisson if

$$\mathsf{P}(\tau^e \geq x) = P_{p,r}(x) = (1-p)\, e^{-\alpha x} + p\delta_r(x)$$

where p is small and

$$\delta_r(x) = \begin{cases} 1, & x\leq r \\ 0, & x>r\,? \end{cases}$$

A similar question can be asked regarding systems "close" to $\langle G_I, 1, E/m, 1\rangle$ (see § 34), when

$$\mathsf{P}(\tau^s \geq x) = P_{p,r}(x).$$

In the first case the answer to the question of stability of the stationary distribution of q^k for small p turns out to be affirmative. In the second case it depends on the number r, more precisely, on the value of the product pr.

We turn first to the simpler systems for which $v^e \equiv 1$ and $q_1=0$.

2. *Stability Theorems for the Systems* $\langle G, 1, G/\infty, 1\rangle$. In order to formulate the posed problem more precisely we introduce the systems $\langle G, 1, G/\infty, 1\rangle^{(r)}$ governed by the stationary sequences $\{\tau_j^{(r)e}, \tau_j^{(r)s}\}$ which depend on the parameter $r=1, 2, \ldots$. All previous notation applying here to the systems $\langle\bullet, \bullet, \bullet, \bullet\rangle^{(r)}$ will be supplied with the superscript (r). Suppose the following conditions hold:

A. There exists a sequence $\{\tau_j^e, \tau_j^s\}$ satisfying the hypotheses of Theorem 1 and such that the finite-dimensional distributions of the sequences $\{\tau_j^{(r)e}, \tau_j^{(r)s}\}$ converge to the corresponding finite dimensional distributions of $\{\tau_j^e, \tau_j^s\}$: We assume here that $\{\tau_j^{(r)e}, \tau_j^{(r)s}\}$ also satisfy the conditions of Theorem 1.

B. $\mathsf{M}\tau^{(r)s} \to \mathsf{M}\tau^s$ as $r\to\infty$ (as usual, we will sometimes suppress the subscript for brevity);

C. The distributions of $\tau_{-j}^s - X_{-j}^e$, where $X_{-j}^e = \sum_{k=-j}^{0} \tau_k^e$, are continuous at the point 0 for all $j \geqslant 0$. In other words, for an arbitrary integer $N>0$ and as $\delta\to 0$

$$\mathsf{P}\left(\bigcup_{j=0}^{N} \{|\tau_j^s - X_{-j}^e| \leqslant \delta\}\right) \to 0\,.$$

Theorem 5. *When Conditions A, B and C are satisfied, the finite-dimensional distributions of the stationary sequences* $\{q^{(r)k}; k\geqslant 0\}$ *converge as* $r\to\infty$ *to the corresponding distributions of* $\{q^k; k\geqslant 0\}$.

Remark. All of the formulated conditions are essential. If at least one of them is dropped we can easily construct an example in which the distributions of $q^{(r)k}$ do not converge.

Proof. We first prove that in the inequality (3) for $q^{(r)0}$ the right side converges to 0 as $N\to\infty$ uniformly in r. Indeed, for arbitrary fixed $b>0$, the last term in (3) is

$$\mathsf{P}\left(\bigcup_{j=N}^{\infty} \{\tau_{-j}^{(r)s} > b(j+1)\}\right) \leqslant \sum_{j=N}^{\infty} \mathsf{P}(\tau_{-j}^{(r)s} > b(j+1))$$

$$\leqslant \int_N^\infty \mathsf{P}(\tau^{(r)s} > bx)\, dx = \frac{1}{b} \int_{bN}^\infty \mathsf{P}(\tau^{(r)s} > x)\, dx\,.$$

But the uniform convergence in r to 0 as $N\to\infty$ of the last integral is necessary and sufficient for the convergence $\mathsf{M}\tau^{(r)s} \to \mathsf{M}\tau^s$ as $r\to\infty$ (Condition B). In fact, if the latter holds, then setting $p_r(x) = \mathsf{P}(\tau^{(r)s} > x)$ and $p(x) = \mathsf{P}(\tau^s > x)$ we find that as $r\to\infty$

$$\int_0^N (p_r - p)\, dx + \int_N^\infty p_r\, dx - \int_N^\infty p\, dx = \int_0^\infty (p_r - p)\, dx \to 0\,.$$

For given $\varepsilon>0$ we can choose an N_ε such that

$$\int_{N_\varepsilon}^\infty p\, dx < \varepsilon/3$$

and thereafter choose $r=r(N_\varepsilon, \varepsilon)$ such that

$$|\int_0^{N_\varepsilon} (p_r-p)\,dx|<\varepsilon/3 \quad \text{and} \quad |\int_0^\infty (p_r-p)\,dx|<\varepsilon/3\,.$$

But this shows that we can find an N_0 such that for all r and $N \geqslant N_0$, $\int_N^\infty p_r\,dx<\varepsilon$. The reader can obtain the converse in an analogous way, although we will not need this here.

We consider now the first term on the right in (3)

$$\mathsf{P}(\textstyle\bigcup_{j=N}^\infty \{\tau_0^{(r)e}+\cdots+\tau_{-j}^{(r)e}<b(j+1)\}) = \mathsf{P}(\sup_{k\geqslant N} Z_k^{(r)}>0)$$

where $Z_k^{(r)}=\sum_{j=0}^k (b-\tau_{-j}^{(r)e})$. Here the required uniform convergence follows from the next lemma, which is proved in § 11 (see the proof of (70) § 11).

Lemma. *Assume* $\{\xi_j\}$ *and* $\{\xi_j^{(r)}\}, r=1,2,\ldots,$ *are strictly stationary sequences such that the finite-dimensional distributions of* $\{\xi_j^{(r)}\}$ *converge as* $r\to\infty$ *to the corresponding distributions of* $\{\xi_j\}$. *Then if* $\{\xi_j\}$ *is metrically transitive,* $\mathsf{M}\xi_j<0$ *and* $\mathsf{M}(\xi_j^{(r)};\,\xi_j^{(r)}>0)\to \mathsf{M}(\xi_j;\,\xi_j>0)$, *we have* $\mathsf{P}(\max_{k\geqslant N}\sum_{j=1}^k \xi_j^{(r)}>0)\to 0$ *for* $N\to\infty$ *uniformly w.r.t.r.*

Now consider for noninteger y the difference

$$\begin{aligned}\mathsf{P}(q^{(r)0}\geqslant y)-\mathsf{P}(q^0\geqslant y) &= \mathsf{P}(\textstyle\sum_{j=0}^N I(\tau_{-j}^{(r)s}>X_{-j}^{(r)e})\geqslant y)\\ &\quad -\mathsf{P}(\textstyle\sum_{j=0}^N I(\tau_{-j}^s>X_{-j}^e)\geqslant y)+\varepsilon(r,N)+\varepsilon(N)\,, \qquad (1.1)\end{aligned}$$

where (by what has already been proved)

$$\varepsilon(N)\leqslant \mathsf{P}(\textstyle\bigcup_{j=N+1}^\infty \{\tau_{-j}^s>X_{-j}^e\})\to 0$$

and

$$\varepsilon(r,N)\leqslant \mathsf{P}(\textstyle\bigcup_{j=N+1}^\infty \{\tau_{-j}^{(r)s}>X_{-j}^{(r)e}\})\to 0$$

as $N\to\infty$ uniformly in r. Let $R(x)$ be the uniform distribution function on [0, 1]. For some $\delta>0$ we put

$$I_{j-}=R\left(\frac{\tau_{-j}^s-X_{-j}^e}{\delta}\right); \qquad I_{j+}=R\left(1+\frac{\tau_{-j}^s-X_{-j}^e}{\delta}\right) \quad \text{and} \quad I_j=I(\tau_{-j}^s>X_{-j}^e)\,.$$

Then clearly

$$I_{j-}\leqslant I_j \leqslant I_{j+}\,.$$

Since $R(x)$ is a continuous function we have (with the obvious notational conventions) as $r\to\infty$

$$\mathsf{P}(I_{j\pm}^{(r)}<x)\Rightarrow \mathsf{P}(I_{j\pm}<x)\,.$$

Hence, we can find an $\varepsilon>0$ so small that

$$\begin{aligned}\limsup_{r\to\infty} \mathsf{P}(\textstyle\sum_{j=0}^N I(\tau_{-j}^{(r)s}>X_{-j}^{(r)e})\geqslant y) &\leqslant \limsup_{r\to\infty} \mathsf{P}(\textstyle\sum_{j=0}^N I_{j+}^{(r)}\geqslant y)\\ &\leqslant \mathsf{P}(\textstyle\sum_{j=0}^N I_{j+}\geqslant y-\varepsilon)\,. \qquad (2.1)\end{aligned}$$

But by Condition C, the probability of the event

$$A_{N,\delta}=\bigcup_{j=0}^{N}\{I_{j+}\neq I_{j-}\} \tag{3.1}$$

equals $\mathsf{P}(\bigcup_{j=0}^{N}\{|\tau_{-j}^{s}-X_{-j}^{e}|\leqslant\delta\})$ and this converges to 0 as $\delta\to 0$. Hence,

$$\begin{aligned}\mathsf{P}(\textstyle\sum_{j=0}^{N} I_{j+}\geqslant y-\varepsilon)-\mathsf{P}(\sum_{j=0}^{N} I_j\geqslant y)\\ &=\mathsf{P}(\textstyle\sum_{j=0}^{N} I_{j+}\geqslant y-\varepsilon;\,\bar{A}_{N,\delta})-\mathsf{P}(\sum_{j=0}^{N} I_j\geqslant y;\,\bar{A}_{N,\delta})+\varepsilon(\delta,N)\\ &=\mathsf{P}(\textstyle\sum_{j=0}^{N} I_j\in[y-\varepsilon,y);\,\bar{A}_{N,\delta})+\varepsilon_1(\delta,N)\end{aligned}$$

where $|\varepsilon_1(\delta,N)|\leqslant\mathsf{P}(A_{N,\delta})$ and $\bar{A}$ is the complement of A. If we choose ε such that the half-open interval $[y-\varepsilon,y)$ contains no integers, then the event $\{\sum_{j=0}^{N} I_j\in[y-\varepsilon,y)\}$ becomes vacuous and by (1.1) and (2.1) we get

$$\limsup_{r\to\infty}(\mathsf{P}(q^{(r)0}\geqslant y)-\mathsf{P}(q^{0}\geqslant y))\leqslant\limsup_{r\to\infty}\varepsilon(r,N)+\varepsilon(N)+\varepsilon_1(\delta,N) \tag{4.1}$$

From the established properties of the functions $\varepsilon(r,N)$, $\varepsilon_1(\delta,N)$ and $\varepsilon(N)$ and the independence of the left side of (4.1) on δ and N it follows that the lim sup to be estimated does not exceed 0. In exactly the same way we can show by means of the inequality

$$\liminf_{r\to\infty}\mathsf{P}(\textstyle\sum_{j=0}^{N} I(\tau_{-j}^{(r)s}>X_{-j}^{(r)e})\geqslant y)\geqslant\mathsf{P}(\sum_{j=0}^{N} I_{j-}\geqslant y+\varepsilon)$$

that

$$\liminf_{r\to\infty}(\mathsf{P}(q^{(r)0}\geqslant y)-\mathsf{P}(q^{0}\geqslant y))\geqslant 0$$

for arbitrary noninteger y. The assertion of the theorem on the convergence of the one-dimensional distributions of $q^{(r)k}$ is thereby proved.

The idea of the proof was rather simple: in the representation

$$q^{(r)k}=\textstyle\sum_{j=0}^{\infty} I(\tau_{k-j}^{(r)e}>X_{k}^{(r)e}-X_{k-j-1}^{(r)e}) \tag{5.1}$$

the contribution of the "tail" $\sum_{j=N}^{\infty}$ was uniformly (w.r.t. r) small for large N. The convergence of the distributions of the finite sums in (5.1) is determined by that of the distributions of the governing sequences when Condition C holds. Hence, it is quite clear that using this same method we can also show the *convergence of arbitrary finite-dimensional distributions of the sequences* $\{q^{(r)k};k\geqslant 0\}$ to the corresponding distributions of the sequence $\{q^{k};k\geqslant 0\}$. The theorem is proved. □

3. *Stronger Continuity Theorems for the Systems* $\langle G, G, G/\infty, 1\rangle$. We turn now to a more general situation in which v_j^e is arbitrary and $q_1\neq 0$, and will consider a more precise characterization of the state of the system by means of the processes

$$\begin{aligned}q_n(x)=\textstyle\sum_{j=1}^{q_1} I(\rho_j>\tau_1^e+\cdots+\tau_{n-1}^e+x)\\ +\textstyle\sum_{j=1}^{v_1^e} I(\tau_{1,j}^s>\tau_1^e+\cdots+\tau_{n-1}^e+x)+\cdots\\ \surd+\textstyle\sum_{j=1}^{v_{n-1}^e} I(\tau_{n-1,j}^s>\tau_{n-1}^e+x)\end{aligned}$$

for which $q_n(0)=q_n$.

The process $\{q_n(x); x \geqslant 0\}$ will clearly have its previous meaning but is defined here for arbitrary initial conditions $(q_1; \rho_1, \ldots, \rho_{q_1})$.

The corresponding stationary sequence of processes $q^k(x)$ will be defined as before by the equality:

$$q^k(x) = \sum_{j=1}^{\nu_k^e} I(\tau_{k,j}^s > \tau_k^e + x) + \sum_{j=1}^{\nu_{k-1}^e} I(\tau_{k-1,j}^s > \tau_{k-1}^e + \tau_k^e + x) + \cdots .$$

Then we find as before that

$$\mathsf{P}\big(q^k(x) > \sum_{j=0}^{N-1} \nu_{-j}^e + q_1\big) \leqslant \mathsf{P}\big(\bigcup_{j=N}^{\infty} \{\tau_0^e + \cdots + \tau_{-j}^e < b(j+1)\}\big)$$
$$+ \mathsf{P}\big(\bigcup_{j=N}^{\infty} \{\max_k \tau_{-j,k}^s > b(j+1)\}\big).$$

Since $\max_k \tau_{1,k}^s \leqslant \lceil \tau_1^s \rceil = \sum_{j=1}^{\nu_1^e} \tau_{1,j}^s$, by assuming that the initial conditions are proper (q_1 and ρ_j are finite) and measurable w.r.t. the extended (see § 3) σ-algebra generated by the governing sequence, we obtain the following generalization of Theorem 3:

Theorem 3A. *The assertion of Theorem 3 remains valid for arbitrary initial values $(q_1; \rho_1, \ldots, \rho_{q_1})$ satisfying the formulated conditions.*

We remark that the required measurability of the initial conditions w.r.t. the extended σ-algebra is not essential for the main part of the assertion of the theorem. If this measurability does not obtain, then $q_{n+k}(x)$ will be represented as $\varepsilon(n, k, x) + q_{n+k}^0(x)$, where q_{n+k}^0 corresponds to zero initial conditions and $\varepsilon(n, k, x) \leqslant \varepsilon(n+k) \to 0$ as $n+k \to \infty$.

We turn now to stability theorems. We will say that the *distributions of the processes* $\{q^{(r)0}(x), x \geqslant 0\}$ *converge weakly as* $r \to \infty$ *to that of* $\{q^0(x); x \geqslant 0\}$ if for some everywhere dense set S the distribution of $(q^{(r)0}(x_1), \ldots, q^{(r)0}(x_m))$ converges weakly to that of $(q^0(x_1), \ldots, q^0(x_m))$ for any $x_1 \in S, \ldots, x_m \in S$.

Denote by S_0 the set of $x \geqslant 0$ which are points of continuity of the distributions of $\tau_{-j,k}^s - X_{-j}^e$ for all $j \geqslant 0$ and k. One has

Theorem 6. *Assume Condition A and Condition B_1:*

$$\mathsf{M}[\tau^{(r)s}] \to \mathsf{M}[\tau^s]$$

hold as $r \to \infty$. Then the distributions of the processes $\{q^{(r)0}(x); x \geqslant 0\}$ converge weakly as $r \to \infty$ to that of $\{q^0(x); x \geqslant 0\}$. Furthermore, the convergence set S coincides with S_0.

We remark that Theorem 5 is a consequence of Theorem 6 and that, by themselves, the conditions of Theorem 6 are more general than those of Theorem 5, where Condition C was present. The reason for this is easily seen from the proof of Theorem 6 which is obtained almost verbatim under the new conditions from the proof of Theorem 1. We get as a result the convergence of the distributions of $q^{(r)0}(x)$ for all values of $x \geqslant 0$ which are continuity points of the distributions of $\tau_{-j,k}^s - X_{-j}^e$ for all j and k (the required convergence holds if

$$\mathsf{P}\big(\bigcup_{j=0}^{N} \bigcup_{l=0}^{\nu_j^e} \{|\tau_{-j,l}^s - X_{-j}^e - x| \leqslant \delta\}\big)$$

converges to zero as $\delta\to 0$ (compare with (3.1)). Since the set of such x coincides with the positive real axis $x\geqslant 0$ from which no more than a countable set has been deleted, the assertion of Theorem 6 can be taken as proved. □

One can obtain quite analogously

Theorem 6A. *Under the assumptions of Theorem 6 the distributions of the processes* $\{q^{(r)k}(x);\, k\geqslant 0,\, x\geqslant 0\}$ *converge.*

That is, for arbitrary $k_1, \ldots, k_m$ *and* $x_1\in S_0, \ldots, x_m\in S_0$ *the distributions of the vectors* $(q^{(r)k_1}(x_1), \ldots, q^{(r)k_m}(x_m))$ *converge.*

Remark. It is also not hard to see that for the convergence as $r\to\infty$ of the distributions of the vectors

$$(q^{(r)k}(0),\, q^{(r)k}(\tau_{k+1}^{(r)e}),\, q^{(r)k}(\tau_{k+1}^{(r)e}+\tau_{k+2}^{(r)e}), \ldots, q^{(r)k}(\tau_{k+1}^{(r)e}+\ldots+\tau_{k+N}^{(r)e})) \tag{6.1}$$

in the systems $\langle G, 1, G/\infty, 1\rangle$ it is sufficient that the distributions of $\tau_{-j}^s-(\tau_{-j}^e+\cdots+\tau_k^e)$ be continuous at 0 for all $j\geqslant 0$ and $N\geqslant k\geqslant 0$. In virtue of the stationarity, this continuity will clearly hold if Condition C is fulfilled. In addition to (6.1), the joint distributions of such vectors will also converge for different k.

§ 32. The Systems $\langle G_I, G_I, G_I/\infty, 1\rangle$

1. We will now find equations which describe the stationary distribution of $q^0(x)$ for the systems $\langle G, 1, G_I/\infty, 1\rangle$. In this case

$$q^0(x)=\textstyle\sum_{k=0}^{\infty} I(\tau_k^s>x+X_k^e)\,, \quad X_k^e=\sum_{j=0}^{k}\tau_j^e\,.$$

Set

$$P_k(x)=\mathsf{P}(q^0(x)=k)\,, \quad k=0, 1, \ldots;$$

$$P(x)=\mathsf{P}(\tau^s>x)\,, \qquad F(x)=\mathsf{P}(\tau^e<x)\,.$$

Theorem 7. *The probabilities* $P_k(x)$ *satisfy the system of integral equations*

$$P_0(x)=\textstyle\int_0^\infty dF(t)(1-P(t+x))P_0(t+x)\,,$$

$$P_k(x)=\textstyle\int_0^\infty dF(t)P(t+x)P_{k-1}(t+x)+\int_0^\infty dF(t)(1-P(t+x))P_k(t+x)\,, \tag{10}$$

$$k=1, 2, \ldots\,.$$

The system consisting of the first $k+1$ *equations has, for arbitrary* k, *a unique solution possessing the following properties:* $P_i(x)$, $0\leqslant i\leqslant k$, *is a function of bounded variation;*

$$P_0(x)\to 1\,, \qquad P_i(x)\to 0\,, \quad 1\leqslant i\leqslant k, \text{ as } x\to\infty\,.$$

Proof. The random variable $q^0(x)$ can be represented as

$$q^0(x)=I(\tau_0^s>x+\tau_0^e)+\tilde{q}(x+\tau_0^e)\,, \tag{11}$$

where $\tilde{q}(x)$ is distributed like $q^0(x)$ and does not depend on τ_0^e and τ_0^s. From this representation and with the aid of the total probability formula w.r.t. the variables τ_0^s and τ_0^e, one easily obtains equations for $P_k(x)$ when $k \geqslant 0$. For the probability $p_0(x) = \mathsf{P}(q^0(x) > 0) = 1 - P_0(x)$ we get

$$p_0(x) = \mathsf{P}(\tau^s - \tau^e > x) + \int_x^\infty d_t F(t-x)(1-P(t))p_0(t) . \tag{12}$$

This is essentially a Volterra equation of the 2nd kind in the class of functions of bounded variation converging to 0 as $x \to \infty$. Its kernel is

$$K(dt, x) = (1 - P(t))\, d_t F(t-x)$$

(as with the standard Volterra equation, Equation (12) is "triangular"). It can be put into the standard Volterra form by the substitutions $x = 1/y$ and $t = 1/u$. As is well known, equations of this type have a unique solution which can be found efficiently by the method of successive approximations.

The same remarks obviously apply to the equation

$$P_1(x) = g_1(x) + \int_x^\infty d_t F(t-x)(1-P(t))P_1(t)$$

as well, where

$$g_1(x) = \int_0^\infty dF(t)P(t+x)(1-p_0(t+x)) ,$$

and likewise for $P_2(x)$, $P_3(x)$, etc. □

2. We now consider the queue length $q(t)$ at time t. To avoid cumbersome computations we will assume that $v^e \equiv 1$ (i.e., we treat the systems $\langle G_I, 1, G_I/\infty, 1\rangle$). In analogy to the preceding, we introduce the processes $q(t, x)$, where $q(t, x)$ is the *number of customers in the system at time t which are still being served at time* $t + x$. Then, if

$$\eta(t) - 1 = \max\{k : \tau_1^e + \cdots + \tau_k^e < t\} ,$$

one has

$$q(t, x) = I(\tau_1^s > t + x) + I(\tau_2^s > t + x - \tau_1^e) + \cdots + I(\tau_{\eta(t)}^s > t + x - \tau_1^e - \cdots - \tau_{\eta(t)-1}^e) ;$$
$$q(t) = q(t, 0) .$$

The "defect" variable here: $\gamma^e(t) = t - \tau_1^e - \cdots - \tau_{\eta(t)-1}^e$, has the following properties (see Appendix 1): If $\mathsf{M}\tau^e < \infty$ and τ^e is nonlattice, then the limiting distribution of $\gamma^e(t)$ exists as $t \to \infty$. Denote by γ^e any random variable for which $\mathsf{P}(\gamma^e < x) = \lim_{t\to\infty} \mathsf{P}(\gamma^e(t) < x)$. Then for arbitrary N the joint limiting distribution of the variables $\gamma^e(t)$; $\tau_{\eta(t)-1}^e, \ldots, \tau_{\eta(t)-N}^e$; $\tau_{\eta(t)}^s, \ldots, \tau_{\eta(t)-N}^s$ exists and coincides with that of the independent random variables γ^e; $\tau_1^e, \ldots, \tau_N^e$; $\tau_0^s, \ldots, \tau_N^s$. Using this fact, it is easy to show that *for* $\mathsf{M}\tau^s < \infty$ *and* $\mathsf{M}\tau^e < \infty$ (τ^e *nonlattice*), *the limiting distribution of* $q(t, x)$ *as* $t \to \infty$ *exists and coincides with the distribution of*

$$q^c(x) = I(\tau_0^s > \gamma^e + x) + I(\tau_1^s > \gamma^e + \tau_1^e + x) + I(\tau_2^s > \gamma^e + \tau_1^e + \tau_2^e + x) + \cdots . \tag{13}$$

It is also not hard to construct the stationary process $q^c(u, x)$ to which $q(t + u, x)$ converges as $t \to \infty$.

Comparing this with (11), we note in addition that the random variable $q^c(x)$ can be written as

$$q^c(x)=I(\tau_0^s>\gamma^e+x)+\tilde{q}(\gamma^e+x),$$

where, as before, $\tilde{q}(x)$ is distributed like $q^0(x)$ and does not depend on τ_0^s and γ^e. Thus, if we put

$$Q_k(x)=\mathsf{P}(q^c(x)=k), \quad k=0, 1, \dots,$$

and recall that the density of the distribution of γ^e equals

$$(\mathsf{M}\tau^e)^{-1}(1-F(t)),$$

we get

$$\begin{aligned}Q_k(x)&=\textstyle\int_0^\infty (\mathsf{M}\tau^e)^{-1}(1-F(t))[\mathsf{P}(I(\tau_0^s>t+x)+\tilde{q}(t+x)=k)]\,dt\\ &=\textstyle\int_0^\infty (\mathsf{M}\tau^e)^{-1}(1-F(t))(1-P(t+x))P_k(t+x)\,dt\\ &\quad+\textstyle\int_0^\infty (\mathsf{M}\tau^e)^{-1}(1-F(t))P(t+x)P_{k-1}(t+x)\,dt; \quad k=0, 1, \dots,\end{aligned}$$

where $P_{k-1}(t+x)$ must be set equal to 0 for $k=0$. Hence, the search for the distribution $Q_k(x)$ reduces to finding the solution of Eq. (10).

Finally, it is also easy to see that if $\mathsf{M}\tau^e=\infty$ and $\mathsf{M}\tau^s<\infty$, *then* $\gamma^e(t)\underset{p}{\to}\infty$ *as* $t\to\infty$ *and* $q(t)\underset{p}{\to}0$.

These results can be extended by the reader to the case of arbitrarily distributed v^e.

§ 33. The Systems $\langle E, 1, G_I/\infty, 1\rangle$

1. For the systems $\langle E, 1, G_I/\infty, 1\rangle$ we have $\mathsf{P}(\tau^e>x)=e^{-\alpha x}$ and the distribution of γ^e coincides with that of τ^e. From this and the representations (13) and (1) it follows that the distributions of $q^0(x)$ and $q^c(x)$ also coincide. In other words, $\mathsf{P}(q_n(x)=k)$ and $\mathsf{P}(q(t, x)=k)$ *converge as* $n\to\infty$ *and* $t\to\infty$ *to the same limit distribution. (This will also hold for arbitrary* v^e.) We will prove that this distribution is Poisson.

Theorem 8. *Set* $\mathsf{M}\tau^s=a$ *and* $P(x)=\mathsf{P}(\tau^s>x)$. *Then*

$$\lim_{n\to\infty}\mathsf{P}(q_n(x)=k)=\lim_{t\to\infty}\mathsf{P}(q(t, x)=k)=\frac{\left(\int_x^\infty \alpha P(t)\,dt\right)^k}{k!}\exp\left(-\textstyle\int_x^\infty \alpha P(t)\,dt\right); \quad k=0, 1, \dots.$$

For $x=0$ *we find that*

$$\mathsf{P}(q^0=k)=\mathsf{P}(q^c=k)=\frac{(a\alpha)^k}{k!}e^{-\alpha a}; \quad k=0, 1, \dots.$$

Proof. We will use Equations (10). In our case they take the form

$$P_k(x)=\alpha e^{\alpha x}\int_x^\infty e^{-\alpha t}P(t)P_{k-1}(t)\,dt+\alpha e^{\alpha x}\int_x^\infty e^{-\alpha t}[1-P(t)]P_k(t)\,dt\,,\quad k\geqslant 1\,.$$

$$P_0(x)=\alpha e^{\alpha x}\int_x^\infty e^{-\alpha t}[1-P(t)]P_0(t)\,dt\,.$$

Differentiating both sides of these equalities w.r.t. x, we get

$$\begin{aligned}P_k'(x)&=\alpha P(x)[P_k(x)-P_{k-1}(x)]\,;\quad k\geqslant 1\,,\\ P_0'(x)&=\alpha P(x)P_0(x)\,.\end{aligned}\tag{14}$$

We make a change of variables, assuming

$$P_k(x)=u_k(x)\exp\int_0^x \alpha P(t)\,dt\,.\tag{15}$$

Inserting this in (14), we have (in abbreviated notation)

$$\begin{aligned}&u_k'\exp\int_0^x+u_k\alpha P\exp\int_0^x=\alpha P[u_k\exp\int_0^x-u_{k-1}\exp\int_0^x]\,,\\ &u_0'\exp\int_0^x+u_0\alpha P\exp\int_0^x=\alpha P u_0\exp\int_0^x\,,\end{aligned}$$

or, what is the same,

$$\begin{cases}u_k'=-\alpha P u_{k-1}\,,\\ u_0'=0\,.\end{cases}$$

Since $P_0(\infty)=1$ and $P_k(\infty)=0$, $k\geqslant 1$, one has

$$u_0(x)=e^{-\alpha a}$$

and

$$u_k(x)=\alpha^k e^{-\alpha a}\int_x^\infty P(t_1)\int_{t_1}^\infty P(t_2)\int_{t_2}^\infty\cdots\int_{t_{k-1}}^\infty P(t_k)\,dt_1\ldots dt_k\,.$$

The latter integral can be written as

$$\int\cdots\int_{x\leqslant t_1\leqslant\ldots\leqslant t_k}P(t_1)P(t_2)\ldots P(t_k)\,dt_1\ldots dt_k=\frac{1}{k!}\Big(\int_x^\infty P(t)\,dt\Big)^k\,.$$

Now use (15). The theorem is proved. □

2. The result for the distribution of q^c can also be obtained directly along with an assertion on the distribution of $q(t)$.

Theorem 9.

$$\mathsf{P}(q(t)=k)=\exp\{-\alpha\int_0^t P(x)\,dx\}[\alpha\int_0^t P(x)\,dx]^k/(k!)\,.$$

Thus $q(t)$ also has a Poisson distribution, but with parameter $\alpha\int_0^t P(x)\,dx$.

Proof. The probability that n customers enter the system during time t equals $e^{-\alpha t}(\alpha t)^n/(n!)$. The disposition of customer arrival times on the interval $[0, t]$, given that n of them appear, is a realization of n independent observations of a random variable distributed uniformly on $[0, t]$. Hence, by the total probability formula

$$\mathsf{P}(q(t)=k)=\sum_{n=k}^\infty e^{-\alpha t}\frac{(\alpha t)^n}{n!}\binom{n}{k}\left[\frac{1}{t}\int_0^t[1-P(x)]\,dx\right]^{n-k}\left[\frac{1}{t}\int_0^t P(x)\,dx\right]^k.\tag{16}$$

Expanding the first square brackets in this expression and changing the order of summation, we get $(a(t)=\int_0^t P(x)\,dx)$

$$\mathsf{P}(q(t)=k)=\sum_{j=0}^{\infty}\sum_{n=j+k}^{\infty} e^{-\alpha t}\frac{\alpha^n}{k!j!(n-k-j)!}a^k(t)t^j a^{n-k-j}(t)(-1)^{n-k-j}$$

$$=e^{-\alpha a(t)}\frac{\alpha^k a(t)^k}{k!}e^{-\alpha t}\sum_{j=0}^{\infty}\frac{\alpha^j t^j}{j!}. \quad \square$$

In an analogous way (compare with (16)) one can also obtain an explicit formula for $\mathsf{P}(q_n=k)$ since the conditional distribution of the arrival times of $n-1$ customers on $[0, t]$, given that $\tau_1^e+\dots+\tau_n^e=t$, is again uniform in this case. Consequently,

$$\mathsf{P}(q_n=k)=\int_0^\infty \mathsf{P}(\tau_1^e+\dots+\tau_n^e\in dt)\binom{n-1}{k}\left[\frac{a(t)}{t}\right]^k\left[1-\frac{a(t)}{t}\right]^{n-k-1}$$

Since

$$\mathsf{P}(\tau_1^e+\dots+\tau_n^e\in dt)=\frac{\alpha^n e^{-\alpha t}t^{n-1}}{(n-1)!}dt,$$

one has

$$\mathsf{P}(q_n=k)=\alpha^n\int_0^\infty e^{-\alpha t}\frac{a^k(t)}{k!}\frac{(t-a(t))^{n-k-1}}{(n-k-1)!}dt.$$

3. *Discrete Systems for which* $\mathsf{P}(\tau^e=k)=p^{k-1}(1-p)$, $k=1, 2, \dots$; τ^s *is integer-valued.* It is necessary to emphasize here that q_n is defined as the number of busy channels in the system before the arrival of the n-th request not counting requests whose service is completed at the time of this arrival. This fact must also be kept in mind when considering the quantities $q_n(x)$ defined for integer $x\geqslant 0$. Setting $P(k)=\mathsf{P}(\tau^s>k)$ as before, we obtain for $k>0$

$$P_k(x)=\mathsf{P}(q^0(x)=k)=(1-p)\sum_{j=1}^{\infty}p^{j-1}P(x+j)P_{k-1}(x+j)$$
$$+(1-p)\sum_{j=1}^{\infty}p^{j-1}(1-P(x+j))P_k(x+j),$$

$$P_0(x)=(1-p)\sum_{j=1}^{\infty}p^{j-1}(1-P(x+j))P_0(x+j).$$

Considering the differences $P_k(x)-pP_k(x+1)$, we obtain

$$\begin{cases}P_k(x)-pP_k(x+1)=(1-p)P(x+1)P_{k-1}(x+1)+(1-p)(1-P(x+1))P_k(x+1),\\ P_0(x)-pP_0(x+1)=(1-p)(1-P(x+1))P_0(x+1).\end{cases}$$

Or, what is the same,

$$\begin{cases}P_k(x)-P_k(x+1)=(1-p)P(x+1)[P_{k-1}(x+1)-P_k(x+1)],\\ P_0(x)-P_0(x+1)=-(1-p)P(x+1)P_0(x+1).\end{cases}$$

We introduce the functions $u_k(x)$, $k=0, 1, \dots$, with the help of the equalities

$$P_k(x)=u_k(x)\Big[\prod_{j=0}^{x}(1-(1-p)P(j))\Big]^{-1}.$$

Note that the product $\prod_{j=0}^{\infty}(1-(1-p)P(j))$ is finite since $\mathsf{M}\tau^s<\infty$. Thus, since $P_0(\infty)=1$ and $P_k(\infty)=0$ for $k\geqslant 1$, we obtain

$$u_0(\infty)=\prod_{j=0}^{\infty}(1-(1-p)P(j)), \qquad u_k(\infty)=0, \qquad k\geqslant 1.$$

For the functions $u_k(x)$ we get the system (after multiplying the equations by $\prod_{j=1}^{x+1}(1-(1-p)P(j)))$

$$\begin{cases} u_k(x)(1-(1-p)P(x+1))-u_k(x+1)=(1-p)P(x+1)[u_{k-1}(x+1)-u_k(x+1)] \\ u_0(x)(1-(1-p)P(x+1))-u_0(x+1)=-(1-p)P(x+1)u_0(x+1). \end{cases}$$

This is equivalent to

$$\begin{cases} u_k(x)-u_k(x+1)=Q(x+1)u_{k-1}(x+1), \quad k\geqslant 1, \\ u_0(x)-u_0(x+1)=0, \end{cases}$$

where $Q(x)=(1-p)P(x)/[1-(1-p)P(x)]$. From this we get $u_0(x)=u_0(\infty)=\prod_{j=0}^{\infty}(1-(1-p)P(j))$. For $k\geqslant 1$ one has

$$u_k(0)-u_k(x+1)=\sum_{j=1}^{x+1}Q(j)u_{k-1}(j),$$
$$u_k(0)=\sum_{j=1}^{\infty}Q(j)u_{k-1}(j).$$

Consequently,

$$u_k(x)=\sum_{j=x+1}^{\infty}Q(j)u_{k-1}(j),$$

whence

$$\begin{aligned} u_k(x)&=\sum_{j_1=x+1}^{\infty}Q(j_1)\sum_{j_2=j_1+1}^{\infty}Q(j_2)\ldots\sum_{j_k=j_{k-1}+1}^{\infty}Q(j_k)u_0(j_k) \\ &=u_0(\infty)\sum_{x<j_1<j_2<\ldots<j_k}\prod_{l=1}^{k}Q(j_l). \end{aligned}$$

Returning to the quantities $P_k(x)$, we get finally

$$P_0(x)=\prod_{j=x+1}^{\infty}(1-(1-p)P(j))$$

and

$$P_k(x)=P_0(x)\sum_{x<j_1<j_2<\ldots<j_k}\prod_{l=1}^{k}Q(j_l).$$

It's not hard to see that for all $x\geqslant 0$ these formulae yield a distribution different from the Poisson. For example, when $x=0$

$$P_1(0)=P_0(0)\sum_{j>0}Q(j)$$

and

$$\begin{aligned} P_2(0)&=P_0(0)\sum_{i>j>0}Q(i)Q(j) \\ &=\frac{P_0(0)}{2}[(\sum_{j>0}Q(j))^2-\sum_{j>0}Q^2(j)]<\frac{P_0(0)}{2}(\sum_{j>0}Q(j))^2. \end{aligned}$$

These comparisons yield inequalities which can be useful: for $k\geqslant 2$

$$P_k(0)<P_0(0)(\sum_{j>0}Q(j))^k/(k!).$$

§ 34. The Systems $\langle G_I, 1, E/\infty, 1\rangle$

1. Here we have $\mathsf{P}(\tau^s \geq x) = e^{-\alpha x}$, $\alpha > 0$. As we have seen in Subsection 1, § 30, in this case $\mathsf{P}(q_n = k)$ converges exponentially as $n \to \infty$ to the limiting distribution $\{p_k\}$.

By (6), for all k and $\lambda > 0$

$$|\mathsf{P}(q_n = k) - p_k| \leq \frac{\alpha\psi^n(-\lambda)}{(\alpha-\lambda)(1-\psi(-\lambda))}, \qquad \psi(\lambda) = \mathsf{M}\exp(\lambda\tau^e).$$

Set

$$\lambda = \alpha - \frac{\psi(-\alpha)}{n\psi'(-\alpha)}.$$

Then the difference to be estimated does not exceed

$$\frac{\alpha e n \psi'(-\alpha)\psi^n(-\alpha)}{\psi(-\alpha)(1-\psi(-\alpha))} c_n,$$

where $c_n \to 1$ as $n \to \infty$.

For the distribution $\{p_k\}$ we have

Theorem 10.

$$p_k = \sum_{j=k}^{\infty} (-1)^{j-k} \binom{j}{k} C_j,$$

where (17)

$$C_j = \prod_{l=1}^{j} \frac{\psi(-l\alpha)}{1-\psi(-l\alpha)}, \qquad C_0 = 1.$$

The proof can be carried out by using (10) again. However, it is apparently simpler to note that in our case, as well as under the conditions of § 29, the quantities q_n form a simple homogeneous Markov chain with transition probabilities (see (30) and (31) § 29)

$$p_{ij} = \mathsf{P}(q_{n+1} = j \mid q_n = i) = \int_0^\infty \pi_{ij}(x)\, d\mathsf{P}(\tau^e < x),$$

where

$$\pi_{ij}(x) = \binom{i+1}{j} e^{-j\alpha x}(1-e^{-\alpha x})^{i-j+1}.$$

This chain is clearly irreducible, so that $\{p_k\}$ is the unique solution of the system of equations

$$p_j = \sum p_i p_{ij}, \quad j = 0, 1, \ldots, \tag{18}$$
$$\sum p_j = 1.$$

One shows by direct substitution that (17) satisfies this system. □

Quite analogously to Theorem 10, Chapter 5 one proves

Theorem 11. *If τ^e is nonlattice, then*

$$\lim_{t\to\infty} \mathsf{P}(q(t)=k)=P_k=\frac{p_{k-1}}{k\alpha\mathsf{M}\tau^e}, \quad k=1,2,\ldots,$$

exist (the existence follows from Subsection 2 § 32).

For the systems $\langle G_I, 1, E/\infty, 1\rangle$ one can also determine the transformations

$$\textstyle\sum_{n=1}^{\infty} z^n \mathsf{P}(q_n=k) \quad \text{and} \quad \int_0^\infty e^{-\lambda t}\mathsf{P}(q(t)=k)\,dt$$

of the distributions $\mathsf{P}(q_n=k)$ and $\mathsf{P}(q(t)=k)$ explicitly. Indeed, for $|z|<1$ and $q(0)=i$

$$\textstyle\sum_{n=1}^{\infty} z^n \mathsf{P}(q_n=k)=\sum_{r=k}^{\infty}(-1)^{r-k}\binom{r}{k}A_r(z); \quad k=0,1,2,\ldots;$$

$$A_r(z)=\textstyle\sum_{j=0}^{r}\binom{i}{j}\prod_{l=j}^{r}\frac{z\psi_l}{1-z\psi_l}, \quad \psi_l=\psi(-\alpha l)=\mathsf{M}\,e^{-\alpha l\tau^e}.$$

For Re $\lambda>0$ and $q(0)=i$

$$\textstyle\int_0^\infty e^{-\lambda t}\mathsf{P}(q(t)=k)\,dt=\sum_{r=k}^{\infty}(-1)^{r-k}\binom{r}{k}\,B_r(\lambda),$$

$$B_r(\lambda)=\frac{1}{(\lambda+r\alpha)}\textstyle\sum_{j=0}^{r}\binom{i}{j}\prod_{l=j}^{r-1}\frac{1-\psi(-\lambda-l\alpha)}{\psi(-\lambda-l\alpha)}$$

(see Takács [71], Chapter 3, § 3).

2. For the *discrete* analogue of the systems $\langle G_I, 1, E/\infty, 1\rangle$ τ^e is integer-valued and $\mathsf{P}(\tau^s=k)=(1-p)p^{k-1}$, $k=1,2,\ldots$.

In this case the transition probabilities of the Markov chain formed by the q_n will have the form

$$p_{ij}=\mathsf{P}(q_{n+1}=j\mid q_n=i)=\textstyle\sum_{k=1}^{\infty}\mathsf{P}(\tau^e=k)\pi_{ij}(k),$$

where the conditional probabilities $\pi_{ij}(k)$ (compare with Subsection 5 § 29) equal

$$\pi_{ij}(k)=\binom{i+1}{j}p^{kj}(1-p^k)^{i-j+1}.$$

If we substitute these p_{ij} into (18) we get a system of equations for the limiting probabilities $p_j=\lim_{n\to\infty}\mathsf{P}(q_n=j)$ which has a unique solution. Put

$$p(z)=\textstyle\sum p_j z^j=\mathsf{M}z^{q^0}.$$

Then using the system just mentioned one gets

$$\begin{aligned} p(z)&=\textstyle\sum_{i=0}^{\infty}\sum_{k=1}^{\infty}p_i\mathsf{P}(\tau^e=k)\sum_{j=0}^{\infty}\binom{i+1}{j}(zp^k)^j(1-p^k)^{i-j+1}\\ &=\textstyle\sum_{k=1}^{\infty}\mathsf{P}(\tau^e=k)(zp^k+1-p^k)\sum_{i=0}^{\infty}p_i(p^kz+1-p^k)^i \qquad (19)\\ &=\textstyle\sum_{k=1}^{\infty}\mathsf{P}(\tau^e=k)(zp^k+1-p^k)p(p^kz+1-p^k).\end{aligned}$$

Write

$$C_j = \frac{1}{j!}\left(\frac{d^j p(z)}{dz^j}\right)_{z=1}, \; C_0 = 1 \quad \text{and} \quad \psi(z) = \mathsf{M} z^{\tau^e}.$$

Then

$$p_k = \frac{1}{k!}\left(\frac{d^k p(z)}{dz^k}\right)_{z=0} = \sum_{j=k}^{\infty} (-1)^{j-k} \binom{j}{k} C_j .$$

Differentiating the equations (19) j times and putting $z=1$, we obtain

$$C_j = \sum_{k=1}^{\infty} \mathsf{P}(\tau^e = k) p^{kj} C_j + \sum_{k=1}^{\infty} p^{kj} \mathsf{P}(\tau^e = k) C_{j-1}$$

or, equivalently,

$$C_j = \frac{\psi(p^j)}{1-\psi(p^j)} C_{j-1} .$$

Since $C_0 = 1$, we have finally

$$C_j = \prod_{l=1}^{j} \frac{\psi(p^l)}{1-\psi(p^l)} \quad \text{and} \quad p_k = \sum_{j=k}^{\infty} (-1)^{j-k} \binom{j}{k} C_j .$$

Chapter 7

Systems with Refusals

A natural characterization of such systems is the number of occupied channels at either the time of arrival of the n-th customer (q_n), or at time t ($q(t)$). The variables q_n and $q(t)$ will be called, as before, the *queue lengths*. As already explained in the introduction, the operation of the systems $\langle G/m\rangle_R$ is analogous to that of $\langle G/m\rangle$ but with the essential difference that here, $q(0)\leqslant m$, and if $q_n=m$ at the moment of arrival of the n-th group of requests, then all requests of this group are refused and leave the system. If, however, $q_n<m$, then those requests of the group are refused which remain after the idle channels have been filled. Thus, we have $q_n\leqslant m$ and $q(t)\leqslant m$ here and the important parameters of the systems are the probabilities that a customer is refused.

As in the case of systems with queues, in the class of systems $\langle G, G, E/m, G\rangle_R$ the sequence $\{q_n\}$ can be described by means of recursion relations. For independent governing sequences (the systems $\langle G_I, G_I, E/m, G_I\rangle_R$) these relations transform the sequence $\{q_n\}$ into a homogeneous, ergodic Markov chain with a finite number of states. The calculation of the stationary distribution of $\{q_n\}$ is carried out by means of the usual equations. In § 38 this will be done for the systems $\langle G_I, 1, E/m, 1\rangle_R$.

We will devote most of our attention in what follows to the investigation of the systems $\langle G, G, G/m, 1\rangle_R$.

The systems $\langle G/\infty\rangle_R$ obviously coincide with the systems $\langle G/\infty\rangle$, which were studied in the previous chapter and will not be considered here. The systems $\langle G\rangle_R$ $(m=1)$ will be treated in this chapter (§ 38).

§ 35. The Systems $\langle G, G, G/m, 1\rangle_R$. General Theorems

1. We consider first the systems $\langle G, 1, G/m, 1\rangle_R$ $(v_j^s\equiv 1)$ with initial condition $q_1=0$. The case in which this agreement is not made (provided that q_1 is a proper random variable) will be treated later since, as before, it does not essentially change the arguments and only complicates the calculations.

Let $q_n(x)$ again be the number of channels occupied by the first $(n-1)$ requests longer than time x after the moment of arrival of the n-th request, so that $q_n=q_n(0)$.

Furthermore, assume ($I(A)$ is the indicator of the event A) that

$$Q_k(x)=I(\tau_k^s>x+\tau_k^e)+I(\tau_{k-1}^s>x+\tau_{k-1}^e+\tau_k^e)+\cdots$$

is the analogous characteristic of the stationary system with an infinite number of channels (see the previous chapter) which is governed by the same sequence $\{\tau_j^e, \tau_j^s; -\infty<j<\infty\}$. Let A_k be the event that *for some* $0\leqslant L\leqslant m-1$ *and* $l_j\geqslant 1$ *such that* $\sum_{j=0}^{L} l_j=m$

$$Q_k(0)\leqslant m-l_0; \quad Q_k(\tau_{k+1}^e)\leqslant m-l_0-l_1; \quad \dots;$$
$$Q_k(\tau_{k+1}^e+\cdots+\tau_{k+L}^e)\leqslant m-l_0-\cdots-l_L=0. \tag{1}$$

holds.

It is clear that $\mathsf{P}(A_k)=\mathsf{P}(A_0)$ if the governing sequence is stationary.

The event A_k means that the stationary queue $Q_k=Q_k(0)$ in the system $\langle G, 1, G/\infty, 1\rangle$ at the moment of arrival of the current request (at time t_0, say) does not exceed $m-1$, so that an incoming request can occupy a vacated channel with number $\leqslant m$ with subsequent requests also occupying the channels with numbers $\leqslant m$ *which have been vacated by requests arriving up to time* t_0, until the last of the requests has left. Hence, there must exist a time interval occupied by the arrival of $L\leqslant m$ successive requests in the course of which there occurs a complete "turnover" of the requests located in the system.

Theorem 1. *Assume that the sequence* $\{\tau_n^e, \tau_n^s; -\infty<n<\infty\}$ *is strictly stationary and ergodic. Further, assume*

$$\mathsf{P}(A_0)>0. \tag{2}$$

Then the distributions of the sequence of processes $\{q_{n+k}(x); k\geqslant 0, x\geqslant 0\}$ *converge for* $n\to\infty$ *to the distribution of the stationary (w.r.t. k) process* $\{q^k(x); k\geqslant 0, x\geqslant 0\}$ *in such a way that* $\mathsf{P}(q^0(0)=m)<1$.

The convergence here is understood in the strong sense of Theorem 3, Chapter 6: there exist processes $\{\tilde{q}_n^k(x); k\geqslant 0, x\geqslant 0\}$, *distributed like* $\{q^k(x); k\geqslant 0, x\geqslant 0\}$, *and such that as* $n\to\infty$

$$\mathsf{P}\Big(\bigcup_{k=0}^{\infty}\bigcup_{x\geqslant 0}\{q_{n+k}(x)\neq\tilde{q}_n^k(x)\}\Big)\to 0.$$

Proof. Put $X_n=\sum_{j=1}^{n}\tau_j^e$ and let $\delta(k)$ be a function equal to 1 for $k<m$ and 0 for $k=m$. Then

$$q_n(x)=I(\tau_1^s>X_{n-1}+x)+\cdots+I(\tau_m^s>X_{n-1}-X_{m-1}+x)$$
$$+I(\tau_{m+1}^s>X_{n-1}-X_m+x)\delta(q_{m+1})+\cdots+I(\tau_{n-1}^s>\tau_{n-1}^e+x)\delta(q_{n-1}).$$

Thus, for the processes $q_n(x)$ we have the recursion relation

$$q_{n+1}(x)=I(\tau_n^s>\tau_n^e+x)\delta(q_n(0))+q_n(x+\tau_n^e). \tag{3}$$

Let U be a random variable shift-transformation corresponding to the shift T of sets from the σ-algebra $\mathfrak{M}$ generated by the governing sequence (see § 3). Then

$$\tau_{n+1}^e(\omega)=U\tau_n^e(\omega) \quad\text{and}\quad \tau_{n-1}^e(\omega)=U^{-1}\tau_n^e(\omega).$$

Put

$$q_n^k(x) = U^{-k} q_n(x) \underset{d}{=} q_n(x) \quad \text{and} \quad q_n^k = U^{-k} q_n \underset{d}{=} q_n .$$

We will prove that for any $-n+1 \leqslant k \leqslant -m+1$ and $r>0$

$$A_k \subset \{q_n^{n-1}(x) = q_{n+r}^{n+r-1}(x)\} \tag{4}$$

(here, the equality in braces is to be understood as an equality of functions, i.e., an equality for all x).

In fact, the inequalities (1) for the given values of k mean that

$$\delta(q_{n+k+j}^{n-1}) = \delta(q_{n+r+k+j}^{n+r-1}) = 1$$

for $j=0, 1, \ldots, L$. On the other hand, they also imply that

$$q_{n+k}^{n-1}(\tau_{n+k}^e + \cdots + \tau_{n+k+L-1}^e) = q_{n+r+k}^{n+r-1}(\tau_{n+k}^e + \cdots + \tau_{n+k+L-1}^e) = 0$$

(if the argument x is random, then $q_n^k(x)$ is also the result of applying U^{-k} to $q_n(x)$). Hence, after L applications of Relation (3) one gets

$$\begin{aligned} q_{n+k+L}^{n-1}(x) &= I(\tau_{k+L}^s > x + \tau_{k+L}^e) + q_{n+k+L-1}^{n-1}(x + \tau_{n+k+L-1}^e) \\ &= I(\tau_{k+L}^s > x + \tau_{k+L}^e) + I(\tau_{k+L-1}^s > x + \tau_{k+L-1}^e + \tau_{k+L}^e) \\ &\quad + q_{n+k+L-2}^{n-1}(x + \tau_{n+k+L-1}^e + \tau_{n+k+L-2}^e) = \cdots \\ &= \textstyle\sum_{j=0}^{L-1} I(\tau_{k+L-j}^s > x + \tau_{k+L-j}^e + \cdots + \tau_{j+L}^e) . \end{aligned}$$

In the same way one obtains

$$q_{n+r+k+L}^{n+r-1}(x) = \textstyle\sum_{j=0}^{L-1} I(\tau_{k+L-j}^s > x + \tau_{k+L-j}^e + \cdots + \tau_{k+L}^e) .$$

Using (3) again, we establish (4).

The essence of these calculations is rather simple: q_n^{n-1} and q_{n+r}^{n+r-1} are functions of the terms of the governing sequences at the moments of time $(-n+1, -n+2, \ldots, 0)$ and $(-n-r+1, \ldots, 0)$, resp. The inequalities (1) mean that all requests entering two systems up to time k which begin operating at times $-n+1$ and $-n-r+1$ leave these systems in both cases during the time interval $(k, k+L)$, and at the same moments of time $L+1$ new requests numbered $k+1, \ldots, k+L+1$ enter them. This means that the subsequent evolutions of the two systems will coincide, so that

$$q_n^{n-1}(x) = q_{n+r}^{n+r-1}(x) .$$

We return now to (4). It implies that

$$\bigcup_{k=-m+1}^{-n+1} A_k \subset \bigcap_{r \geqslant 1} \{q_n^{n-1}(x) = q_{n+r}^{n+r-1}(x)\} . \tag{5}$$

Thus, if

$$\mathsf{P}\Big(\bigcup_{k \leqslant -m+1} A_k\Big) = 1 , \tag{6}$$

then the sequence of processes $q_n^{n-1}(x)$ is fundamental in the sense of a.s. convergence since as $n \to \infty$

$$\mathsf{P}\Big(\bigcup_{r \geqslant 1} \{q_n^{n-1}(x) \neq q_{n+r+1}^{n+r}(x)\}\Big) \leqslant 1 - \mathsf{P}\Big(\bigcup_{k=-m+1}^{-n+1} A_k\Big) \to 0 .$$

Hence, when (6) holds, $q_n^{n-1}(x)$ converges almost everywhere to some limit $q^0(x)$ (which implies the existence of the limiting distribution of $q_n(x)$). It is clear that the coincidence of $q_n^{n-1}(x)$ and $q_{n+r}^{n+r-1}(x)$ implies that of $q_{n+j}^{n-1}(x)$ and $q_{n+r+j}^{n+r-1}(x)$ for all $j=0, 1, \ldots$. Thus, $\lim_{n\to\infty} q_{n+j}^{n-1}(x)$ exist, and we will denote them by $q^j(x)$. The process $\{q^j(x); j\geqslant 0, x\geqslant 0\}$ possesses the required properties. The processes $\tilde{q}_n^j(x)$ appearing in the formulation of the theorem can obviously be obtained from the $q^j(x)$ by the shift transformation U^n.

It remains to show (6). Assume $B=\bigcup_{k\leqslant -m+1} A_k$. Since $A_k=T^k A_0$, $TB\supset B$. Moreover, $\mathsf{P}(TB)=\mathsf{P}(B)$. Consequently, $TB=B$ up to a set of measure 0 and B is invariant w.r.t. the transformation T. Because of the metric transitivity of our sequence this means that $\mathsf{P}(B)$ is either 0 or 1. Since $\mathsf{P}(B)\geqslant \mathsf{P}(A_0)>0$, $\mathsf{P}(B)=1$. The theorem is proved. □

Remark 1. For the systems $\langle G_I, 1, G_I/m, 1\rangle_R$ it is sufficient for the validity of the condition $\mathsf{P}(A_0)>0$ that

$$\mathsf{P}(\tau^s\leqslant m\tau^e)>0 \quad\text{and}\quad \mathsf{M}\tau^s<\infty\,. \tag{7}$$

Indeed, let us prove that in this case,

$$\mathsf{P}(Q_0(0)\leqslant m-1\,, Q_0(\tau_1^e)\leqslant m-2, \ldots, Q_0(\tau_1^e+\cdots+\tau_{m-1}^e)=0)>0 \tag{8}$$

holds, which implies the fulfillment of Condition (2) of Theorem 1.

Using the formula for $Q_0(x)$, we find that the event following the probability symbol in (8) occurs if the event

$$A=\bigcap_{j=m-1}^{\infty}\{\tau_{-j}^s\leqslant\tau_{-j}^e+\cdots+\tau_0^e\}\bigcap_{j=m-2}^{\infty}\{\tau_{-j}^s\leqslant\tau_{-j}^e+\cdots+\tau_1^e\}\cdots$$
$$\bigcap_{j=0}^{\infty}\{\tau_{-j}^s\leqslant\tau_{-j}^e+\cdots+\tau_{m-1}^e\}$$

occurs. The latter obviously coincides with the event

$$\bigcap_{j=m}^{\infty}\{\tau_{-j}^s\leqslant\tau_{-j}^e+\cdots+\tau_0^e\}\bigcap_{k=0}^{m-1}\{\tau_{-k}^s\leqslant\tau_{-k}^e+\cdots+\tau_{-k+m-1}^e\}\,.$$

But Condition (7) means that there exists an $a>0$ such that

$$\mathsf{P}(\tau^e\geqslant a)=p>0 \text{ and } \mathsf{P}(\tau^s\leqslant ma)=q>0\,.$$

Since the event

$$A_1=\bigcap_{k=0}^{m-1+N}\{\tau_{-k}^s\leqslant ma\}\bigcap_{k=-m+1}^{m-1+N}\{\tau_{-k}^e\geqslant a\}\bigcap_{j=m+N}^{\infty}\{\tau_{-j}^s\leqslant\tau_{-j}^e+\cdots+\tau_0^e\}$$

implies the event A, one has

$$\mathsf{P}(A)\geqslant\mathsf{P}(A_1)\geqslant q^{m+N}p^{2m+N-1}\,\mathsf{P}\Big(\bigcap_{j=m+N}^{\infty}\{\tau_{-j}^s\leqslant\tau_{-j}^e+\cdots+\tau_0^e\}\Big)\,.$$

The positivity of the last factor for large enough N follows from the fact that the probability of the complementary event $\bigcup_{j=m+N}^{\infty}\{\tau_{-j}^s>\tau_{-j}^e+\cdots+\tau_0^e\}$, as we already know, can be made arbitrarily small (see (3), § 30). Hence, $\mathsf{P}(A)>0$ and the inequality at (8) is proved. □

Using the derived estimates it is easy to see that in the general case the conditions

$$\mathsf{M}\tau^s<\infty$$

and

$$\mathsf{P}(\bigcap_{k=0}^{m-1} \{\tau^s_{-k} \leqslant \tau^e_{-k} + \cdots + \tau^e_{-k+m-1}\} \bigcap_{k=m}^{N} \{\tau^s_{-k} \leqslant \tau^e_{-k} + \cdots + \tau^e_0\}) > 0$$

for arbitrary finite $N \geqslant m$ will be sufficient for (8) to hold.

Remark 2. Condition (2) is much more than is needed to ensure the convergence to one as $n \to \infty$ of the probability that for all $r > 0$ and $x \geqslant 0$

$$q_n^{n-1}(x) = q_{n+r}^{n+r-1}(x). \tag{9}$$

However, as we will see in Theorem 11, one cannot completely renounce additional restrictions on the governing sequences.

One can establish other sufficient conditions guaranteeing (9). Using the results of the last chapter on the distribution of $q^k(x)$, it is not hard to establish, for example, that for the systems $\langle G_I, 1, G_I/m, 1\rangle_R$, the following condition is sufficient, being in a certain sense supplementary to (7): *for some* x_0 *and all* $x \geqslant x_0$, $\Delta > 0$

$$\mathsf{P}(\tau^s \in (x, x+\Delta)) > 0, \qquad \mathsf{M}\tau^s < \infty.$$

This condition implies convergence to one as $n \to \infty$ of the probability that $q_{k-1}^{n-1} = q_{r+k-1}^{n+r-1} = m$ and $q_k^{n-1} = q_{r+k}^{n+r-1} = 0$ for at least one k, $-n+1 < k \leqslant 0$, and any $r > 0$.

2. We consider now the *systems* $\langle G, G, G/m, 1\rangle_R$ *with arbitrary initial conditions*. Let $(q_1;\, \rho_1, \ldots, \rho_{q_1})$, $q_1 \leqslant m$, be a vector defined on the basic probability space, where $\rho_1, \ldots, \rho_{q_1}$ are the service times of the initial requests.

It is convenient to give here the governing sequence $\{\tau_j^e, \nu_j^e, \tau_j^s\}$ in another form: either, as was done in Chapter 6, as the sequence $\{\tau_j^e, \nu_j^e, \boldsymbol{\tau}_j^s\}$, where $\boldsymbol{\tau}_n^s$ is the vector of service times of the n-th lot of customers, or by considering the system $\langle G, 1, G/m, 1\rangle_R$, equivalent to the original, with a governing sequence $\{\tilde{\tau}_j^e, \tau_j^s\}$, where $\tilde{\tau}_j^e$ can assume zero values. If, for example, $\tau_1^e = r_1$, $\nu_1^e = k_1$, $\tau_2^e = r_2$ and $\nu_2^e = k_2$, then we must assume that

$$\tilde{\tau}_1^e = \cdots = \tilde{\tau}_{k_1-1}^e = 0,\ \tilde{\tau}_{k_1}^e = r_1,\ \tilde{\tau}_{k_1+1}^e = \cdots = \tilde{\tau}_{k_1+k_2-1}^e = 0,\ \tilde{\tau}_{k_1+k_2}^e = r_2\text{, etc.}[1]$$

However, the possibility of zero values for the random variables τ_n^e changes nothing in the previous considerations. For such systems with arbitrary initial conditions

$$q_n(x) = \sum_{j=1}^{q_1} I(\rho_j > X_{n-1} + x) + \sum_{k=1}^{n-1} I(\tau_k^s > X_{n-1} - X_{k-1} + x)\,\delta(q_k).$$

The first sum converges to 0 as $n \to \infty$ for arbitrary proper $\rho_1, \ldots, \rho_{q_1}$. To retain the proof of Theorem 1 for this case we need merely replace (5) by (let n be even)

$$\bigcup_{k=-m+1}^{-n/2} A_k B \subset \bigcap_{r>0} \{q_n^{n-1}(x) = q_{n+r}^{n+r-1}(x)\},$$

---

[1] A certain inconvenience associated with such substitutions is due to the fact that the stationary original sequence does not always go over into a stationary sequence in the new form (even when account is taken of corrections which can be introduced by initial conditions). However, if the g.c.d. of the values of ν^e is 1 and $\mathsf{M}\nu^e < \infty$, then the governing sequence $\{\tilde{\tau}_j^e, \tau_j^s\}$ can be made stationary.

where

$$B=\bigcap_{j=1}^{U^{-n+1}q_1}\{U^{-n+1}\rho_j<U^{-n+1}X_{n/2}\}\bigcap_{j=1}^{U^{-n+1}q_1}\{U^{-n+1}\rho_j<U^{-n+1}X_{n/2+r}\}$$

and

$$\mathsf{P}(B)\to 1$$

as $n\to\infty$. Here we assume that $(q_1;\rho_1,\ldots,\rho_{q_1})$ is a random variable measurable w.r.t. the σ-algebra $\mathfrak{M}$ generated by $\{\tilde{\tau}_j^e,\tau_j^s;\ -\infty<j<\infty\}$ or w.r.t. some larger σ-algebra $\mathfrak{M}^*\subset\mathfrak{B}$ onto whose sets we can extend the definition of the measure-preserving shift-transformation T. If, for example, $(q_1,\rho_1,\ldots,\rho_{q_1})$ does not depend on the governing sequence, then, as we saw in § 3, such an extension of $\mathfrak{M}$ to an $\mathfrak{M}^*$ w.r.t. which $(q_1,\rho_1,\ldots,\rho_{q_1})$ is measurable, is always possible.

The rest of the calculations of the proof of Theorem 1 remain unchanged.

Hence, *for the system $\langle G, G, G/m, 1\rangle_R$ with arbitrary initial conditions measurable w.r.t. $\mathfrak{M}^*$, the assertion of Theorem 1 remains valid if its assumptions are satisfied by the sequence $\{\tilde{\tau}_j^e,\tau_j^s\}$. The σ-algebra $\mathfrak{M}^*$ and the sequence $\{\tilde{\tau}_j^e,\tau_j^s\}$ are defined above.*

3. We turn now to the behavior of $q(t)$—the number of busy channels at time t. In addition to $q(t)$ we will again consider the process $\{q(t,x);\,t\geqslant 0,\,x\geqslant 0\}$, where *$q(t,x)$ is the number of requests remaining in the system up to time $t+x$ out of those which were there at time t.* In other words, $q(t,x)$ is the number of requests in the system at time $t+x$ whose service continues longer than x.

We treat first the systems $\langle G, 1, G/m, 1\rangle_R$. With respect to the governing of these systems we can consider here a *formulation of the problem differing somewhat from the previous one.* It can be given with the help of the continuous time input process $\{e(t);\,-\infty<t<\infty\}$ which gives the number of customers who have entered the system up to time t. The input process $e(t)$ is related to the governing sequence as follows: If $t_1,t_2,\ldots$ are the jump times of $e(t)$ on $t>0$, then $t_2-t_1=\tau_1^e$, $t_3-t_2=\tau_2^e,\ldots$. The service time of a request arriving at the jump time t_j (i.e., of the j-th request), is τ_j^s.

We will assume in what follows that $\{e(t)\}$ is a *process with strictly stationary increments.* We saw in Chapter 1 that if the τ_j^e are i.i.d., then it is sufficient for the existence of such a process that τ_j^e be nonlattice and $\mathsf{M}\tau^e<\infty$. If t is an integer, then it is sufficient that the g.c.d. of τ_j^e's values equal 1 and that $\mathsf{M}\tau^e<\infty$. The distribution of t_1 coincides with that of the size of the first overshoot χ of a barrier at infinity in a random walk with jumps $\tau_1^e,\tau_2^e,\ldots$.

Let $\ldots,t_{-2},t_{-1},t_0$ be the jump times of $e(t)$ on $t\leqslant 0$ and

$$\eta(t)-1=\max\{k\colon t_k<t\}\,,\qquad \chi(t)=t_{\eta(t)}-t\,;\quad -\infty<t<\infty$$

(for a process $e(t)$ with stationary increments, $\chi(t)$ is distributed like t_1 for arbitrary $t>0$). Then for $q(0)=0$

$$\begin{aligned}q(t,x)&=I(t_1+\tau_1^s>t+x)+\cdots+I(t_1+X_{m-1}+\tau_m^s>t+x)\\&\quad+I(t_1+X_m+\tau_{m+1}^s>t+x)\,\delta(q_{m+1})+\cdots\\&\quad+I(t_1+X_{\eta(t)-2}+\tau_{\eta(t)-1}^s>t+x)\,\delta(q_{\eta(t)-1})\\&=\textstyle\sum_{j=1}^{\eta(t)-1}I(t_j+\tau_j^s>t+x)\,\delta(q_j)\,,\end{aligned}\tag{10}$$

where q_j is, as before, the queue length before the arrival of the j-th customer.

If some nonzero initial conditions $(q(0), \rho_1, \ldots, \rho_{q(0)})$ are given (ρ_j is the service time of the first customer in the j-th channel), then

$$q(t, x) = q(t, x; \rho_1, \ldots, \rho_{q(0)}) = \sum_{j=1}^{q(0)} I(\rho_j > t + x) + \sum_{j=1}^{\eta(t)-1} I(t_j + \tau_j^s > t + x)\,\delta(q_j). \tag{11}$$

Let $\mathfrak{N}$ be the σ-algebra generated by the increments of the process $e(t)$. Then we can associate with this process a group of measure-preserving shift-transformations: $T^t : T^t T^s = T^{t+s}$ defined on the elements of $\mathfrak{N}$ and the corresponding group of transformations U^t on $\mathfrak{N}$-measurable random variables. Hence, the difference $e(u_2 + t) - e(u_1 + t)$ can be represented as

$$e(u_2 + t) - e(u_1 + t) = U^t(e(u_2) - e(u_1)).$$

It is clear that the random variables t_k, $k = \ldots, -1, 0, 1, \ldots$; $\chi(t)$, $\eta(t)$ in (10), τ_j^e and $\tau_{\eta(t)-j}^e$ are all $\mathfrak{N}$-measurable. Here, $U^t\eta(0) = \eta(t)$ and $U^t\chi(0) = \chi(t)$.

In place of σ-algebra $\mathfrak{N}$ it is sometimes more convenient to consider a larger σ-algebra $\mathfrak{N}^* \subset \mathfrak{B}$ (such, for example, that the initial conditions are $\mathfrak{N}^*$-measurable) to whose elements one can extend a measure-preserving transformation T^t. Such σ-algebras $\mathfrak{N}^*$ will be said to be *extended.*

Theorem 2. *Assume the sequence $\{\tau_n^e; \tau_n^s\}$ satisfies the assumptions of Theorem 1 but that the entering stream of customers is described by the process $\{e(t); -\infty < t < \infty\}$ with strictly stationary increments. Then for arbitrary initial conditions $(q(0); \rho_1, \ldots, \rho_{q(0)})$ measurable w.r.t. some extended σ-algebra $\mathfrak{N}^*$, the distribution of the "processes" $\{q(t+u, x); u \geq 0, x \geq 0\}$ converges as $t \to \infty$ to the distribution of the stationary (w.r.t. u) process $\{q^c(u, x); u \geq 0, x \geq 0\}$.*

(More precisely: there exist processes $\tilde{q}^c(u, x) = \tilde{q}_t^c(u, x)$ distributed like $q^c(u, x)$ for all t and such that as $t \to \infty$

$$\mathsf{P}\Big(\bigcup_{u \geq 0} \bigcup_{x \geq 0} \{q(t+u, x) \neq \tilde{q}_t^c(u, x)\}\Big) \to 0).$$

The *proof* of the theorem is completely analogous to that of Theorem 1. Let $q(0) = 0$ for simplicity of notation. Then

$$\begin{aligned} U^{-t}q(t, x) = {} & I(\chi(-t) + \tau_{\eta(-t)}^s > t + x) \\ & + I(\chi(-t) + \tau_{\eta(-t)}^e + \tau_{\eta(-t)+1}^s > t + x) + \cdots \\ & + I(\chi(-t) + \tau_{\eta(-t)}^e + \cdots + \tau_{-1}^e + \tau_0^s > t + x)\,\delta(q_0^{-t}), \end{aligned}$$

where $q_j^{-t} = U^{-1}q_{\eta(t)-1+j}$ is the queue length at time t_j (the arrival time of the j-th customer) if the system begins operating at time $-t$.

In a similar way we can write down the value of $U^{-t-u}q(t+u, x)$.

Using again the majorants $Q(u, x)$ (corresponding to the stationary distribution of the queue length in the system $\langle G, 1, G/\infty, 1\rangle$ with the same governing scheme):

$$Q(u, x) = I(\tau_{\eta(u)-1}^s > \gamma(u) + x) + I(\tau_{\eta(u)-2}^s > \tau_{\eta(u)-2}^e + \gamma(u) + x) + \cdots,$$

where $\gamma(u) = u - t_{\eta(u)-1}$ is the defect up to the level u for the process $e(t)$, and noting that $\eta(-t)\to-\infty$ as $t\to\infty$ w.p.1, we can show as in Theorem 1 that there exists an interval of the form (t_j, t_{j+L}) in $(-t, 0)$ on which a complete "renewal" of the process $Q(u, x)$ occurs. Hence, the quantities $U^{-t}q(v, x)$ and $U^{-t-u}q(u+v, x)$, as functions of v, $0\leqslant v\leqslant t$, will coincide from some time onwards with probability converging to 1 as $t\to\infty$. This implies the a.s. convergence of $U^{-t}q(t, x)$.

As we have already seen in Subsection 2, the introduction of arbitrary initial conditions does not change the situation. □

The remarks made above on the nature of the systems $\langle G, G, G/m, 1\rangle_R$ and on the nature of other sufficient conditions guaranteeing the convergence of $U^{-t}q(t, x)$ remain in force.

An assertion analogous to Theorem 2 clearly holds when τ_j^e is integer-valued.

We return now to the case of an ordinary governing scheme in which requests enter the system at times $0, \tau_1^e, \tau_1^e+\tau_2^e, \cdots$. We denote now by $q_{\rho_1,\dots,\rho_{q(0)}}(t, x)$ the number of channels occupied at time $t+x$ by customers whose service lasts longer than x if the initial values were $(q(0); \rho_1, \dots, \rho_{q(0)})$. Then it is easy to see that the relation between this process and the process $q(t, x)=q(t, x; \rho_1, \dots, \rho_{q(0)})$ defined above (see (11)) can be written as

$$q_{\rho_1,\dots,\rho_{q(0)}}(t, x)=q(t+t_1, x; \rho_1+t_1, \dots, \rho_{q(0)}+t_1),$$

where t_1 is the time of first jump of the process $e(t)$. From the proof of Theorem 2 we see that $q(t, x)$ is asymptotically independent of t_1, as well as of the initial conditions, as $t\to\infty$. Hence, if there exists a process $e(t)$ with strictly stationary increments and the assumptions of Theorem 1 are fulfilled, then the limiting distribution of $q_{\rho_1,\dots,\rho_{q(0)}}(t, x)$ as $t\to\infty$ coincides with that of $q(t, x)$ and is independent of the initial conditions $q(0); \rho_1, \dots, \rho_{q(0)}$.

§ 36. Stability Theorems

To simplify the exposition we will limit ourselves in this section to consideration of the systems $\langle G, 1, G/m, 1\rangle_R$ with zero initial conditions. As in the case of systems with an infinite number of service channels, the stability theorems obtained below can be extended without difficulty to the systems $\langle G, G, G/m, 1\rangle_R$ when v_j^e is arbitrary and $q_1\neq 0$.

Assume given a family of sequences $\{\tau_j^{(r)e}, \tau_j^{(r)s}\}$, $r=1, 2, \dots$, governing systems which we denote by $\langle G, 1, G/m, 1\rangle_R^{(r)}$.

A. We assume that the *finite-dimensional distributions of these sequences converge as $r\to\infty$ to the distribution of the sequence $\{\tau_j^e, \tau_j^s\}$ and that all of the indicated sequences satisfy the assumptions of Theorem 1.*

We now ask whether or not the distributions of the corresponding processes $\{q^{(r)0}(x); x\geqslant 0\}$ will converge weakly. (The index (r) will correspond throughout to the system $\langle G, 1, G/m, 1\rangle_R^{(r)}$; weak convergence is understood as in Subsection 3 § 31.)

In order to formulate the basic result we impose the following conditions, which figured in § 31:

B. $\mathsf{M}\tau^{(r)s} \to \mathsf{M}\tau^s$ as $r \to \infty$;

C. The distribution of $\tau^s_{-j} - X^e_{-j}$, where $X^e_{-j} = \sum_{k=-j}^{0} \tau^e_k$, is continuous at 0 for all $j \geqslant 0$. In other words, for arbitrary integer N and $\delta \to 0$,

$$\mathsf{P}\left(\bigcup_{j=0}^{N} \{|\tau^s_{-j} - X^e_{-j}| \leqslant \delta\}\right) \to 0 .$$

Because of the stationarity it follows from this that the distributions of

$$\tau^s_{-j} - (\tau^e_{-j} + \cdots + \tau^e_k)$$

will also be continuous at 0 for all $j \geqslant 0$, $k \geqslant 0$.

Theorem 3. *Assume the governing sequences $\{\tau_j^{(r)e}, \tau_j^{(r)s}\}$ and $\{\tau_j^e, \tau_j^s\}$ satisfy Conditions A, B and C. Then the distributions of the processes $\{q^{(r)0}(x); x \geqslant 0\}$ converge weakly as $r \to \infty$ to that of $\{q^0(x); x \geqslant 0\}$; the convergence set S coincides with the set S_0 of points x at which the distributions of $\tau^s_{-j} - (\tau^e_{-j} + \cdots + \tau^e_k)$ are continuous for all $j \geqslant 0$, $k \geqslant 0$.*

Proof. Consider the random variable ν equal to the largest $k \leqslant -m+1$ for which the event A_k occurs. In other words, $\nu(\omega) = k$ if $\omega \in A_k \bar{A}_{k+1} \cdots \bar{A}_{-m+1}$, $k = -m+1, -m, \ldots$, where $\bar{A} = \text{compl.}(A)$. Since $\mathsf{P}(\bigcup_{k \leqslant -m+1} A_k) = 1$ (see the proof of Theorem 1), ν is a proper random variable and for prescribed $\varepsilon > 0$ we can find an N such that

$$\mathsf{P}(\nu < -N) < \varepsilon/2 .$$

But by the remark following Theorem 6A § 31 and Condition C, the distributions of the vectors (6.1) § 31 will converge, so that those of $\nu^{(r)}$ also converge to the distribution of ν. This means that we can find an r_ε such that for all $r \geqslant r_\varepsilon$

$$\mathsf{P}(\nu^{(r)} < -N) < \varepsilon .$$

We now note that the $q^k(x)$ satisfy the recursion relation (apply the shift transformation U^{n-k} in (3) and let $n \to \infty$)

$$q^k(x) = q^{k-1}(x + \tau^e_k) + I(\tau^s_k > \tau^e_k + x)\, \delta(q^{k-1}(0)) ,$$

where $\delta(l) = 0$ for $l = m$ and $\delta(l) = 1$ for $l < m$.

Using this rule, let us define the processes $\tilde{q}^k_N(x)$ as follows: For $0 \leqslant l \leqslant m-1$ define the initial values

$$\tilde{q}_N^{-N+l}(x) = \sum_{j=0}^{l} I(\tau^s_{-N+j} > \tau^e_{-N+j} + \cdots + \tau^e_{-N} + x) . \qquad (1.1)$$

For $l \geqslant m$ set

$$\tilde{q}_N^{-N+l}(x) = \tilde{q}_N^{-N+l-1}(x + \tau^e_{-N+l}) + I(\tau^s_{-N+l} > \tau^e_{-N+l} + x)\, \delta(\tilde{q}_N^{-N-l-1}(0)) . \quad (2.1)$$

As we have already remarked, the event A_k for $-N \leqslant k \leqslant -m+1$ implies the "renewal" or complete turnover described above in the sequence $Q_k(x)$ and hence, also in the sequences $q^k(x)$ and $\tilde{q}^k_N(x)$. This implies that on the set $\omega \in \{\nu \geqslant -N\}$ $\tilde{q}^k_N(x) = q^k(x)$ for all $k \geqslant 0$. Moreover, $\tilde{q}^0_N(x)$ is a function whose values depend only on the elements of the governing sequences numbered from $-N$ to 0. The form of this function can be obtained by using (1.1) and performing $N-m$ iterations with the help of (2.1). The result is an explicit expression containing sums of the indicators $I(\tau^s_{-N+j}>\tau^e_{-N+j}+\dots+\tau^e_{-N+j-k}+x)$ for different j and k and values of the function δ of those sums for which the number of iterations is $\leqslant N-m$. But the distribution of the sums of the indicators defined for the sequences $\{\tau_j^{(r)e}, \tau_j^{(r)s}\}$ as well as the distribution of the values of the function δ (whose arguments are integers) converge as $r\to\infty$ for $x\in S_0$ to the distributions of the corresponding expressions for the sequences $\{\tau_j^e, \tau_j^s\}$. Hence, if $x\in S_0$, then

$$\mathsf{P}(q^{(r)0}(x)\geqslant y)=\mathsf{P}(\tilde{q}_N^{(r)0}(x)\geqslant y;\, \nu^{(r)}\geqslant -N)+\mathsf{P}(q^{(r)0}(x)\geqslant y;\, \nu^{(r)}<-N);$$

$$\limsup_{r\to\infty}\mathsf{P}(q^{(r)0}(x)\geqslant y)\leqslant\varepsilon+\mathsf{P}(\tilde{q}_N^0(x)\geqslant y)\leqslant\tfrac{3}{2}\varepsilon+\mathsf{P}(\tilde{q}_N^0(x)\geqslant y;\, \nu\geqslant -N)$$

$$\leqslant\tfrac{3}{2}\varepsilon+\mathsf{P}(q^0(x)\geqslant y)\,.$$

An analogous estimate for the $\liminf_{r\to\infty}$ completes the proof for the one-dimensional distributions of $q^{(r)0}(x)$. The proof of the convergence of the multi-dimensional distributions is essentially the same. □

§ 37. The Systems $\langle G_I, 1, G_I/m, 1\rangle_R$

1. As we have already seen, Theorem 1 implies

Theorem 4. *If*

$$\mathsf{P}(\tau^s\leqslant m\tau^e)>0 \quad\text{and}\quad \mathsf{M}\tau^s<\infty\,, \tag{12}$$

then for arbitrary initial conditions $(q_1, \rho_1, \dots, \rho_{q_1})$ *measurable w.r.t. some extended σ-algebra* $\mathfrak{M}^*$ (see Subsection 2 § 35), *the distributions of the processes* $\{q_{n+k}(x);\, k\geqslant 0, x\geqslant 0\}$ *converge for* $n\to\infty$ *to that of some stationary (w.r.t. k) process* $\{q^k(x);\, k\geqslant 0, x\geqslant 0\}$.

This assertion remains valid if instead of (12) one requires that for some x_0 and all $x>x_0$, $\Delta>0$

$$\mathsf{P}(\tau^s\in(x, x+\Delta))>0 \quad\text{and}\quad \mathsf{M}\tau^s<\infty\,.$$

We now find an equation satisfied by the stationary distribution of $q^0(x)$. We have already noted that the process $q^0(x)$ is completely determined by the position of its jumps. Denote by

$$\mu(\underbrace{0, \dots, 0}_{j\text{ times}}, dx_{j+1}, \dots, dx_m) \tag{13}$$

the probability that $q^0(0)=q^0=m-j$, and $m-j$ of the jumps of the process $q^0(x)$ are included in the intervals $dx_k=(x_k, x_k+dx_k)$, $k=j+1, \dots, m$, resp. (The use of the symbol dx_k simultaneously as a scalar and as an interval should cause no misunderstanding.)

Theorem 5. *The stationary distribution* $\mu(0, \ldots, 0, dx_{j+1}, \ldots, dx_m)$, $j=0, 1, \ldots, m$, *satisfies*

$$\mu(0, \ldots, 0, dx_{j+1}, \ldots, dx_m)=\sum_{k=j+1}^{m} \int_0^\infty \mathsf{P}(\tau^e \in dt)\mathsf{P}(\tau^s \in t+dx_k)M_{1,k} + \int_0^\infty \mathsf{P}(\tau^e \in dt)\mathsf{P}(\tau^s \leqslant t)M_2 + \int_0^\infty \mathsf{P}(\tau^e \in dt)M_3, \tag{14}$$

where

$$M_{1,k}=\int\cdots\int_{0\leqslant y_2\leqslant y_3\leqslant\ldots\leqslant y_{j+1}\leqslant t} \mu(0, dy_2, \ldots, dy_{j+1}, dx_{j+1}+t, \ldots, dx_{k-1} + t, dx_{k+1}+t, \ldots, dx_m+t),$$

$$M_2=\int\cdots\int_{0\leqslant y_2\leqslant y_3\leqslant\ldots\leqslant y_j\leqslant t} \mu(0, dy_2, \ldots, dy_j, dx_{j+1}+t, \ldots, dx_m+t)$$

and

$$M_3=\int\cdots\int_{0< y_1\leqslant y_2\leqslant\ldots\leqslant y_j\leqslant t} \mu(dy_1, \ldots, dy_j, dx_{j+1}+t, \ldots, dx_m+t).$$

(*Remarks*. The interval dy is to be understood for $y=0$ as the point 0. Hence, "discrete" values such as $\mu(0, \ldots, 0, dx_{j+1}+t, \ldots, dx_m+t)$ (in M_2), etc. also enter into the integrals $M_{1,k}$ and M_2. They do not enter into M_3.

On the right side of the equation for $\mu(0, \ldots, 0)$, the terms containing $M_{1,k}$ will be missing. In the equation for $\mu(dx_1, \ldots, dx_m)$ with $x_1>0, \ldots, x_m>0$, the term containing M_2 will be missing. The integral M_3 in this equation goes over into $\mu(dx_1+t, \ldots, dx_m+t)$.)

Equation (14) *has a unique solution satisfying*

$$\int\cdots\int_{0\leqslant x_1\leqslant\ldots\leqslant x_m} \mu(dx_1, \ldots, dx_m)=1.$$

Proof. For the systems $\langle G_I, 1, G_I/m, 1\rangle_R$, the sequence of processes $\{q_n(x); x\geqslant 0\}$, $n=1, 2, \ldots$, forms a homogeneous Markov chain, as indicated by (3) (the state space of this chain can be identified with the Euclidean space R^m).

If we go to the limit in (3) (applying first the transformation U^{-n}), then for the distribution of the stationary process $q^k(x)$ we get

$$q^0(x) \underset{d}{=} q^1(x)=I(\tau_0^s>\tau_0^e+x)\,\delta(q^0(0))+q^0(x+\tau_0^e).$$

If we write this equality in terms of the measure μ using on the right the total probability formula w.r.t. the variables τ_0^e, τ_0^s and $q^0(0)$, we get (14). The terms containing $M_{1,k}$ and M_2 correspond to $q^0(0)<m$; that containing M_3 to $q^0(0)=m$. The uniqueness of the solution follows from the fact that the sequence $q_n(x)$ is Markov and from Theorem 3 on the convergence of the distributions.

2. Theorem 6. *If* $\mathsf{P}(\tau^e>x)=e^{-\alpha x}$ *and* $a=\mathsf{M}\tau^s$ (*the systems* $\langle E, 1, G_I/m, 1\rangle_R$), *then*

$$\mu(\underbrace{0, \ldots, 0}_{j\text{ times}}, dx_{j+1}, \ldots, dx_m)=c\alpha^{m-j}P(x_{j+1})\ldots P(x_m)\,dx_{j+1}\ldots dx_m, \tag{15}$$

where

$$P(x)=\mathsf{P}(\tau^s\geqslant x) \text{ and } c=\left[\sum_{k=0}^{m}\frac{(a\alpha)^k}{k!}\right]^{-1}.$$

Proof. Since (12) is fulfilled here, by Theorem 3 it is enough to show that the distribution (15) satisfies (14). Writing $I=I(t)=\int_0^t P(x)\,dx$, we have with accuracy up to the constant c

$$M_{1,k}=\left[\frac{\alpha^{m-1}}{j!}I^j+\frac{\alpha^{m-2}}{(j-1)!}I^{j-1}+\cdots+\alpha^{m-j-1}\right]\prod_{l=j+1,\,l\neq k}^m P(x_l+t)\,dx_l\,,$$

$$M_2=\left[\frac{\alpha^{m-1}}{(j-1)!}I^{j-1}+\frac{\alpha^{m-2}}{(j-2)!}I^{j-2}+\cdots+\alpha^{m-j}\right]\prod_{l=j+1}^m P(x_l+t)\,dx_l\,,$$

$$M_3=\frac{\alpha^m}{j!}I^j\prod_{l=j+1}^m P(x_l+t)\,dx_l\,.$$

Integrating by parts we find that the sum $\sum_{k=j+1}^m$ on the right in (14) equals

$$-\sum_{k=j+1}^m\int_0^\infty \alpha e^{-\alpha t}\,dt\,d_{x_k}P(x_k+t)M_{1,k}$$

$$=-\prod_{l=j+1}^m dx_l\int_0^\infty \alpha e^{-\alpha t}\left[\sum_{s=1}^{j+1}\frac{\alpha^{m-s}}{(j+1-s)!}I^{j+1-s}\right]d_t\prod_{k=j+1}^m P(x_k+t)$$

$$=\alpha^{m-j}\prod_{l=j+1}^m P(x_l)\,dx_l$$

$$+\prod_{l=j+1}^m dx_l\int_0^\infty\prod_{l=j+1}^m P(x_l+t)\alpha d\left\{e^{-\alpha t}\left[\sum_{s=1}^{j+1}\frac{\alpha^{m-s}}{(j+1-s)!}I^{j+1-s}\right]\right\}.$$

Put $\Pi(t)=\prod_{j=j+1}^m P(x_l+t)$. Then the right side of (14), leaving out the product $\prod_{l=j+1}^m dx_l$ and the constant c, equals

$$\alpha^{m-j}\Pi(0)+\int_0^\infty \alpha e^{-\alpha t}\Pi(t)\left[-\alpha\sum_{s=1}^{j+1}\frac{\alpha^{m-s}}{(j+1-s)!}I^{j+1-s}\right.$$

$$\left.+\sum_{s=1}^{j}\frac{\alpha^{m-s}}{(j-s)!}I^{j-s}P(t)+(1-P(t))\sum_{s=1}^{j}\frac{\alpha^{m-s}}{(j-s)!}I^{j-s}+\frac{\alpha^m I^j}{j!}\right]dt\,.$$

The sum in square brackets equals 0. Since $c\alpha^{m-j}\Pi(0)\prod_{l=j+1}^m dx_l$ stands on the left in (14), the theorem is proved. □

Corollary. *For* $0\leqslant k\leqslant m$, $k\geqslant l\geqslant 0$

$$\mathsf{P}(q^0(0)=k,\,q^0(x)=k-l)=c\frac{(a\alpha)^k}{k!}\binom{k}{l}\left(\frac{I(x)}{a}\right)^l\left(1-\frac{I(x)}{a}\right)^{k-l},$$
$$\mathsf{P}(q^0(0)=k)=c\frac{(a\alpha)^k}{k!}\cdot \tag{16}$$

These formulae follow directly from the equality

$$\mathsf{P}(q^0(0)=k,\,q^0(x)=k-l)$$
$$=\int\cdots\int_{\substack{0<x_{m-k+1}\leqslant\cdots\leqslant x_{m-k+l}\leqslant x\\ x\leqslant x_{m-k+l+1}\leqslant\cdots\leqslant x_m}}\mu(\underbrace{0,\ldots,0}_{m-k},dx_{m-k+1},\ldots,dx_m)\,,$$

when the obtained value of μ is substituted.

From (16) it becomes clear that the conditional probability that in the stationary regime a fixed channel will continue serving a request longer than time x after the arrival of the current request, given that at this arrival time service has already been requested (the channel is occupied), is equal to $1-(I(x)/a)$. That is, this conditional service duration is distributed like the overshoot of a barrier at infinity in a random walk with jumps $\tau_1^s, \tau_2^s, \ldots$.

3. We now consider the stationary process $q^c(u, x)$ which is the limit "in distribution" of the sequence $q(t+u, x)$ as $t\to\infty$. (As we saw at the end of Subsection 3 § 35, the limiting distribution of $q(t+u, x)$ will be the same for both varieties of stationary governing schemes considered in § 35 and does not depend on the initial conditions.)

The distribution of $q^c(0, x)$, as well as that of $q^0(x)$, is determined by a certain measure $\lambda(\underbrace{0, \ldots, 0}_{j\text{ times}}, dx_{j+1}, \ldots, dx_m)$, which is defined in analogy with the measure $\mu(\underbrace{0, \ldots, 0}_{j}, dx_{j+1}, \ldots, dx_m)$ in (14).

Theorem 7. *For the systems $\langle E, 1, G_I/m, 1\rangle_R$, the stationary distributions λ and μ coincide.*

Hence, for the distribution of $q^c(u, x)$ all of the explicit formulae obtained for $q^0(x)$ are valid. From this follow, in particular, Erlang's formulae for the stationary distribution of $q(t, x)$.

Proof. In the case $\mathsf{P}(\tau^e \geqslant x)=e^{-\alpha x}$ the process $\mathbf{q}(t)=(q(t, x), x\geqslant 0)$ is Markovian. By the theorem on convergence to a stationary distribution the measure λ is the unique solution of the corresponding equation for the invariant measure constructed w.r.t. the transition function for $\mathbf{q}(t)$. Since this transition function can be uniquely recovered from the infinitesimal operator, we can consider an equation constructed for small time intervals as the equation for λ.

Indeed, by the total probability formula for the distributions of $q^c(\Delta, x)$ and $q^c(0, x)$, we find that

$$\begin{aligned}\lambda(\underbrace{0, \ldots, 0}_{j}, dx_{j+1}, \ldots, dx_m) &= e^{-\alpha\Delta}\lambda(\underbrace{0, \ldots, 0}_{j}, dx_{j+1}+\Delta, \ldots, dx_m+\Delta)\\ &\quad + e^{-\alpha\Delta}\lambda(\underbrace{0, \ldots, 0}_{j-1}, (0, \Delta), dx_{j+1}+\Delta, \ldots, dx_m+\Delta)\\ &\quad + \alpha\Delta \textstyle\sum_{k=j+1}^{m} \lambda(\underbrace{0, \ldots, 0}_{j+1}, dx_{j+1}+\Delta, \ldots, dx_{k-1}+\Delta,\\ &\qquad dx_{k+1}+\Delta, \ldots, dx_m+\Delta)\\ &\quad \times \mathsf{P}(\tau^s \in dx_k+\theta\Delta)+o(\Delta); \quad \theta\leqslant 1, j=1, \ldots, m.\end{aligned} \tag{17}$$

To estimate the remainder here we have assumed that λ has a density. It is surely sufficient to show that the measure λ, definable by the densities

$$\frac{\lambda(0, \ldots, 0, dx_{j+1}, \ldots, dx_m)}{dx_{j+1}\ldots dx_m} = p_{m-j}(x_{j+1}, \ldots, x_m) = \alpha^{m-j}\prod_{k=j+1}^{m} P(x_k), \tag{18}$$

where $P(x) = \mathsf{P}(\tau^s \geqslant x)$, satisfies (17). In order to make this verification more intuitive we assume that τ^s has a right-continuous density

$$p(x) = -\frac{dP(x)}{dx}.$$

Then, as is clear from (17), the derivatives $\partial p_k/\partial x_l$ exist and our equation takes the form

$$-\sum_{l=j+1}^{m}\frac{\partial p_{m-j}(x_{j+1}, \ldots, x_m)}{\partial x_l} = -\alpha p_{m-j}(x_{j+1}, \ldots, x_m) + p_{m-j+1}(0, x_{j+1}, \ldots, x_m)$$
$$+\alpha\sum_{k=j+1}^{m} p_{m-j-1}(x_{j+1}, \ldots, x_{k-1}, \ldots, x_m)p(x_k). \tag{19}$$

Substituting (18) here we obtain an identity since the first two terms on the right in (19) cancel one another and the left side equals

$$\alpha^{m-j}\sum_{l=j+1}^{m}\prod_{k=j+1,\, k\neq l}^{m} P(x_k)p(x_l).$$

We leave to the reader the verification of the fact that (18) satisfies (17) in the general case (in which there is no $p(x)$). In order to get a differential equation in this case we can introduce the functions

$$\tilde{p}_{m-j}(x_{j+1}, \ldots, x_m) = p_{m-j}(x_{j+1}, \ldots, x_m)/\prod_{k=j+1}^{m} P(x_k). \quad \square$$

4. We also entrust to the reader the verification of the fact that assertions similar to Theorems 6 and 7 remain valid for the discrete analogue of the systems $\langle E, 1, G_I/m, 1\rangle_R$, where $\mathsf{P}(\tau^e = k) = p^{k-1}(1-p)$, $k = 1, 2, \ldots$, and τ^s is integer-valued. The distribution of $q^0(0)$ in this case will not be Poisson (see Subsection 2 § 34).

5. *The rate of convergence of the distributions of $q_{n+k}(x)$ and $q^k(x)$ for the systems $\langle G_I, 1, G_I/m, 1\rangle_R$.* We assume for simplicity that $\mathsf{P}(\tau^s \leqslant \tau^e) > 0$. We then have

Theorem 8. *If $\mathsf{M}\exp(\lambda\tau^s) < \infty$ for some $\lambda > 0$, then there exist $\varepsilon > 0$ and $c < \infty$ and processes $\tilde{q}^k(x)$ distributed like $q^k(x)$ such that*

$$\mathsf{P}\Big(\bigcup_{k\geqslant 0}\bigcup_{x\geqslant 0}\{q_{n+k}(x) \neq \tilde{q}^k(x)\}\Big) \leqslant ce^{-\varepsilon n}. \tag{20}$$

If $\mathsf{M}\exp(\lambda\tau^s) = \infty$ for arbitrary $\lambda > 0$, but

$$P_n = \mathsf{P}(\tau_1^e + \cdots + \tau_n^e < \tau_1^s) \leqslant c_1 n^{-\alpha}\varepsilon(n), \quad \alpha > 2,$$

where $\varepsilon(n)$ *is a slowly-varying function and* c_1 *is a constant, then for some* $c_2<\infty$ *the left side of* (20) *does not exceed* $c_2 n^{-\alpha+1}\varepsilon(n)$.

At the end of the Subsection and at the end of the proof of the theorem we will refer to methods of constructing various effective estimates for the left side of (20) as well as other types of conditions useful in the search for estimates of the form (20).

Proof. Let B_k denote the event $\{Q_k(0)=0\}$ (see (1)). Then, as in (4), it is easy to establish that

$$\bigcup_{k=-n+1}^{0} B_k \subset \bigcap_{r\geq 1} \bigcap_{x\geq 0} \{q_n^{n-1}(x)=q_{n+r}^{n+r-1}(x)\}.$$

This implies that ($q^0(x)=\lim_{n\to\infty} q_n^{n-1}(x)$)

$$\mathsf{P}(\bigcup_{x\geq 0}\bigcup_{r\geq 0}\{q_{n+r}^{n+r-1}(x)\neq q^0(x)\})\leqslant \mathsf{P}(\overline{\bigcup_{k=-n+1}^{0} B_k})=\mathsf{P}(\bigcap_{k=-n+1}^{0}\bar{B}_k).$$

Let $p=\mathsf{P}(B_0)$ and $p^{(k)}$, $k=1,2,\ldots$, be the probability that the event B_j, $j=-n+1$, $-n+2,\ldots,0,1,\ldots$ occurs for the first time when $j=-n+k$. Our problem then consists of estimating $\sum_{k=n+1}^{\infty} p^{(k)}$. We have

$$\begin{aligned} p=\mathsf{P}(B_0)=p^{(n)}&+p^{(n-1)}\mathsf{P}(\tau_0^s\leqslant\tau_0^e)\\ &+p^{(n-2)}\mathsf{P}(\tau_0^s\leqslant\tau_0^e,\ \tau_{-1}^s\leqslant\tau_{-1}^e+\tau_0^e)+\cdots\\ &+p^{(1)}\mathsf{P}(\tau_0^s\leqslant\tau_0^e,\ldots,\tau_{-n+1}^s\leqslant\tau_{-n+1}^e+\cdots+\tau_0^e). \end{aligned} \tag{21}$$

Set

$$D_k=\{\tau_k^s\leqslant\tau_k^e+\cdots+\tau_0^e\},\quad k=0,-1,-2,\ldots;$$

$$C_n=\bigcap_{k=0}^{-n+1} D_k,\qquad r_n=\mathsf{P}(C_n\bar{C}_\infty),\qquad r_0=1-p,$$

$$r(z)=\sum_{n=0}^{\infty} r_n z^n \quad\text{and}\quad p(z)=\sum_{n=1}^{\infty} p^{(n)}z^n.$$

Then $\mathsf{P}(C_n)=p+r_n$ and (21) implies that for $|z|<1$

$$\frac{pz}{1-z}=p(z)\left(\frac{p}{1-z}+r(z)\right),\qquad p(z)=\frac{pz}{p+(1-z)r(z)}=\frac{pz}{1-\Delta(z)}, \tag{22}$$

where $\Delta(z)=\sum_{k=1}^{\infty} z^k\Delta_k$, $\Delta_k=r_{k-1}-r_k\geqslant 0$, $k=1,2,\ldots$. Now note that $\Delta(1)=r_0=1-p<1$ and

$$\Delta_n=\mathsf{P}(C_{n-1}\bar{D}_{-n+1})\leqslant\mathsf{P}(\bar{D}_{-n+1})=\mathsf{P}(\tau_{-n+1}^e+\cdots+\tau_0^e<\tau_{-n+1}^s). \tag{23}$$

The asymptotic behavior of $p^{(n)}$, as is clear from (22), will vary considerably depending on the relation between the two numbers

$$b_1=\inf\{z:\Delta(z)\geqslant 1\}\geqslant 1$$

and

$$b_+=\sup\{z:\Delta(z)<\infty\}\geqslant b_1\geqslant 1;$$

b_+ is the first singularity of the function $\Delta(z)$ and b_1 is the first zero of the function $1-\Delta(z)$, ($\Delta(z)\uparrow$, $\Delta(1)<1$).

If $b_+ = b_1$, $\Delta(b_1) < 1$ and for some $a \geqslant 1$ $P_n a^n < cn^{-\alpha}\varepsilon(n)$, $\alpha > 2$, then by Theorem 1 of Appendix 3 there exists a constant c_1 such that

$$p^{(n)} < c_1 n^{-\alpha}\varepsilon(n) a^{-n} .$$

If, however, $b_1 < b_+$, then $b_1 > 1$, and picking out the first singularity of the function $p(z)$, we obtain

$$p^{(n)} = \frac{p b_1^{-n}}{\Delta'(b_1)}(1 + o(h^n)), \quad h < 1 . \tag{24}$$

Hence, the estimates for $p^{(n)}$ and so also for $\sum_{n+1}^{\infty} p^{(k)}$, are exponential if $b_+ > 1$. It remains to remark that $b_+ > 1$ when $\mathsf{M}\exp(\lambda\tau^s) < \infty$ for some $\lambda > 0$. □

In the case $b_+ > b_1$ the relation (24) will be useful only when the known properties of the coefficients of Δ_n allow computation of the values of b_1 and $\Delta'(b_1)$.

We now indicate some methods for the construction of rougher but simpler estimates. Using, for example, the inequality ($\lambda \geqslant 0$)

$$\mathsf{P}(\tau_1^e + \cdots + \tau_n^e < \tau_1^s) \leqslant \psi_s(\lambda)\psi_e^n(-\lambda) ,$$

$$\psi_s(\lambda) = \mathsf{M}\exp(\lambda\tau^s) , \qquad \psi_e(\lambda) = \mathsf{M}\exp(\lambda\tau^e) ,$$

we find that $p^{(n)}$ is majorized for each fixed N by the coefficients of the series

$$\begin{aligned} pz&\left[1 - \Delta_1 z - \Delta_2 z^2 - \cdots - \Delta_{N-1} z^{N-1} - \frac{\psi_s(\lambda)\psi_e^N(-\lambda)z^N}{1 - z\psi_e(-\lambda)}\right]^{-1} \\ &= (1 - z\psi_e(-\lambda)) \\ &\quad \times pz[(1 - z\psi_e(-\lambda))(1 - \Delta_1 z - \cdots - \Delta_{N-1}z^{N-1}) - \psi_s(\lambda)\psi_e^N(-\lambda)z^N]^{-1} , \end{aligned}$$

which can be found in explicit form if we expand the right side into partial fractions and then choose $\lambda > 0$ in the best way.

Another type of condition which allows us to find estimates of the rate of convergence is determined by the following relation: If $\mathfrak{M}_n$ is the σ-algebra generated by the sequence $\{\tau_j^e, \tau_j^s; -\infty < j \leqslant n\}$ and the initial conditions, then for some n_0 and all $n \geqslant 0$ one has almost everywhere

$$\mathsf{P}_{\mathfrak{M}_n}(q_{n+n_0} = 0) > c_0 > 0 .$$

When this condition holds we can apply the methods we used in Theorems 6 and 7, Chapter 5.

§ 38. The Systems $\langle G_I, G_I, E/m, G_I\rangle_R$

1. Here the sequence of random variables q_n forms a simple homogeneous Markov chain with a finite number of states $E_0, \ldots, E_m$ which are defined in a natural way. If we assume that $\mathsf{P}(\nu^s = 1) > 0$, then its possible jumps are $E_k \to E_{\min(m,\,k-j+1)}$ for any k and $k+l \geqslant j \geqslant 0$, where l is the smallest value of v^e. As is easy to see, in this case the chain has a single class of communicating states with no subclasses. Consequently,

$$\lim_{n\to\infty} \mathsf{P}(q_n=j)=p_j>0$$

exist and

$$|\mathsf{P}(q_n=j)-p_j|<ce^{-\varepsilon n}$$

for some $c<\infty$ and $\varepsilon>0$.

2. *The System* $\langle G_I, 1, E/m, 1\rangle_R$.

Theorem 9. *If* $\mathsf{P}(\tau^s>x)=e^{-\alpha x}$, *then*

$$p_k=\sum_{r=k}^{m}(-1)^{r-k}\binom{r}{k}B_r\,, \quad k=0,1,\ldots,m\,, \tag{25}$$

where

$$B_r=C_r\frac{\sum_{j=r}^{m}\binom{m}{j}\frac{1}{C_j}}{\sum_{j=0}^{m}\binom{m}{j}\frac{1}{C_j}}, \quad C_0=1,\ C_r=\prod_{j=1}^{r}\frac{\psi_j}{1-\psi_j}$$

and

$$\psi_j=\mathsf{M}\exp(-\alpha\tau^e j)\,.$$

Proof. The probabilities p_k are the unique solution of a system of equations for the invariant measure of the form (32) § 29 in which the transition probabilities p_{ij} are defined by (30) and (31) § 29 for $i\leqslant m$ and $j\leqslant m$ if we put $p_{m,m+1}=0$ and understand $p_{m,m}$ as the sum $p_{m,m}+p_{m,m+1}$ of probabilities calculated for the system with a queue. This change closes a system of the first $m+1$ equations for $p_0,\ldots,p_m$. It remains to verify directly that the substitution of (25) into this system converts it into an identity. We leave this to the reader. □

The following theorem can be proved just like Theorem 10 of Chapter 5.

Theorem 10. *If* τ^e *is nonlattice and* $\mathsf{M}\tau^e<\infty$, *then for arbitrary initial conditions*

$$\lim_{t\to\infty}\mathsf{P}(q(t)=k)=P_k=\frac{p_{k-1}}{k\alpha\mathsf{M}\tau^e}, \quad k=1,2,\ldots,m\,,$$

exist.

For the systems considered in this section we can also find

$$\sum_{n=1}^{\infty} z^n\mathsf{P}(q_n=k) \quad\text{and}\quad \int_0^\infty e^{-\lambda t}\mathsf{P}(q(t)=k)\,dt$$

in explicit form.

We will not do this here as the results are rather cumbersome (see Takács [71] p. 180).

§ 39. The Systems $\langle G\rangle_R$

1. From the standpoint of behavior of the processes $q_n(x)$ and $q(t, x)$, the systems $\langle G, G, G, G\rangle_R$ are clearly equivalent to the systems $\langle G, 1, G, 1\rangle_R$. From this and the theorems of § 35 we obtain the following assertion for the systems $\langle G\rangle_R$:

Theorem 11. *If the sequence* $\{\tau_j^e, \tau_j^s; -\infty<j<\infty\}$ *is strictly stationary, ergodic and*

$$\mathsf{P}(\tau_0^s \leqslant \tau_0^e, \tau_{-1}^s \leqslant \tau_{-1}^e + \tau_0^e, \tau_{-2}^s \leqslant \tau_{-2}^e + \tau_{-1}^e + \tau_0^e, \dots) > 0,$$

then

$$\lim_{n\to\infty} \mathsf{P}(q_n(x)=1) = p(x).$$

exists.

An analogous statement holds for $\mathsf{P}(q(t, x)=1)$ when $t\to\infty$. There also exist processes $\{q^k(x)\}$ and $\{q^c(u, x)$, stationary w.r.t. k and u, to which $q_{n+k}(x)$ and $q(t+u, x)$ converge in distribution as $n\to\infty$ and $t\to\infty$. If, in addition, the sequence $\{v_i^e, v_i^s\}$ is also stationary and does not depend on $\{\tau_j^e, \tau_j^s\}$, and π_n is the probability that a request taken at random (with uniform distribution) from the n-th lot is refused, then

$$\begin{aligned}\lim_{n\to\infty} \pi_n &= p(0) + (1-p(0))\left(1 - \mathsf{M}\min\left(1, \frac{v^s}{v^e}\right)\right)\\ &= 1-(1-p(0))\mathsf{M}\min\left(1, \frac{v^s}{v^e}\right).\end{aligned} \tag{26}$$

exists.

We turn now to the systems $\langle G, 1, G, 1\rangle_R$ for which the pairs (τ_j^e, τ_j^s) form a sequence of *i.i.d. vectors* independent of the initial conditions (q_1, ρ_1) (ρ_1 is the service time of the initial request). We turn our attention to the fact that here, as everywhere else in this chapter, the variable τ_n^s is understood as the service time of the nth request to enter the system *provided this request is accepted for service*. Otherwise, the element τ_n^s of the governing sequence is not used. For systems with refusals the element τ_n^s can also be understood in another way. We can assume, for example, that τ_n^s is the service time of the n-th request to be accepted for service (the number under which it enters will be as a rule greater than n; in this case the assumption on the dependence of the elements of the vector (τ_n^e, τ_n^s) seems to be unnatural). If in the sequel the sequence $\{\tau_j^s\}$ is understood in this different way we will be point this out.

Hence, we assume that the vectors (τ_j^e, τ_j^s) are independent and set

$$X_k = \textstyle\sum_{j=1}^k \tau_j^e, \qquad X_0=0; \qquad r_k = \mathsf{P}(\tau_1^s \in (X_{k-1}, X_k]), \quad k=1, 2, \dots,$$

so that $\sum_{k=1}^\infty r_k = 1$. The probabilities r_k define the distribution of the first passage time η of the level τ_1^s in a random walk with jumps $\tau_1^e, \tau_2^e, \dots$:

$$\eta = \min\{k: X_k \geqslant \tau_1^s\}.$$

If χ is the overshoot $X_\eta - \tau_1^s$, then $\mathsf{M}\chi$ and $\mathsf{M}\eta = \sum kr_k$ are connected by Wald's identity in the relation

$$\mathsf{M}\tau^s + \mathsf{M}\chi = \mathsf{M}\tau^e \mathsf{M}\eta . \tag{27}$$

For the indicated systems we have

Theorem 12. *For the existence with arbitrary initial condition* (q_1, ρ_1) *of*

$$\lim_{n\to\infty} \mathsf{P}(q_n(x)=1)=p(x)$$

it is necessary and sufficient that the set of indices k for which $r_k>0$ *(or, what is the same, the set of possible values of* η*) have g.c.d. equal to 1. When this holds,*

$$p(x)=\frac{1}{\mathsf{M}\eta}\sum_{j=1}^{\infty} \mathsf{P}(\tau_1^s>\tau_1^e+\cdots\tau_j^e+x) .$$

If τ^e *is nonlattice (or integral with lattice spacing 1 for discrete time), then*

$$\lim_{t\to\infty} \mathsf{P}(q(t,x)=1)=P(x)=\frac{1}{\mathsf{M}\tau^e\mathsf{M}\eta}\int_x^\infty \mathsf{P}(\tau^s\geqslant u)\,du .$$

always exists. In these formulae

$$p(0)=1-\frac{1}{\mathsf{M}\eta} \quad\text{and}\quad P(0)=\frac{\mathsf{M}\tau^s}{\mathsf{M}\tau^e\mathsf{M}\eta} . \tag{28}$$

Proof. First assume $q_1=0$. We split up the sequence $\tau_1^e, \tau_2^e, \ldots$ into independent cycles $(\tau_1^e,\ldots,\tau_{\eta_1}^e)$, $(\tau_{\eta_1+1}^e,\ldots,\tau_{\eta_1+\eta_2}^e)$, ... generated by passage of the levels $\tau_1^s, \tau_{\eta_1+1}^s, \ldots$, by successive sums of terms of the sequence $\{\tau_j^e\}$ so that

$$\eta_1=\min\{k: \tau_1^e+\cdots+\tau_k^e\geqslant\tau_1^s\}, \qquad \eta_2=\min\{k: \tau_{\eta_1+1}^e+\cdots+\tau_{\eta_1+k}^e\geqslant\tau_{\eta_1+1}^s\}, \quad \text{etc.}$$

Since the η_k are Markovian, they form a sequence of independent random variables distributed like η. The event $\{q_n=0\}$ is clearly equivalent to one of the sums

$$X_k^\eta=\sum_{j=1}^k \eta_j$$

being exactly equal to n. But it is well known (see also Appendix 1) that the limit of the probability of this event exists as $n\to\infty$ (and equals $1/\mathsf{M}\eta=1-p(0)$) iff the g.c.d. of the values of η is 1. This proves the necessity.

Now by the total probability formula with $n\to\infty$

$$\begin{aligned}\mathsf{P}(q_n(x)=1)&=\sum_{k=0}^{\infty}\sum_{j=0}^{n-1} \mathsf{P}(X_k^\eta=j)\mathsf{P}(\tau_j^s-\tau_j^e-\cdots-\tau_{n-1}^e>x)\\&=\sum_{j=0}^{n-1} h(j)\mathsf{P}(\tau_1^s>\tau_1^e+\cdots+\tau_{n-j}^e+x)\\&\to\frac{1}{\mathsf{M}\eta}\sum_{j=1}^{\infty} \mathsf{P}(\tau_1^s>\tau_1^e+\cdots+\tau_j^e+x) ,\end{aligned}$$

where $h(n)$ is the renewal density of the random variable η.

If τ^e is nonlattice, then the limiting distribution of the defect $\gamma(t)$ up to the level t in a random walk with jumps

$$\zeta_1 = \tau_1^e + \cdots + \tau_{\eta_1}^e, \qquad \zeta_2 = \tau_{\eta_1+1}^e + \cdots + \tau_{\eta_1+\eta_2}^e, \ldots$$

exists. Hence, using the renewal theorem again and putting $X_k^\zeta = \sum_{j=1}^k \zeta_j$, with $H_\zeta(t)$ the renewal function of ζ_1, we get

$$\begin{aligned}\mathsf{P}(q(t,x)=1) &= \textstyle\sum_{k=0}^{\infty} \int_0^t \mathsf{P}(X_k^\zeta \in du)\mathsf{P}(\tau^s > t-u+x)\\ &= \textstyle\int_0^t dH_\zeta(u)\mathsf{P}(\tau^s > t-u-x) \to \int_0^\infty \frac{du}{\mathsf{M}\zeta}\mathsf{P}(\tau^s > u+x)\\ &= \frac{1}{\mathsf{M}\tau^e \mathsf{M}\eta} \textstyle\int_x^\infty \mathsf{P}(\tau^s > u)\, du\,.\end{aligned}$$

It is clear that the introduction of nonzero initial conditions causes no essential changes in the previous arguments. □

2. We have thus seen that under the hypotheses of Theorem 12, for the systems $\langle G, 1, G, 1 \rangle_R$ the limiting probability π_n that the n-th request is refused exists and equals $1-(1/\mathsf{M}\eta)$.

However, *if we interpret the probability of refusal in a broader sense*, for example, as the limit of the ratio $(n-r_n)/n$, where r_n is the number served among the first n requests, then the assertion of Theorem 12 will hold under more general assumptions.

Theorem 13. *Let $\{\tau_j^e, \tau_j^s\}$ be an arbitrary stationary sequence of independent vectors. Then*

$$1-\pi_n = \frac{r_n}{n} \xrightarrow[\text{a.s.}]{} \frac{1}{\mathsf{M}\eta}\,.$$

Proof. As we have seen, the sequence $\eta_1, \eta_2, \cdots$ will consist of independent, identically distributed random variables. The number r_n of served customers is the same as the first passage time of the level n in a random walk with jumps $\eta_1, \eta_2, \ldots$ (i.e., the number of sums $\eta_1, \eta_1+\eta_2, \eta_1+\eta_2+\eta_3, \ldots$ which are less than n). By the strong law of large numbers

$$\frac{1}{k}\sum_{j=1}^k \eta_j \xrightarrow[\text{a.s.}]{} \mathsf{M}\eta \quad \text{so that} \quad (r_n/n) \xrightarrow[\text{a.s.}]{} (1/\mathsf{M}\eta)\,. \quad □$$

We remark that under our assumptions $\lim_{n\to\infty}(r_n/n)$ evidently coincides with $\lim_{n\to\infty}(1/n)\sum_{k=1}^n \mathsf{P}(q_k=0)$.

Somewhat different conditions can also be found for the assertion of Theorem 13. Let τ_j^* be the service time of the j-th customer to be accepted for service (as already remarked, $\tau_j^* = \tau_\theta^s$ for some $\theta \geqslant j$).

Theorem 13A. *Suppose the sequence $\{\tau_j^e\}$ consists of i.i.d. variables and that the sequence $\{\tau_j^*\}$ does not depend on $\{\tau_j^e\}$ and is stationary and metrically transitive.*

Then

$$1-\pi_n=\frac{r_n}{n}\xrightarrow[\text{a.s.}]{}\frac{1}{\mathsf{M}\eta}.$$

Proof. As before, we form cycles of lengths $\eta_1, \eta_2, \ldots$ generated by passage of the levels $\tau_1^*, \tau_2^*, \ldots$. To prove the theorem it is enough to show (see the proof of Theorem 12) that the sequence $\{\eta_k\}$ is stationary and metrically transitive. The stationarity of η_k follows from that of τ_k^* and the fact that the sequences $(\tau_1^e, \tau_2^e, \ldots)$, $(\tau_{\eta_1+1}^e, \tau_{\eta_1+2}^e, \ldots), \ldots$ are identically distributed. By the last remark the sequence $\{\eta_n\}$ will have this same distribution if we put

$$\eta_k=f(\tau_k^*, \{\tau_j^{(k)e}\})=\min\{n: \tau_1^{(k)e}+\cdots+\tau_n^{(k)e}\geqslant\tau_k^*\},$$

when the sequences $\tau_j^{(k)e}\}$ are independent of each other and of $\{\tau_j^*\}$, and are distributed like $\{\tau_j^e\}$. From this representation of η_k it is now easy to obtain the metric transitivity since the sequence $\{\tau_k^*, \{\tau_j^{(k)e}\}\}$ is metrically transitive and $\eta_k=f(\tau_k^*, \{\tau_j^{(k)e}\})=U\eta_{k-1}$, where U is the corresponding shift for our sequence. □

If $\{\tau_j^e, \tau_j^s\}$ is an arbitrary stationary, metrically transitive sequence, then the sequence $\eta_1, \eta_2, \ldots$ will, generally speaking, still not be stationary. For example, we can set $\Omega=(\omega_1, \omega_2)$, $\mathsf{P}(\omega_1)=\mathsf{P}(\omega_2)=\frac{1}{2}$, $T\omega_1=\omega_2$, $T\omega_2=\omega_1$, $\tau_1^s(\omega_1)=\tau_1^s(\omega_2)=2$, $\tau_1^e(\omega_1)=1$ and $\tau_1^e(\omega_2)=3$. Then $\eta_1(\omega_1)=2\neq\eta_1(\omega_2)=1$ although $\eta_2(\omega_1)=\eta_2(\omega_2)=2$. In this connection, the problem of the existence of a limit of the ratio r_n/n for an arbitrary metrically transitive sequence remains unsolved.

We turn now to the systems $\langle G\rangle_R$. Assume that the hypotheses of Theorem 13 hold. Then, if in addition to these hypotheses we assume that the sequence $\{v_j^e, v_j^s\}$, say, does not depend on $\{\tau_j^e, \tau_j^s\}$ and that it consists also of i.i.d. vectors, then r_n can be written as the sum of a random number R_n of random variables not depending on R_n and distributed like $\min(v_n^s, v_n^e)$. By Theorem 13 we have in this case

$$1-\pi_n=\frac{r_n}{\sum_{j=1}^n v_j^e}=\frac{r_n}{R_n}\cdot\frac{R_n}{n}\cdot\frac{n}{\sum_{j=1}^n v_j^e}\xrightarrow[\text{a.s.}]{}\frac{\mathsf{M}\min(v_0^s, v_0^e)}{\mathsf{M}v_0^e\mathsf{M}\eta}.$$

A similar observation holds under the conditions of Theorem 13A.

We remark that this limit for π_n is generally speaking different from (26) where, in place of $p(0)$ we must insert $1-(1/\mathsf{M}\eta)$. Hence, for the systems $\langle G\rangle_R$ the possible definitions of the limiting probability of refusal differ not only in form but also in substance.

Analogous theorems and remarks can also be formulated for a broader conception of $\lim_{t\to\infty}\mathsf{P}(q(t)=1)$. For example, we can take it as the limiting value of the fraction of time in $[0, t]$ during which the system is busy.

3. Let us consider the following variational problem. Let the distribution of τ^e in the system $\langle G_I, 1, G_I, 1\rangle_R$ be fixed and assume that we can choose the distribution of τ^s arbitrarily from the class of distributions $\mathscr{F}_a$ for which $\mathsf{M}\tau^s=a$. What distribution $F\in\mathscr{F}_a$ minimizes the value of the refusal probability $p(0)$ (or $P(0)$)?

We encounter here a curious example of how a small, and at first glance insignificant deviation from the precise posing of a problem can lead to results directly opposite to those we are seeking. Indeed, if in a problem on the minimization of $p(0)$ we try to minimize $P(0)$ instead of $p(0)$ (agreeing that these are similar characteristics), we arrive at a distribution for which $p(0)$ is *maximized*! In fact, as we have already seen, (28) holds, where $\mathsf{M}\tau^s$ and $\mathsf{M}\tau^e$ are fixed, so that

$$p(0)+P(0)\frac{\mathsf{M}\tau^e}{\mathsf{M}\tau^s}=1\,,$$

and the smaller $P(0)$, the greater $p(0)$.

In accordance with the above, for an exponential distribution of τ^e, $p(0)=P(0)$ and doesn't depend on the distribution of τ^s from $\mathscr{F}_a$ (in this case, the distribution of χ in (27) coincides with that of τ^e).

We return to the posed problem. Let $\mathscr{F}_a^2\subset\mathscr{F}_a$ be the subclass of "two-point" distributions $F(x)$ (of the random variables τ^s) whose variations are concentrated at two points. Then we have

Theorem 14.

$$\inf_{F\in\mathscr{F}_a} p(0)=\inf_{F\in\mathscr{F}_a^2} p(0)\,,\qquad \inf_{F\in\mathscr{F}_a} P(0)=\inf_{F\in\mathscr{F}_a^2} P(0)\,.$$

By (28), the problem reduces to finding distributions on which

$$\mathsf{M}\eta=\int dF(x)H(x)$$

is minimized and maximized resp. where $F(x)=\mathsf{P}(\tau^s<x)\in\mathscr{F}_a$ and $H(x)$ is the renewal function of τ^e. From this representation it follows, in particular, that if $H(x)$ is concave ($dH(x)/dx$ is nonincreasing for $x>0$), then $\inf p(0)$ is attained on the one-point distribution: $\tau^s\equiv a$ and $\inf P(0)(\sup p(0))$ is attained on the sequence of "expanding" distributions:

$$\mathsf{P}\left(\tau^s=\frac{1}{n}\right)=1-\mathsf{P}(\tau^s=n)\,,\qquad \mathsf{P}(\tau^s=n)=\frac{a}{n}-\frac{n-a}{n(n-1)}\,,\qquad n\to\infty\,.$$

If, however, $H(x)$ is convex, we have the reversed situation: on the first distribution $\inf P(0)$ is attained ($\sup p(0)$) and on the second $\inf p(0)$.

In the general case, as Theorem 14 shows, we can approximate $\inf p(0)$ and $\inf P(0)$ as closely as we like by choosing two-point distributions for τ^s. The value of the infinium itself, as we have just seen, cannot be attained on proper (non-degenerate distributions.

Denote by $\mathscr{F}_a^0\subset\mathscr{F}_a$ the set of step-functions from $\mathscr{F}_a$ with a finite number of jumps. The proof of the theorem will follow from the next two lemmas (dual assertions w.r.t. $\inf\mathsf{M}\eta$ will be omitted for the sake of brevity).

Lemma 1.

$$\sup_{F\in\mathscr{F}_a}\mathsf{M}\eta=\sup_{F\in\mathscr{F}_a^0}\mathsf{M}\eta\,.$$

Lemma 2.

$$\sup_{F\in\mathscr{F}_a^0}\mathsf{M}\eta=\sup_{F\in\mathscr{F}_a^2}\mathsf{M}\eta\,.$$

Proof of Lemma 1. Noting that for some $\infty > c > 0$ and $\infty > b > 0$ (see Appendix 1) $H(x) < bx + c$, for given $\varepsilon > 0$ we can find an N such that

$$\mathsf{M}(\tau^s; \tau^s \geqslant N) < \varepsilon, \qquad \int_{x \geqslant N} dF(x)H(x) < \varepsilon$$

and

$$|H(x_0)\mathsf{P}(\tau^s > N) - \int_{x \geqslant N} dF(x)H(x)| < \varepsilon,$$

where $x_0 \geqslant N$ is defined by

$$x_0\mathsf{P}(\tau^s > N) = \mathsf{M}(\tau^s; \tau^s > N).$$

Hence, the value of $\mathsf{M}\eta$ for an arbitrary distribution $F \in \mathscr{F}_a$ can be approximated by a distribution from $\mathscr{F}_a$ concentrated on $[0, N]$ and at the point x_0.

It remains to choose on $[0, N]$ a step "distribution" F_n,

$$F_n(N) = F(N), \qquad \int_0^N x dF_n(x) = \int_0^N x dF(x), \tag{29}$$

having $n < \infty$ jumps of sizes $p_1, p_2, \ldots, p_n$ at the points $x_1, x_2, \ldots, x_n$ and such that

$$|\sum_{k=1}^n p_k H(x_k) - \int_{x \leqslant N} dF(x)H(x)| < \varepsilon. \tag{30}$$

Without loss of generality we can assume that all jumps of $F(x)$ on $[0, N]$ are smaller than $\varepsilon/2$. Taking into account the fact that $H(x)$ is left-continuous, we partition $[0, N]$ into intervals $\Delta_i = (b_i, b_{i+1}]$, $b_1 = 0$, $b_{n+1} = N$, such that:

1) The maximal increment of $H(x)$ on Δ_i, $i = 1, \ldots, n+1$, does not exceed ε. For this it is sufficient to assign all jumps greater in magnitude than $\varepsilon/2$ to the endpoint b_i and then decrease the maximal length of Δ_i to the necessary dimensions;

2) The maximal increment of $F(x)$ on Δ_i also does not exceed ε; this can be effected in the same way.

Now assume that

$$p_i = \mathsf{P}(\tau^s \in \Delta_i) \quad \text{and} \quad x_i = \frac{1}{p_i}\mathsf{M}(\tau^s; \tau^s \in \Delta_i).$$

Then F_n, generated by the jump sizes p_i at the points x_i, enjoys the property (29) and the difference (30) does not exceed

$$\sum p_i \max_{x \in \Delta_i} |H(x) - H(x_i)| \leqslant \varepsilon. \quad \square$$

Proof of Lemma 2. Let $\mathscr{F}_a^m \subset \mathscr{F}_a$ be the set of distributions concentrated at m points. For the proof of the lemma it is enough to show that for any $F \in \mathscr{F}_a^{m+1}$ with $m \geqslant 2$ there exists a $G \in \mathscr{F}_a^m$ such that

$$I(G) \equiv \int dG(x)H(x) \geqslant \int dF(x)H(x).$$

Denote by $x_0, \ldots, x_m$ the jump points of the distribution F and by $p_0, \ldots, p_m$ the sizes of these jumps. Thus, $p_i > 0$, $\sum_{i=0}^m p_i = 1$ and $\sum_{i=0}^m x_i p_i = a$. We construct a distribution G_t which places probabilities $P_i = p_i + td_i$ at the points x_i. For G_t to be in $\mathscr{F}_a$ it is necessary that t and d_i be chosen so that

$$p_i + t\,d_i \geqslant 0, \tag{31}$$

$$\sum_{i=0}^m d_i = 0 \quad \text{and} \quad \sum_{i=0}^m x_i\,d_i = 0. \tag{32}$$

For such a G_t we have

$$I(G_t)-I(F)=t\sum_{i=0}^{m} H(x_i)\,d_i\,.$$

Set $z=1$ or $z=0$ depending on whether or not the rank of the matrix

$$\begin{pmatrix} 1 & 1 & \dots & 1 \\ x_0 & x_1 & \dots & x_m \\ H(x_0) & H(x_1) & \dots & H(x_m) \end{pmatrix}$$

is equal to or less than three. Then the system of equations (32) and

$$\sum H(x_i)\,d_i=z$$

has a nonzero solution $(d_0^0, \dots, d_m^0)$. Since $\sum d_i^0=0$, at least one $d_{i_0}^0<0$. Choose t_0 such that $p_{i_0}+t_0\,d_{i_0}^0=0$. It is clear that $t_0>0$ so that

$$G_{t_0}\in\mathscr{F}_a^{\,m} \quad \text{and} \quad I(G_{t_0})-I(F)=t_0\sum H(x_i)\,d_i^0=t_0 z\geqslant 0\,. \quad \square$$

It is not hard to see from the proof of Lemma 2 that by fixing not one, but k of the moments of τ^s we can show that sup $\mathsf{M}\eta$ (inf $p(0)$) is attained for $(k+1)$-point distributions of τ^s.

§ 40. Asymptotic Analysis of Multi-Channel Systems

1. For the queueing systems $\langle G_I, 1, E/m, 1\rangle$ and the systems with refusals $\langle G_I, 1, E/m, 1\rangle_R$ we have obtained explicit formulae for the limiting distributions of the number of busy servers. However, if the number of servers m is large and a large fraction of them is occupied as a rule, then the usefulness of these formulae decreases due to their complexity. In such cases, asymptotic methods turn out to be more useful. The advantage of these methods lies also in the fact that they are applicable under considerably more general conditions than those yielding any explicit formulae. At the conclusion of this section we will quote as illustrations some assertions in this vein; in particular, theorems related to the systems $\langle G_I/m\rangle$ operating under conditions of heavy traffic. The proofs of these assertions go beyond the scope of this book and they are special cases of even more general laws.

2. The contents of this subsection will be the asymptotic analysis of the probability of refusal for the systems $\langle G_I, 1, E/m, 1\rangle_R$ based on the use of the exact formulae of § 38.

We will consider the systems $\langle G_I, 1, E/m, 1\rangle_R$ with high customer arrival rate and a large number m of service channels. In order to pose the problem more precisely it is necessary to introduce a family of sequences

$$\{\tau_{j,1}^e, \tau_{j,1}^s\,;\ -\infty<j<\infty\}\,, \qquad \{\tau_{j,2}^e, \tau_{j,2}^s\,;\ -\infty<j<\infty\}\,, \dots,$$

(double sequence) governing systems with $m_1, m_2, \dots$ channels, resp., and to assume that

$$\mathsf{P}(\tau^s_{j,n} > x) = \exp(-\alpha_n x) \qquad \left(\mathsf{M}\tau^s_{j,n} = \frac{1}{\alpha_n}\right),$$
$$\frac{\mathsf{M}\tau^e_{j,n}}{\mathsf{M}\tau^s_{j,n}} = a_n\alpha_n \to 0 \quad \text{and} \quad m_n \to \infty \tag{33}$$

as $n \to \infty$. In the sequel, the index n will be suppressed for brevity and in place of $n \to \infty$ we will write $m \to \infty$ (one can interpret n as the number of channels from the outset).

The ratio $\mathsf{M}\tau^s/\mathsf{M}\tau^e = 1/a\alpha$ in (33) indicates how many times the input stream (the number of requests arriving per unit time) exceeds the capacity of a single server working continuously. Hence, it is clear that if the customer losses are not to be too large, it is necessary that the number of channels m be near $1/a\alpha$. We introduce a parameter ρ characterizing the closeness of m and $1/a\alpha$ by means of the relation $\rho = 1 - ma\alpha$, so that ρ is always <1. The case $\rho > 0$ means that the system is usually filled and $\rho < 0$ that it is not being used to capacity.

We introduce in addition the random variable $\zeta = m\alpha\tau^e$, so that $\mathsf{M}\zeta = 1 - \rho$ and put

$$M_\beta = \mathsf{M}\zeta^\beta, \quad 1 < \beta \leqslant 2,$$
$$\varphi(t) = \mathsf{M}\, e^{-t\zeta} = \int e^{-tm\alpha x}\mathsf{P}(\tau^e \in dx)$$

and

$$r(t) = \frac{(1-t)(1-\varphi(t))}{t\varphi(t)}.$$

In what follows the letters c and ε will denote constants not depending on n. Finally, we let

$$p = p_m = \lim_{k\to\infty} \mathsf{P}(q_k = m)$$

be the probability of refusal.

Theorem 15. 1) *If $\rho > \varepsilon$ as $m \to \infty$ for some $\varepsilon > 0$, then $p \sim \rho$.*
If $M_\beta < c$, then this relation also holds when $\rho \to 0$ in such a way that

$$\rho m^{(\beta-1)/\beta} \to \infty, \quad 1 < \beta \leqslant 2.$$

2) *If $\rho\sqrt{m} \sim v$ and $M = M_2/2 < c$, then*

$$p \sim \sqrt{M/2\pi m}\exp(-v^2/2M)\Phi^{-1}(-v/\sqrt{M})$$
$$\sim \rho\Phi'(-v/\sqrt{M})[(v/\sqrt{M})\Phi(-v/\sqrt{M})]^{-1},$$

where Φ is the normal-(0, 1) distribution function.

3) *If $\rho < 0$, then there exists a unique positive solution $\mu > 0$ of the equation $\varphi(\mu) = 1 - \mu$. If, in addition, $M < c$, then*

(a) *for $\rho < -\varepsilon < 0$*

$$p \sim (1-\mu)\sqrt{\frac{|I''(\mu)|}{2\pi m}}\, e^{-mI(\mu)},$$

where

$$I(\mu)=\int_0^\mu \ln r(t)\,dt\,.$$

(The number μ can also be considered as the value of the argument at which the integrand $\ln r(t)$ *changes sign.)*;

(b) *as* $\rho\to 0$ *and* $-\rho\sqrt{m}\to\infty$

$$p\sim\sqrt{M/2\pi m}\,e^{-mI(\mu)}\quad and\quad \mu=(-\rho/M)+o(\rho)\,.$$

Remark 1. If $\rho>\varepsilon$ then, as already mentioned, the system is in heavy traffic and the servers are almost never idle. Hence, during a long time t they will be able to serve about $mt/\mathsf{M}\tau^s$ requests and the fraction of unserved customers will be near

$$\frac{(t/\mathsf{M}\tau^e)-(tm/\mathsf{M}\tau^s)}{t/\mathsf{M}\tau^e}=1-\frac{m\mathsf{M}\tau^e}{\mathsf{M}\tau^s}=1-m\alpha\alpha=\rho\,.$$

This is, roughly speaking, the meaning of the first assertion of the theorem.

The second assertion is related to the fact that for $\rho\sqrt{m}\sim v$ the number of free channels, normalized by $\sqrt{m}$, behaves like a diffusion process with reflection for whose one-dimensional distributions explicit formulae can be found in our case. Part 2 of the theorem is simply a local limit theorem for the number of idle channels at the "reflection point". The content of 3) is of interest in the study of high-reliability systems in which customer losses must be kept negligible.

Remark 2. Assuming the existence of $\mathsf{M}\zeta^3$ or of moments of higher order, we can find estimates for convergence rates or asymptotic expansions for all of the relations quoted.

Remark 3. It is natural to expect that if $\rho\to\rho_0>\varepsilon$ as $n\to\infty$ and the distribution of ζ converges to some limit, then there exists a proper limiting distribution $\{p_j\}$ of the number $m-q_n$ of idle servers. It follows from the results of § 38 that

$$\mathsf{P}(m-q^0=j)=p\sum_{s=m-j}^{m}(-1)^{s-m+j}\binom{s}{m-j}$$

$$\times\sum_{l=s}^{m}\binom{m}{l}\prod_{i=s+1}^{l}\frac{1-\varphi(i/m)}{\varphi(i/m)},$$

whence, setting $R(t)=(1-\varphi(t))/\varphi(t)$, we find that

$$\mathsf{P}(m-q^0=1)=pR(1)\,,$$

$$\mathsf{P}(m-q^0=2)=p\left[R(1)+R(1)R\left(1-\frac{1}{m}\right)-m\left(R(1)-R\left(1-\frac{1}{m}\right)\right)\right],\ \ldots.$$

If the limiting values are given the subscript 0, then

$$p_1=\rho_0R_0(1)\,,\qquad p_2=\rho_0[R_0(1)+R_0^2(1)-R_0'(1)],\ \ldots.$$

The probability p_k will depend on $k-1$ derivatives of R_0 at the point 1.

Thus, in this case the distribution of the number of idle channels will not, generally speaking, have a summarizing character. This can also be said of the value of p in the region of "large deviations" with $\rho < -\varepsilon$ (Statement 3 of the theorem).

Proof of Theorem 15. We will write the refusal probability p as (see Theorem 9)

$$p = \left(\sum_{j=0}^{m} A_j\right)^{-1}, \tag{34}$$

where

$$A_0 = 1, \qquad A_j = \binom{m}{j} \prod_{k=1}^{j} \frac{1-\varphi(k/m)}{\varphi(k/m)}; \quad j = 1, 2, \ldots, m.$$

Consider the ratio r_j, $j = 1, \ldots, m$, of two successive terms in (34):

$$r_j = \frac{A_j}{A_{j-1}} = \frac{m-j+1}{j} \cdot \frac{1-\varphi(j/m)}{\varphi(j/m)} = r\left(\frac{j}{m}\right)\left(1 + \frac{1}{m-j}\right).$$

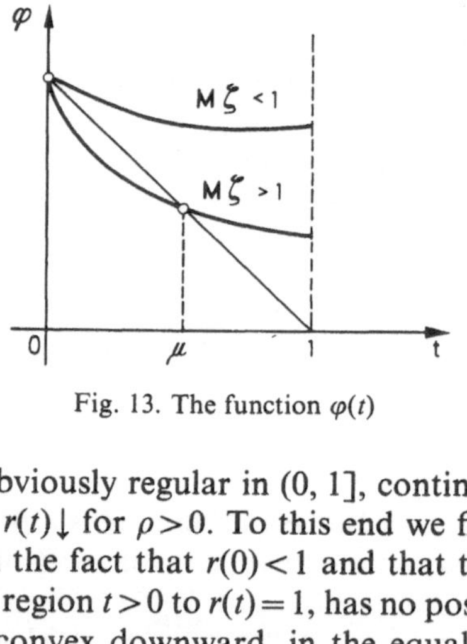

Fig. 13. The function $\varphi(t)$

The function $r(t)$ is obviously regular in (0, 1], continuous at $t=0$ and $r(0) = 1-\rho$. We will show that $r(t)\downarrow$ for $\rho > 0$. To this end we first note that $r(t) < 1$ on (0, 1]. This follows from the fact that $r(0) < 1$ and that the equation $\varphi(t) = 1-t$, which is equivalent in the region $t > 0$ to $r(t) = 1$, has no positive solutions (Fig. 13). Moreover, since $\varphi(t)$ is convex downward, in the equality $(1-\varphi(t))/t = -\varphi'(\tilde{t})$, $0 < \tilde{t} < t$ we have $\tilde{t} = \tilde{t}(t)\uparrow$ and $-\varphi'(\tilde{t})\downarrow$. Further .

$$-\varphi'(t) < -\varphi'(\tilde{t}) = \frac{1-\varphi(t)}{t} < \frac{\varphi(t)}{1-t}$$

and

$$\left(\frac{1-t}{\varphi(t)}\right)' = -\frac{\varphi(t) + (1-t)\varphi'(t)}{\varphi^2(t)} < 0.$$

Thus, $r(t)$ is the product of two monotonically decreasing factors. The same thing obviously holds for the function $r_1(t) = (1-t)(1-\varphi_1(t))/t\varphi_1(t)$ when $\rho_1 = 1 - \mathsf{M}\zeta_1 > 0$, where $\zeta_1 = (m+1)\zeta/m$ and $\varphi_1(t) = \mathsf{M}\, e^{-t\zeta_1}$. In terms of the function $r_1(t)$, the ratio r_j can be written as $r_j = r_1(j/(m+1))$.

We proceed to the determination of the sum (34). First let

$$M_\beta = \mathsf{M}\zeta^\beta < c, \qquad 1 < \beta \leqslant 2, \qquad \rho m^{(\beta-1)/\beta} \to \infty.$$

Then, clearly,

$$\mathsf{M}\zeta_1^\beta < c\,, \qquad \rho_1 m^{(\beta-1)/\beta} \to \infty\,, \qquad r_1(t) = 1 - \rho_1 + t^{\beta-1} M(t)\,,$$

where $0 \geqslant M(t) \geqslant -c$ for $0 \leqslant t \leqslant 1$. We have

$$\ln A_j = \sum_{k=1}^{j} \ln r_1\left(\frac{k}{m+1}\right) = j \ln(1-\rho_1)$$

$$+ \sum_{k=1}^{j} \ln\left(1 + \left(\frac{k}{m+1}\right)^{\beta-1} \frac{M(k/(m+1))}{1-\rho_1}\right)$$

$$= j \ln(1-\rho_1) + \varepsilon_{j,m}\,, \quad \varepsilon_{j,m} < 0\,.$$

Further, for $j \leqslant j_0 = [(m^{(\beta-1)/\beta} \rho_1^{-1})^{1/2}]$ (the case $\limsup_{m\to\infty} \rho = 1$ is excluded)

$$|\varepsilon_{j,m}| < c\frac{j^\beta}{m^{\beta-1}} \leqslant c \frac{(m^{\beta-1} \rho_1^{-\beta})^{1/2}}{m^{\beta-1}} = c(m^{(\beta-1)/\beta} \rho_1)^{-\beta/2} \to 0\,.$$

Hence,

$$\sum_{j=j_0+1}^{m} A_j < \sum_{j=j_0+1}^{m} (1-\rho_1)^j < \frac{(1-\rho_1)^{j_0}}{\rho_1}\,, \tag{35}$$

$$\sum_{j=0}^{j_0} A_j = \sum_{j=0}^{j_0} (1-\rho_1)^j e^{\varepsilon_{j,m}} = \frac{1-(1-\rho_1)^{1+j_0}}{\rho_1} (1+o(1))\,.$$

Noting that

$$j_0 \ln(1-\rho_1) < -j_0\rho_1 = -(m^{(\beta-1)/\beta}\rho_1)^{1/2} \to \infty$$

we find that

$$\sum_{j=0}^{m} A_j = \frac{1}{\rho_1}(1+o(1)) = \frac{1}{\rho}(1+o(1))\,. \tag{36}$$

For $\rho > \varepsilon$ the equality (36) can be obtained as a simple consequence of (35) and the fact that $r_j \sim 1-\rho$ when $j/m \to 0$. The first assertion of the theorem is proved.

In what follows we will find useful an integral representation of $\sum A_j$. One has

$$\ln\binom{m}{j} = \ln\sqrt{\frac{m}{2\pi j(m-j)}} - j\ln\frac{j}{m} - (m-j)\ln\left(1-\frac{j}{m}\right)$$

$$+ R_m - R_j - R_{m-j}$$

$$= \ln\sqrt{\frac{m}{2\pi j(m-j)}} + m\int_0^{j/m} \ln\frac{1-t}{t}\,dt + R_m - R_j - R_{m-j}\,, \tag{37}$$

where R_n is the remainder term in Stirling's formula for $\ln n!$ Furthermore,

$$\ln \prod_{k=1}^{j} \frac{1-\varphi(k/m)}{\varphi(k/m)} = \sum_{k=1}^{j} \ln\frac{1-\varphi(k/m)}{(k/m)\varphi(k/m)} + \sum_{k=1}^{j} \ln\frac{k}{m}\,.$$

Since by assumption the function $B(t) = \ln[(1-\varphi(t))/t\varphi(t)]$ has a continuous and uniformly bounded derivative on $[0, 1]$ ($\varphi''(0) < c$),

$$m\int_{(k-1)/m}^{k/m} B(t)\,dt = B\left(\frac{k}{m}\right) - \frac{1}{2m} B'\left(\frac{k}{m}\right) + o\left(\frac{1}{m}\right),$$

uniformly in k. This allows us to write

$$\ln \prod_{k=1}^{j} \frac{1-\varphi(k/m)}{\varphi(k/m)} = m\int_0^{j/m} \ln\frac{1-\varphi(t)}{\varphi(t)}\,dt + \left(\sum_{k=1}^{j} \ln\frac{k}{m} - m\int_0^{j/m} \ln t\,dt\right)$$
$$+\frac{1}{2}\left(B\left(\frac{j}{m}\right) - B(0)\right) + o(1),$$

uniformly in j. Taking (37) into account along with the relations

$$B(0) = \ln \mathsf{M}\zeta \quad\text{and}\quad \ln(j!m^{-j}) - m\int_0^{j/m} \ln t\,dt = \ln\sqrt{2\pi j} + R_j,$$

we find that

$$\ln A_j = \tfrac{1}{2}\ln\frac{1-\varphi(j/m)}{(j/m)(1-j/m)\varphi(j/m)\mathsf{M}\zeta} + m\int_0^{j/m} \ln r(t)\,dt + o(1) - R_{m-j},$$

or, what is the same,

$$A_j = C\left(\frac{j}{m}\right) e^{mI(j/m)}(1 + o(1) - R_{m-j}), \tag{38}$$

uniformly in $j = 0, 1, \ldots, m-1$, where

$$C(t) = 1/(1-t)\sqrt{r(t)/\mathsf{M}\zeta} \quad\text{and}\quad I(t) = \int_0^t \ln r(u)\,du.$$

For $j = m$ one gets immediately

$$A_m = \sqrt{2\pi m\frac{1-\varphi(1)}{\varphi(1)\mathsf{M}\zeta}}\, e^{mI(1)}(1 + o(1)). \tag{39}$$

We now return to (34) and assume that $\mathsf{M}\zeta = 1 - \rho > 1$. In this case $\varphi'(0) < -1$ and the equations $\varphi(t) = 1 - t$ and $r(t) = 1$ have a unique positive root $\mu > 0$ (see Fig. 13). The point μ is obviously the location of a maximum of the function $I(t)$ since

$$I'(\mu) = \ln r(\mu) = 0$$

and

$$I''(\mu) = r'(\mu) = -\varphi'(\mu)\frac{1-\mu}{\mu\varphi^2(\mu)} - \frac{1-\varphi(\mu)}{\mu^2\varphi(\mu)} = -\frac{1+\varphi'(\mu)}{\mu\varphi(\mu)} < 0.$$

Here, $\varphi'(\mu)$ will be close to -1 only when ρ is small. But as $\rho \to 0$,

$$\mu = -\frac{2\rho}{\mathsf{M}\zeta^2} + o(\rho) = -\frac{\rho}{M} + o(\rho) \text{ and } I''(\mu) = -M + o(1), \tag{40}$$

so that $I''(\mu)$ is uniformly bounded from above by a negative constant. Since $I(t)$ has no other extrema on $(0,1)$, we see that for some $\varepsilon_1, \varepsilon > 0$

$$I(t) < I(\mu) - \varepsilon_1$$

outside the interval $(\mu - \varepsilon, \mu + \varepsilon) \cap [0, 1]$. Furthermore, the boundedness of M implies that for some $c > 0$, $\varphi(1) > c$ and $\mu < 1 - c$. Thus, for arbitrary b and B satisfying $0 \leqslant b < \mu - \varepsilon$, $\mu + \varepsilon < B < 1$, we will have (see (38), (39))

$$\sum_{j=0}^{m} A_j = \sum_{j=bm}^{Bm} C\left(\frac{j}{m}\right) e^{mI(j/m)}(1+o(1)).$$

To compute this sum we will use the method of Laplace. First note that, uniformly in j,

$$m \int_{j/m}^{(j+1)/m} h(t)C(t)\, e^{mI(t)}\, dt = C\left(\frac{j}{m}\right) e^{mI(j/m)}(1+o(1)),$$

where $h(t) = I'(t)(e^{I'(t)} - 1)^{-1}$, so that

$$\sum_{j=bm}^{Bm} A_j = m \int_b^B h(t)C(t)\, e^{mI(t)}\, dt(1+o(1)).$$

From this it follows immediately that for $\mu > \varepsilon$

$$p^{-1} = mC(\mu)\, e^{mI(\mu)} \sqrt{\frac{2\pi}{m|I''(\mu)|}}(1+o(1)).$$

As $\mu \to 0$ $(\rho \to 0)$

$$p^{-1} = mC(0)\, e^{mI(\mu)} \sqrt{\frac{2\pi}{m|I''(\mu)|}}\, \Phi(\mu\sqrt{m|I''(\mu)|})(1+o(1)).$$

Using (40) and the facts that

$$C(0) = 1 \quad \text{and} \quad I(\mu) = \mu \ln(1-\rho) - \frac{\mu^2 M}{2} + o(\mu^2) = \frac{\rho^2}{2M} + o(\rho^2),$$

we find, in particular, that for $\rho \sim v/\sqrt{m}$

$$p^{-1} = \sqrt{\frac{2\pi m}{M}}\, e^{v^2/2M} \Phi\left(-\frac{v}{\sqrt{M}}\right)(1+o(1)).$$

If $-\rho\sqrt{m} \to \infty$, then

$$p^{-1} = \sqrt{2\pi m/M}\, e^{mI(\mu)}(1+o(1)).$$

It remains to treat the case $\mathsf{M}\zeta \leqslant 1$, $\mathsf{M}\zeta \to 1$. Here, obviously,

$$p^{-1} = m \int_0^{\varepsilon} h(t)C(t)\, e^{mI(t)}\, dt(1+o(1)).$$

Writing $I''(0) = -D^2$ and $I'(0) = -dD^2 < 0$ we obtain

$$mI(t) = -\frac{m}{2}(D^2t^2 - 2I'(0)t) + mt^2\varepsilon(t)$$

$$= -\frac{mD^2}{2}(t+d)^2 + \frac{mD^2d^2}{2} + mt^2\varepsilon(t),$$

where $\varepsilon(t) \to 0$ as $t \to 0$;

$$p^{-1} = \sqrt{m} \exp(mD^2d^2/2)$$
$$\times \int_{d\sqrt{m}}^{(d+\varepsilon)\sqrt{m}} \exp\left\{-\frac{D^2u^2}{2} + \varepsilon\left(\frac{u}{\sqrt{m}} - d\right)(u - d\sqrt{m})^2\right\} du(1+o(1)).$$

Let $N \to \infty$ so slowly that $N^2\varepsilon(N/\sqrt{m}) \to 0$ as $m \to \infty$. Then

$$p^{-1} = \sqrt{m}\exp(mD^2d^2/2)$$
$$\times\left[\int_{d\sqrt{m}}^{\infty}\exp\left\{-\frac{D^2u^2}{2}\right\}du\right.$$
$$\left.+\int_{d\sqrt{m}+2N}^{(d+\varepsilon)\sqrt{m}}\exp\left\{-\frac{D^2u^2}{2}+\varepsilon\left(\frac{u}{\sqrt{m}}-d\right)(u-d\sqrt{m})^2\right\}du\right]$$
$$\times(1+o(1)),$$

where in the second integral $(u > d\sqrt{m}+2N)$

$$-\frac{D^2u^2}{2}-\varepsilon\left(\frac{u}{\sqrt{m}}-d\right)(u-d\sqrt{m})^2$$
$$< -\frac{D^2}{2}(d\sqrt{m}+N)^2-\frac{D^2}{4}(u-d\sqrt{m}-2N)^2\,.$$

Since

$$\exp\left\{-\frac{D^2}{2}(d\sqrt{m}+N)^2\right\} = o\left(\frac{1}{d\sqrt{m}}\exp\left\{-\frac{mD^2d^2}{2}\right\}\right),$$

we get finally

$$p^{-1} = \frac{\sqrt{2\pi m}}{D}\exp(mD^2d^2/2)\Phi(-Dd\sqrt{m})(1+o(1))$$
$$= \sqrt{\frac{2\pi m}{|I''(0)|}}\exp\left(-\frac{m(I'(0))^2}{2I''(0)}\right)\Phi\left(\frac{I'(0)\sqrt{m}}{\sqrt{|I''(0)|}}\right)(1+o(1))\,.$$

In particular, for $\rho \sim v/\sqrt{m}$, $v \geqslant 0$ $(I'(0) = \ln(1-\rho)$, $I''(0) = -M+o(1))$ we find

$$p^{-1} = \sqrt{\frac{2\pi m}{M}}\exp(v^2/2M)\Phi\left(-\frac{v}{\sqrt{M}}\right)(1+o(1))\,.$$

As $\rho\sqrt{m} \to \infty$, we have in accordance with the preceding $(md^2 \to \infty)$

$$p^{-1} \sim \frac{1}{dD^2} \sim \frac{1}{\rho}\,.$$

The theorem is proved. □

We also quote here an assertion relating to the case $\rho\sqrt{m} \sim v$, $-\infty \leqslant v < c$. Let q^0, as before, be the number of busy channels in a stationary system.

Theorem 16. *As $m \to \infty$ and for $M < c$*

$$\mathsf{P}\left(q^0 - \frac{1}{a\alpha} < x\sqrt{mM}\right) \to \begin{cases}\Phi(x)\Phi^{-1}(-v/\sqrt{M}), & x \leqslant -v/\sqrt{M},\\ 1, & x > -v/\sqrt{M}.\end{cases}$$

3. In conclusion we formulate a result obtained in [12] and relating[2] to the multi-channel queueing systems

$$\langle G_I/m\rangle \quad \text{and} \quad \langle G_I/m\rangle_A$$

(see Chapter 8).

We again treat a double sequence, i.e., a family of governing sequences

$$\{\tau_j^e, v_j^e, \tau_j^s, v_j^s; \ -\infty<j<\infty\}$$

depending on some parameter. This parameter can be taken as the number

$$\delta=(\mathsf{M}v^e/\mathsf{M}\tau^e)-(m\mathsf{M}v^s/\mathsf{M}\tau^s).$$

This is clearly the difference between the mean number of requests arriving in the system per unit time and the mean number of requests which the system can serve per unit time when the queue is infinite. The parameter δ can be considered as a characterization of the traffic in the system. If δ is small, accumulations in the queue will be dispelled quite slowly when $\delta<0$ and won't be dispelled at all when $\delta\geqslant 0$, so that the length of the queue $q(t)$ can become large. Under these conditions, as in the single-channel case (see § 25), the question naturally arises as to the existence of simple approximate relations describing the distribution of, say, $q(t)$ for large t.

The assertion presented below describes the behavior of $q(t)=q(t,\delta)$ when $t\to\infty$ and $\delta\to 0$ simultaneously (the indices and arguments referring to the dependence of the parameters of the system on δ will be suppressed as a rule for the sake of brevity).

Write

$$P(v,x,b)=\mathsf{P}(\omega(u)<x+(u/b),\ 0\leqslant u\leqslant v), \quad v>0,$$

where $\omega(u)$ is a standard Wiener process. We gave the explicit form of this expression in (67) § 25. In the important special case in which $v=\infty$ and $b>0$

$$P(\infty,x,b)=1-\exp\{-2x/b^2\}.$$

Theorem 17. *Assume that as $\delta\to 0$*

$$\sigma^2(\delta)=\frac{\mathsf{D}v^e}{\mathsf{M}\tau^e}+\frac{(\mathsf{M}v^e)^2\mathsf{D}\tau^e}{(\mathsf{M}\tau^e)^3}+\left(\frac{\mathsf{D}v^s}{\mathsf{M}\tau^s}+\frac{(\mathsf{M}v^s)^2\mathsf{D}\tau^s}{(\mathsf{M}\tau^s)^3}\right)\to\sigma^2>0$$

and for some $\gamma>0$ the moments

$$\mathsf{M}\tau^i,\quad \mathsf{M}\left|\frac{\tau^i-\mathsf{M}\tau^i}{\sqrt{\mathsf{D}\tau^i}}\right|^{2+\gamma} \quad \text{and} \quad \frac{\mathsf{M}|v^i-\mathsf{M}v^i|^{2+\gamma}}{\mathsf{D}v^i}, \quad i=e,s,$$

are uniformly bounded. Assume in addition that the initial conditions $q(0)$ and times $\rho_1,\dots,\rho_m$ of commencement of operation of the 1st, 2nd, …, m-th *channel, resp. satisfy the conditions*

$$(\delta t)^{-t}\max_j\left(q(0),\frac{\rho_j}{\mathsf{M}\tau^e}\right)\xrightarrow[\mathsf{P}]{}0 \quad \text{if } \delta\sqrt{t}\geqslant 1$$

[2] In [12] are considered systems in which different service channels are governed by sequences of random variables with, generally speaking, different distributions.

and

$$t^{-1/2} \max_j \left(q(0), \frac{\rho_j}{\mathsf{M}\tau^e} \right) \xrightarrow[\mathsf{P}]{} 0 \quad \text{if } \delta\sqrt{t} < 1 .$$

Then for the systems $\langle G_I/m \rangle$ *and* $\langle G_I/m \rangle_A$ *the limiting distribution of* $q(t)$ *as* $\delta \to 0$ *and* $t \to \infty$ *is described by the following three relations:*

A. *As* $\delta\sqrt{t} \to v,\, v \leqslant \infty$,

$$\lim_{\delta \to 0} \mathsf{P}\left(q(t) < \frac{x}{|\delta|} \right) = P\left(v^2, \frac{x}{\sigma}, \sigma \operatorname{sign} \delta \right).$$

B. *As* $\delta\sqrt{t} \to 0$

$$\lim_{t \to \infty} \mathsf{P}(q(t) < x\sqrt{t}) = P\left(1, \frac{x}{\sigma}, \infty \right) = \sqrt{\frac{2}{\pi}} \int_0^{x/\sigma} e^{-u^2/2}\, du .$$

C. *As* $\delta\sqrt{t} \to -\infty$

$$\lim_{t \to \infty} \mathsf{P}(q(t) < -\delta t + x\sqrt{t}) = \frac{1}{\sqrt{2\pi}} \int_{-\infty}^{x/\sigma} e^{-u^2/2}\, du .$$

A similar theorem can be formulated for the waiting time.

Thus, if we want to investigate the distribution of $q(t)$ for large t for systems in heavy traffic with small parameter δ, we can use as approximations for $\mathsf{P}(q(t) < y)$ the right sides of A, B and C, depending on the magnitude of $\delta\sqrt{t}$.

Chapter 8

Systems with Autonomous Service

§ 41. General Properties

1. The governing of the systems $\langle G\rangle_A$ (see § 1), as well as other types of systems, is effected by means of the sequence of random vectors

$$\{\tau_j^e, v_j^e, \tau_j^s, v_j^s; j \geqslant 1\} \tag{1}$$

with some initial conditions. These systems differ from the usual ones in that service begins at times 0, τ_1^s, $\tau_1^s+\tau_2^s$, $\tau_1^s+\tau_2^s+\tau_3^s$, ..., independently of the input stream. The most interesting systems of this kind are systems with queues and we will treat only them here. Service proceeds by groups of sizes $v_1^s, v_2^s, \ldots$, or of smaller size if the queue is not long enough.

Consider the processes $\{e(t)\}$ and $\{s(t)\}$. The process $e(t)$ describes the input stream: the value of $e(t)$ is the number of requests arriving up to time t. The value of $s(t)$ is defined analogously as the sum

$$s(t) = v_1^s + \cdots + v_{\eta(t)-1}^s,$$

where $\eta(t)$ is the first passage time to the threshold t in a random walk with jumps $\tau_1^s, \tau_2^s, \ldots$. Roughly speaking, $s(t)$ is the number of requests from an infinite customer queue which the system serves by time t.

Write $X(t)=e(t)-s(t)$ and let $q(t)$ be the length of the queue at time t, not counting requests already being served (here it is natural to separate the busy channels from the "pure" queue since several requests can be located in each channel). The processes $e(t)$, $s(t)$ and $q(t)$ have the form of jumps; we will assume that they are right-continuous. For example, requests whose service begins at t are assumed to have left the queue at time t (these will also include requests entering the system at time t.)

The distribution of $\{q(t)\}$ will be completely defined if on the same probability space on which the sequence (1) is given (or, what is the same, the processes $\{e(t)\}, \{s(t)\}$ are given), one also defines the random variable $q(0)$—the queue length at time $t=0$. Then

$$q(t) = q(0) + X(t) - \inf_{u \in [0,t]} (0, X(u) + q(0)). \tag{2}$$

Indeed, if for all $u \in [0, t]$ we had

$$X(u)+q(0) \geqslant 0\,,$$

then clearly

$$q(t)=q(0)+X(t)\,.$$

If the inequality is not satisfied, then $q(t)$ exceeds $q(0)+X(t)$ by the sum of the unused "capacities" of the service system, which equals

$$-\inf(0, X(u)+q(0))\,.$$

The equality (2) can also be obtained as the unique solution of the equation

$$\begin{aligned} dq(t) &= de(t)-\min(ds(t), q(t-0)+de(t)) \\ &= dX(t)-\min(0, q(t-0)+dX(t))\,. \end{aligned}$$

We remark that from the point of view of the properties of $q(t)$, picking out the subclass of multi-channel systems from those under consideration makes little sense as such systems are equivalent to single-channel systems for which the service process $s(t)$ is the sum of the corresponding processes for the single channels. Here the situation is the same as that in which it was unnecessary to separate subclasses of systems with multiple input channels from the systems $\langle G \rangle$.

The reader will also have noticed that the basic equality (2) describing the behavior of $q(t)$, by which

$$q(t)=\sup_{u\in[0,t]}(q(0)+X(t), X(t)-X(u))\,, \tag{3}$$

is completely analogous to the relations we investigated in Chapter 1.

In this sense the systems $\langle G \rangle_A$ are the simplest—they admit an explicit representation for $q(t)$ in terms of governing processes with no restrictions on the elements of the governing sequences.

A comparison with the results of Chapter 1 leads immediately to

Theorem 1. *If $X(t)$ is an ergodic process with strictly stationary increments and*

$$\mathsf{M}(X(1)-X(0))<0\,,$$

then for an arbitrary initial condition $q(0)$ the distributions of the processes

$$\{q_t(u)=q(t+u); u \geqslant 0\}$$

converge to that of the process

$$\overline{X}(u)=\sup_{v\leqslant u}(X(u)-X(v))\,.$$

That is, there exist processes

$$\{\tilde{X}(u)=\tilde{X}_t(u); u \geqslant 0\}\,,$$

distributed for all t like $\{\overline{X}(u), u \geqslant 0\}$ and such that

$$\mathsf{P}(\textstyle\bigcup_{u\geqslant 0}\{q_t(u) \neq \tilde{X}(u)\}) \to 0$$

as $t \to \infty$.

In virtue of (3), it is easy to carry over the other results of Chapter 1. For example, one can obtain estimates for the rate of convergence in Theorem 1, and so on.

2. *Relation between the systems $\langle G, G, E, 1\rangle_A$ and $\langle G, G, E, 1\rangle$.* We consider the systems $\langle G, G, E, 1\rangle_A$ satisfying the hypotheses of Theorem 1. For these systems the entering stream $e(t)$ has stationary increments and is ergodic, and $s(t)$ is a homogenous Poisson process independent of $e(t)$.

Along with these we consider the "ordinary" (see Chapter 1) systems $\langle G, G, E, 1\rangle$ with the same initial conditions, the same entering stream $e(t)$ and exponentially distributed service times. For such systems we write $q_*(t)$ for the queue length at time t.

We will prove that the *distribution of $q(t)$ for the systems $\langle G, G, E, 1\rangle_A$ and the distribution of $q_*(t)$ for the systems $\langle G, G, E, 1\rangle$ coincide.*

From this and Theorem 1 it will follow, in particular, that *under the hypotheses of Theorem* 1 *the stationary queue $q_*^c(u)$ for the systems $\langle G, G, E, 1\rangle$ can be represented as*

$$q_*^c(u) = \sup_{v \leqslant u} (e(u) - e(v) - s(u) + s(v)) .$$

To prove this we construct with the aid of the processes $e(t)$ and $s(t)$ a new queueing system in which a customer arriving at time t and finding the system free is immediately accepted for service which runs during

$$\chi_s(t) = \inf\{u \geqslant 0 : s(t+u) = s(t) + 1\} ,$$

the time remaining from the interval between jumps of the process $s(t)$ which "covers" the level t.

If the system is busy then the request takes its place in line and is subsequently served during a time equal to the next interval between jumps of the process $s(t)$.

The variable $\chi_s(t)$ is the overshoot of the level t in a random walk with exponentially distributed jumps τ_1^s, τ_2^s, ... Hence, the distributions of $\chi_s(t)$ and τ^s coincide. It is also clear that this situation is not changed if t is a random variable independent of the future of the process $s(t)$. Since the customer arrival times $\tau_1^e, \tau_1^e + \tau_2^e, \ldots$, possess this property (they are independent of $s(t)$), we find that *the distribution of the queue length $q^*(t)$ for the new system which we have introduced coincides with the distribution of $q_*(t)$.*

We now compare our new system with the system $\langle G, G, E, 1\rangle_A$ governed by the same processes $e(t)$ and $s(t)$. In the latter, a customer arriving at time t and finding the system free stays in line for a time $\chi_s(r)$ and then leaves the "pure" queue to be served during a time distributed like τ^s. Hence, the state $\{q(t)=0\}$ in the system $\langle G, G, E, 1\rangle_A$ coincides with $\{q^*(t)=0\}$ in our new system. The behavior of the systems near these states is also the same. Since the formation of the queues does not otherwise differ, we find that the *distribution of $q(t)$ for the systems $\langle G, G, E, 1\rangle_A$ coincides with that of $q^*(t)$ in the auxiliary system.*

3. *Relation between the systems $\langle G, G, E_k, 1\rangle_A$* and *$\langle G, G, E_k, 1\rangle$.* We turn now to the systems $\langle G, G, G_I, 1\rangle_A$ and $\langle G, G, G_I, 1\rangle$, where the service times τ_j^s

are independent and representable as

$$\tau_j^s \underset{d}{=} \gamma_1 + \cdots + \gamma_m$$

with the $\gamma_1, \ldots, \gamma_m$ independent and exponentially distributed:

$$\mathsf{P}(\gamma_j > x) = e^{-\alpha_j x}, \alpha_j > 0, x \geqslant 0 .$$

The distributions of these τ_j^s are designated as *Erlang* or *type* E_m distributions.

We define the renewal process $\{\zeta(t); -\infty < t < \infty\}$ as a process with stationary increments for which the time intervals between unit jumps are independent and have "alternating" distributions. These distributions form cycles: if a certain interval is distributed like γ_j, then the following one will be distributed like γ_l, $l = j + 1 \pmod m$. To finish the characterization of the process $\zeta(t)$ we set $\zeta(0) = 0$ and define the position of its jumps "near" $t = 0$ as follows (see Subsection 4 Appn. 1): Let ν be a random variable equal to j with probability

$$\frac{\mathsf{M}\gamma_j}{\mathsf{M}(\gamma_1 + \cdots + \gamma_m)} = \frac{1}{\alpha_j}\left(\sum_{l=1}^m \frac{1}{\alpha_l}\right)^{-1}, \quad j = 1, \ldots, m .$$

Then we will assume that for *each fixed* $\nu = j$ the distances χ and γ from the point $t = 0$ to the first positive and negative jump times, resp. of $\zeta(t)$ are distributed like γ_j and are independent, so that the interval including $t = 0$ is distributed like $\chi + \gamma$. The following (in the sense of increasing time) intervals are set equal to $\gamma_{j+1}, \gamma_{j+2}, \ldots$, where $\{\gamma_k; k = 1, 2, \ldots\}$ is a sequence of independent random variables with γ_{lm+j} distributed like γ_j. $1 \leqslant j \leqslant m$, $l = 0, 1, \ldots$; the $\gamma_{j+1}, \gamma_{j+2}, \ldots$ do not depend on $\chi + \gamma$ (conditionally for *fixed* ν!). This means that the time of the r-th jump after 0 of the process $\zeta(t)$ is the end of the interval $\gamma_{\nu+r-1}$.

Using Subsection 4 of Appn. 1 it is easy to show that defining $\zeta(t)$ in this way yields a *process with stationary increments*.

Furthermore, if we set

$$s(t) = \left[\frac{\zeta(t) + \nu - 1}{m}\right]$$

($[x]$ is the integral part of x; we assume that the first time $s(t) = 1$ is after the first appearance of an interval with the distribution of γ_1), then the process $s(t)$ will also have stationary increments and will be the renewal process for the random variable τ^s. The distance $\chi + \gamma_{\nu+1} + \cdots + \gamma_m$ from 0 to the first jump of $s(t)$ has the same distribution as the overshoot of an infinite barrier in a random walk with jumps $\tau_1^s, \tau_2^s, \ldots$, since it has the characteristic function

$$\left(\sum_{j=1}^m \frac{1}{\alpha_j}\right)^{-1} \sum_{j=1}^m \frac{1}{\alpha_j} \prod_{l=j}^m \frac{1}{\left(1 - \dfrac{i\lambda}{\alpha_j}\right)} = \left(i\lambda \sum_{j=1}^m \frac{1}{\alpha_j}\right)^{-1} \left[\prod_{j=1}^m \left(1 - \frac{i\lambda}{\alpha_j}\right)^{-1} - 1\right].$$

This function, as we would naturally expect, is symmetric in the α_j, so that we could just as well set

$$s(t) = \left[\frac{\zeta(t) + (\nu - 1 + j)_m}{m}\right]$$

for arbitrary $0 \leqslant j \leqslant m-1$, where

$$(v-1+j)_m = \begin{cases} v-1+j & \text{if } v-1+j \leqslant m-1 \\ v-1+j-m & \text{if } v-1+j > m. \end{cases}$$

Returning now to service problems, along with the system $\langle G, G, E_k, 1\rangle$ (as before, its queue will be denoted by $q_*(t)$) we consider an auxiliary system having the previous input stream but a somewhat different service regime given by means of the process $\zeta(t)$. If a customer arriving at time t finds the system free he is accepted and undergoes service during a time $\tau^* = \chi(t) + \gamma_{\mu+1} + \dots + \gamma_{\mu+m-1}$, where $\chi(t) = \inf\{u > 0 : \zeta(t+u) = \zeta(t) + 1\}$ is the time from t to the time of the first jump after t of the process $\zeta(t)$; $\mu = \mu(t)$ is the number of the interval from the sequence $\{\gamma_k; k = 1, 2, \dots\}$ which includes the level t: $\mu = \inf\{k \geqslant v : \chi + \sum_{j=v+1}^{k} \gamma_j > t\}$. By the properties of the exponential distribution, τ^* will be distributed like τ^s (for $\mu = j$, $\chi(t)$ is distributed like γ_j) and this does not change if t is a random variable not depending on the future of the process $\zeta(t)$.

If the system is busy, then an arriving customer takes his place in the queue and is subsequently served for a time equal to the sum of the corresponding m intervals of the process $\zeta(t)$ which follow each other in succession.

Since the arrival times $\tau_1^e, \tau_1^e + \tau_2^e, \dots$ do not depend on the process $\zeta(t)$, we find that the distribution of the queue length $q^*(t)$ for the system we have introduced coincides with that of $q_*(t)$.

We consider now a second auxiliary system (whose characteristics will be provided with the superscript**) which has autonomous service, input stream

$$e^{**}(t) = m\, e(t)$$

(for the new system $v_j^{e**} = m v_j^e$), initial value $q^{**}(0) = m q^*(0)$ and service stream $s^{**}(t) = \zeta(t)$. For this system, the assertions of Subsection 1 hold: in particular, one has the representation (3) for $q^{**}(t)$ and for the distribution of the stationary queue,

$$q_c^{**} \underset{d}{=} \sup_{t \leqslant 0} \{e^{**}(t) - s^{**}(t)\} = \sup\{m\, e(t) - \zeta(t)\}.$$

However, as is easily seen by the construction of our auxiliary systems,

$$q^*(t) = \begin{cases} \dfrac{q^{**}(t)}{m}, & \text{if } q^{**}(t) \text{ is a multiple of } m, \\ \left[\dfrac{q^{**}(t)}{m}\right] + 1, & \text{otherwise,} \end{cases}$$

since exit from the "pure" queue of the k-th customer in the second system corresponds to the end of service of the k-th customer in the first.

We obtain as a result a representation of the form (3) *for the queue length* $q_*(t)$ *in the systems* $\langle G, G, E_k, 1\rangle$ *and an assertion on the stationary process* $q_*^c(u)$ for systems governed by an arrival stream $e(t)$ with stationary increments which satisfies the condition

$$\mathsf{M}(e(1) - e(0)) - \left(\sum_{j=1}^{m} \frac{1}{\alpha_j}\right)^{-1} < 0.$$

Indeed, *under the formulated conditions the distributions of the processes*

$$\{q_*(t+u);\, u\geqslant 0\}$$

for the systems $\langle G, G, E_k, 1\rangle$ *converge as* $t\to\infty$ *to that of the process*

$$q_*^c(u)=\sup_{v\leqslant u}(X_*(u)-X_*(v)),$$

where

$$X_*(u)=e(u)-\left[\frac{\zeta(u)}{m}\right].$$

We can now compare the obtained relations with those for the stationary queue $q^c(u)$ in the systems $\langle G, G, E_k, 1\rangle_A$, governed by the processes $e(t)$ and $s(t)$. Since

$$\left[\frac{\zeta(t)}{m}\right]\leqslant s(t)=\left[\frac{\zeta(t)+v-1}{m}\right]<\left[\frac{\zeta(t)}{m}\right]+1,$$

we find that

$$q_*^c(u)-1\underset{d}{\leqslant} q^c(u)\underset{d}{\leqslant} q_*^c(u).$$

§ 42. Methods of Calculating the Stationary Distributions

1. *The systems* $\langle G_I, G_I, E, G_I\rangle_A$. Conditions for the existence of a governing process $X(t)$ with stationary increments for this case are given in § 9 Chapter 1. The nature of the governing sequences here allows us to write

$$\overline{X}(0)\underset{d}{=}\sup_{t\geqslant 0}X(t).$$

If $\tau_1^e, \tau_2^e, \ldots$ are points on the positive axis where $e(t)$ has jumps, then the random variables

$$x_k=X(\tau_k^e+0)-X(\tau_{k-1}^e+0),\quad k=2,3,\ldots,$$

are independent and distributed according to the law

$$\mathsf{P}(x_k=j)=\int d\mathsf{P}(\tau^e<y)\mathsf{P}(v_1^e-v_1^s-\cdots-v_\eta^s=j),\quad k\geqslant 2,$$

where η does not depend on v_1^e and $v_1^s, v_2^s, \ldots$, and has a Poisson distribution with parameter αy (α is the parameter of the distribution of τ^s):

$$\mathsf{P}(\eta=k)=e^{-\alpha y}\frac{(\alpha y)^k}{k!}.$$

We find analogously that $x_1=X(\tau_1^e+0)$ is distributed according to

$$\mathsf{P}(x_1=j)=\int d\mathsf{P}(\chi^e<y)\mathsf{P}(v_1^e-v_1^s-\cdots-v_\eta^s=j),$$

where χ^e is the size of the overshoot of an infinite barrier in a random walk with jumps $\tau_1^e, \tau_2^e, \ldots$. If we write

$$\varphi_i(z)=\mathsf{M}z^{v^i}\quad\text{and}\quad\psi_i(\lambda)=\mathsf{M}\,e^{\lambda\tau^i},\quad i=e,s, \tag{4}$$

then for $k \geqslant 2$

$$\begin{aligned} \mathsf{M} z^{x_k} &= \int d\mathsf{P}(\tau^e < y)\varphi_e(z)\exp\{\alpha y(\varphi_s(z^{-1})-1)\} \\ &= \varphi_e(z)\mathsf{M}\exp\{\alpha\tau^e(\varphi_s(z^{-1})-1)\} = \varphi_e(z)\psi_e(\alpha[\varphi_s(z^{-1})-1]) . \end{aligned} \tag{5}$$

Using the fact that

$$\mathsf{M}\, e^{\lambda \chi^e} = \frac{\psi_e(\lambda)-1}{\lambda \mathsf{M}\tau^e},$$

we find in a similar way that

$$\mathsf{M} z^{x_1} = \varphi_e(z)\frac{\psi_e(\alpha[\varphi_s(z^{-1})-1])-1}{\mathsf{M}\tau^e\alpha[\varphi_s(z^{-1})-1]} .$$

We note now that

$$\begin{aligned} \overline{X}(0) &\underset{d}{=} \max(0, x_1, x_1+x_2, x_1+x_2+x_3, \ldots) \\ &= \max(0, x_1 + \overline{X}) , \end{aligned}$$

where $\overline{X} = \sup_{k \geqslant 0} X_k$, $X_0 = 0$ and X_k, $k = 1, 2, \ldots$, are successive sums, not depending on x_1, of i.i.d. random variables with generating function equal to the right side of (5). Methods of finding the distribution of $\overline{X}$ (and thus also of $\overline{X}(0)$) are studied in Chapters 3 and 4.

It's not hard to see that if τ^e is exponentially distributed, then $\chi^e \underset{d}{=} \tau^e$ and

$$\overline{X}(0) \underset{d}{=} \overline{X} .$$

In this case we also find easily that the limiting distribution of $q(t)$ as $t \to \infty$ coincides with that of the queue length q_n at the time of arrival of the n-th lot as $n \to \infty$. For the systems $\langle G_I, G_I, E, G_I \rangle_A$ the latter obviously satisfies

$$\lim_{n\to\infty} \mathsf{P}(q_n = j) = \mathsf{P}(\overline{X} = j) .$$

For the systems $\langle E, G_1, E, G_1 \rangle_A$, $X(t)$ turns out to be a generalized Poisson process and to find the distribution of $\overline{X}$ we can also use the results of Chapter 2.

2. *The systems* $\langle E, G_I, G_I, G_I \rangle_A$. If we put

$$x_1 = X(\tau_1^s - 0) , \qquad x_k = X(\tau_k^s - 0) - X(\tau_{k-1}^s - 0) , \quad k \geqslant 2 ,$$

where τ_k^s are the jump times of $s(t)$, then, as in the preceding subsection, we get

$$\overline{X}(0) \underset{d}{=} \max(0, x_1 + \overline{X}) .$$

Here $x_1 \geqslant 0$ and $\overline{X}$, which does not depend on x_1, is the supremum of successive sums of random variables distributed like x_2. In the notation of (4)

$$\mathsf{M} z^{x_2} = \varphi_s(z^{-1})\psi_s(\alpha[\varphi_e(z)-1])$$

and

$$\mathsf{M} z^{x_1} = \frac{\psi_s(\alpha[\varphi_e(z)-1])-1}{\alpha \mathsf{M}\tau^s[\varphi_e(z)-1]} .$$

3. *The systems* $\langle G_I \rangle_A$. In the general case

$$\mathsf{P}(\bar{X}(0)=k)=\mathsf{M}p(k,\chi^e,\chi^s)\,,$$

where $p(k, u, v)$ is the conditional distribution of $X(0)$ given that the first jumps of $e(t)$ and $s(t)$ occur at times u and v, resp. The random variable χ^s is defined like χ^e in Subsection 1, but corresponds here to the sequence $\tau_1^s, \tau_2^s, \ldots$. The function $p(k, u, v)$ obviously satisfies

$$\begin{aligned} p(k,u,v)&=0 && \text{for } k<0,\, u\geqslant 0,\, v\geqslant 0\,,\\ p(k,u,v)&=\mathsf{M}p(k-v^e,\tau^e,v-u) && \text{for } u<v\,,\\ p(k,u,v)&=\mathsf{M}p(k+v^s,u-v,\tau^s) && \text{for } u>v \end{aligned}$$

and

$$p(k,u,v)=\mathsf{M}p(k-v^e+v^s,\tau^e,\tau^s) \quad \text{for } u=v\,.$$

Appendices

Appendix 1. Some Theorems from Renewal Theory

1. Let $\tau_1, \tau_2, \ldots$ be a sequence of i.i.d., nonnegative random variables with

$$F(x)=\mathsf{P}(\tau<x), \qquad P(x)=1-F(x), \qquad a=\mathsf{M}\tau \quad \text{and} \quad \sigma^2=\mathsf{D}\tau .$$

Put $X_0=0$, $X_k=\sum_{j=1}^{k}\tau_j$ and $H(x)=\sum_{k=0}^{\infty}\mathsf{P}(X_k<x)$. The function $H(x)$ is the renewal function of the random variable τ. Clearly, $H(x)<\infty$ for any x provided that $a\neq 0$.

Assume also, in agreement with the notation of Chapter 3, that $\eta(x)=\min\{k\colon X_k\geqslant x\}$ is the first passage time to the level x, corresponding to the random variable τ, and that $\chi(x)=X_{\eta(x)}-x$, $\gamma(x)=x-X_{\eta(x)-1}$ are, resp., the overshoot and defect variables w.r.t. the level $x\geqslant 0$.

Then, it's easy to see that $\mathsf{M}\eta(x)=H(x)$ and since $\eta(x)$ does not depend on the future,

$$\mathsf{M}\chi(x)+x=\mathsf{M}\tau\mathsf{M}\eta(x)\,; \qquad H(x)=\frac{x}{a}+\frac{\mathsf{M}\chi(x)}{a}\cdot \tag{1}$$

Here and throughout the sequel we will assume, if nothing is said to the contrary, that $a<\infty$. From (1) we then have the so-called *integral renewal theorem*:

$$\lim_{x\to\infty}\frac{H(x)}{x}=\frac{1}{a}\cdot$$

Indeed, since $\chi(x)\geqslant 0$, one always has

$$\frac{H(x)}{x}\geqslant\frac{1}{a}\cdot$$

Hence, it is enough to show that

$$\limsup_{x\to\infty}\frac{H(x)}{x}\leqslant\frac{1}{a}\cdot$$

Let $\eta^*(x)$ be the first passage time corresponding to the random variable $\tau^*=\min\{\tau, N\}$. For prescribed $\varepsilon>0$ we choose an N such that $a^*=\mathsf{M}\tau^*\geqslant a-\varepsilon$. Then, since $\tau^* \leqslant \tau$ and $\tau^* \leqslant N$, we have $\eta(x) \leqslant \eta^*(x)$, and by (1)

$$H(x) \leqslant \mathsf{M}\eta^*(x) \leqslant \frac{x+N}{a^*},$$

$$\limsup_{x\to\infty} \frac{H(x)}{x} \leqslant \frac{1}{a^*} \leqslant \frac{1}{a-\varepsilon}.$$

Since ε is arbitrary, we conclude that

$$\lim_{x\to\infty} \frac{H(x)}{x} = \frac{1}{a}. \quad \square$$

A more precise assertion is the *local renewal theorem* (Blackwell, Feller): *If τ is nonlattice, then for arbitrary $h>0$*

$$\lim_{x\to\infty} \frac{H(x+h)-H(x)}{h} = \frac{1}{a}.$$

(D) *If τ is integer-valued with the g.c.d. of its values equal to* 1, *then*

$$\lim_{x\to\infty} (H(x+1)-H(x)) = \frac{1}{a}.$$

From these statements we get (see, for example, Takács [70], p. 227)

The Fundamental Renewal Theorem. *If $g(u)$ is a function of bounded variation on* $(0, \infty)$ *and τ is nonlattice, then*

$$\lim_{t\to\infty} \int_0^t g(t-u)\, dH(u) = \frac{1}{a} \int_0^\infty g(u)\, du \tag{2}$$

when the integral on the right exists.

An analogous result will hold when (D) *is satisfied, but the sequence $t \to \infty$ must be integer-valued: on the right in* (2) *we then have*

$$\frac{1}{a} \sum_{k=1}^\infty g(k).$$

We now obtain some estimates related to the renewal function.

Theorem 1. *For an arbitrary distribution function $F(x)$ (the case $a=\infty$ is not excluded) we have for all $x \geqslant 0$*

$$H(x) \leqslant 2 + \frac{2x}{m_0},$$

where m_0 is the median[1] *of the distribution F.*

---

[1] If τ can also assume the value 0 then it can happen that $m_0=0$. In this case we can take the $(\frac{1}{2}+p)$-th order quantile m_p instead of m_0. Then

$$H(x) \leqslant \frac{2}{1-2p} + \frac{2x}{(1-2p)m_p}, \quad -\tfrac{1}{2} < p < \tfrac{1}{2}.$$

Proof. Let τ_j^* be random variables assuming the two values 0 and m_0 with probabilities $\frac{1}{2}$. Then, if $X_n^* = \sum_{j=1}^n \tau_j^*$, we obviously have $X_n \underset{d}{\geqslant} X_n^*$, $n = 1, 2, \ldots$, and by (1)

$$H(x) \leqslant H^*(x) \leqslant \frac{2x}{m_0} + 2 . \quad \square$$

Theorem 2. *If* $\mathsf{M}\tau^2 < \infty$, *then*

$$H(x) = \frac{x}{a} + r(x) ,$$

where

$$0 < r(x) < 2 + \frac{2}{m_0 a} \int_0^x G(t)\, dt , \tag{3}$$

and $G(x)$ *is the "double tail" of the distribution, equal to*

$$\int_x^\infty P(t)\, dt .$$

From the estimate obtained it follows, in particular, that

$$\sup_{x \geqslant 0} \left| H(x) - \frac{x}{a} \right| \leqslant c < \infty , \qquad \text{where } c \leqslant 2 + \frac{\mathsf{M}\tau^2}{a m_0} .$$

Proof. We have

$$\begin{aligned} \mathsf{P}(\chi(x) > y) &= \textstyle\sum_{k=0}^\infty \int_0^x \mathsf{P}(X_k \in dt) \mathsf{P}(\tau \geqslant x + y - t) \\ &= \textstyle\int_0^x \mathsf{P}(x + y - t)\, dH(t) . \end{aligned} \tag{4}$$

From (4) it follows that

$$\begin{aligned} \mathsf{M}\chi(x) &= \textstyle\int_0^x G(x-t)\, dH(t) = aH(x) - \int_0^x H(t) P(x-t)\, dt \\ &= G(x)H(x) + \textstyle\int_0^x (H(x) - H(t)) P(x-t)\, dt \\ &\leqslant G(x)H(x) + \textstyle\int_0^x H(t) P(t)\, dt . \end{aligned}$$

The last inequality is valid because of the fact that $H(x+y) - H(x) \leqslant H(y)$ since $H(x+y) - H(x)$ is the average of $H(\max(0, y - \chi(x)))$. Hence, if $H_1(t) \geqslant H(t)$, then

$$\mathsf{M}\chi(x) \leqslant G(x) H_1(x) + \textstyle\int_0^x H_1(t) P(t)\, dt = aH_1(0) + \int_0^x G(t)\, dH_1(t) .$$

Substituting here $H_1(t) = 2 + (2t/m_0)$ we obtain the assertion of the theorem using Theorem 1 and (1). $\square$

We remark that $H(x+y) - H(x) \leqslant H(y)$ also implies the following inequality for $H(x)$: for arbitrary integer k $H(x) \leqslant kH(x/k)$, so that $H(x) \leqslant xH(1)$ for integer x.

If $\mathsf{M}\tau^2 < \infty$, then by the fundamental renewal theorem with $x \to \infty$

$$\mathsf{M}\chi(x) \to \frac{1}{a} \int_0^\infty G(t)\, dt = \frac{\mathsf{M}\tau^2}{2a}$$

and

$$H(x)=\frac{x}{a}+\frac{\mathsf{M}\tau^2}{2a^2}+o(1)$$

(in the discrete case $x\to\infty$ through integer values).

What is the rate of convergence to 0 of the difference

$$R(x)=H(x)-\frac{x}{a}-\frac{\mathsf{M}\tau^2}{2a^2}\ ?$$

It is known (see [66]) that if $\mathsf{M}\tau^{k+2}<\infty$ and

$$\liminf_{|\lambda|\to\infty}|1-f(\lambda)|>0\,,\quad \text{where } f(\lambda)=\mathsf{M}\,e^{i\lambda\tau}\,,$$

then

$$R(x)=-\frac{1}{a^2}\int_x^\infty G(t)\,dt+o\left(\frac{\ln x}{x^{k+1}}\right).$$

From this it follows that under our assumptions

$$R(x)=o(x^{-k})\,.$$

In the discrete case when (D) is fulfilled and $\mathsf{M}\tau^{k+2}<\infty$ for integer x we have

$$R(x)=-\frac{1}{a^2}\sum_{j\geq k}\sum_{k>j}P(k+1)+o(x^{k+1})\,.$$

If $P(x)$ decreases exponentially we get

Theorem 3. *If the distribution $F(x)$ has an absolutely continuous component and* $\mathsf{M}\,e^{\mu\tau}<\infty$ *for some* $\mu>0$, *then*

$$H(x)=\frac{x}{a}+\frac{\sigma^2+a^2}{2a^2}+R(x)\,, \tag{5}$$

where

$$|R(x)|<ce^{-\varepsilon x}\quad \textit{and}\quad \underset{(x,\,\infty)}{\text{Variation}}\,R(t)<ce^{-\varepsilon x} \tag{6}$$

for some $c<\infty$ *and* $\varepsilon>0$.

This remains valid in the discrete case if the requirement of an absolutely continuous component is replaced by (D).

Proof. For $\operatorname{Im}\lambda\geqslant 0$

$$\int_0^\infty e^{i\lambda x}\,dH(x)=\frac{1}{1-f(\lambda)}\,,$$

and

$$\int_0^\infty e^{i\lambda x}\,dR_1(x)=\frac{1}{1-f(\lambda)}+\frac{1}{i\lambda a}=\frac{1}{i\lambda a}\left(1+\frac{i\lambda a}{1-f(\lambda)}\right)$$

$$=\frac{i\lambda-1}{i\lambda a}\left(\frac{1}{i\lambda-1}+\frac{i\lambda a}{(1-f(\lambda))(i\lambda-1)}\right),$$

where $R_1(x)$ differs from $R(x)$ for $x>0$ only by an additive constant.

Using Theorem 8 of Appn. 2 (see also the remark following Theorem 8) we find that along with the function

$$\mathfrak{v}(\lambda)=\frac{1-f(\lambda)}{i\lambda a}(i\lambda-1)$$

the ring $\mathfrak{B}$ (see the notation of Appn. 2) also contains $\mathfrak{v}^{-1}(\lambda)$.

But the set of elements of the ring $\mathfrak{B}$ for which inverses exist is open. Since $\|\mathfrak{v}(i\varepsilon+\lambda)-\mathfrak{v}(\lambda)\|\to 0$ as $\varepsilon\to 0$, for small enough ε, $\mathfrak{v}^{-1}(i\varepsilon+\lambda)$ will belong to the ring $\mathfrak{B}$ along with $\mathfrak{v}^{-1}(\lambda)$. In other words, $\mathfrak{v}^{-1}\in\mathfrak{B}(-\varepsilon,\varepsilon)$. Under our conditions this means that

$$\mathfrak{v}^{-1}\in\mathfrak{B}(-\varepsilon,\infty)\,. \tag{7}$$

By Theorem 2 of Appn. 2 this implies that

$$\frac{1}{i\lambda}\left(\frac{1}{i\lambda-1}+\frac{i\lambda a}{(1-f(\lambda))(i\lambda-1)}\right)\in\mathfrak{B}(-\varepsilon,\infty)\,. \tag{8}$$

It remains to use Theorems 1 and 2 of Appendix 2 by virtue of which (7) and (8) imply (6).

The proof in the discrete case follows easily from known theorems on power series. □

It is not difficult to see that for ε in the statement of the theorem we can choose any number from the interval $(0, \sup(\mu\colon \mathsf{M}\, e^{\mu\tau}<\infty))$.

2. Let us find the joint limiting distribution of $\chi(x)$ and $\gamma(x)$ as $x\to\infty$. By (4) one has

$$\begin{aligned}\mathsf{P}(\gamma(x)\geqslant t, \chi(x)\geqslant u)&=\mathsf{P}(\chi(x-t)\geqslant u+t)\\&=\textstyle\int_0^{x-t} dH(v)\mathsf{P}(\tau\geqslant x-v+u)\,.\end{aligned}$$

Thus, by the fundamental renewal theorem for nonlattice τ

$$\lim_{x\to\infty}\mathsf{P}(\gamma(x)\geqslant t, \chi(x)\geqslant u)=\frac{1}{a}\textstyle\int_{u+t}^{\infty}\mathsf{P}(v)\,dv\,.$$

A similar result obviously holds for an integer-valued τ with g.c.d. of its values equal to 1.

3. *The joint limiting distribution of* $\gamma(x)$, $\tau_{\eta(x)-1}, \tau_{\eta(x)-2}, \ldots, \tau_{\eta(x)-N}$. We have

$$\begin{aligned}&\mathsf{P}(\gamma(x)\geqslant t,\, \tau_{\eta(x)-1}\in dx_1,\ldots,\tau_{\eta(x)-N}\in dx_N)\\&\quad=\textstyle\sum_{k=0}^{\infty}\int_0^{x-x_1-\cdots-x_N-t}\mathsf{P}(X_k\in dv)\mathsf{P}(\tau_{k+1}\in dx_N)\ldots\mathsf{P}(\tau_{k+N}\in dx_1)\\&\qquad\times\mathsf{P}(\tau_{k+N+1}\geqslant x-v-x_1-\cdots-x_N)\\&\quad=\textstyle\prod_{j=1}^{N}\mathsf{P}(\tau_j\in dx_j)\int_0^{x-z-t}dH(v)\mathsf{P}(x-v-z)\,,\end{aligned}$$

where $z=x_1+\cdots+x_N$.

By the fundamental renewal theorem the last integral converges to

$$\frac{1}{a}\textstyle\int_t^{\infty}\mathsf{P}(v)\,dv\,. \tag{9}$$

Thus, the joint limiting distribution for $x\to\infty$ of $\gamma(x)$, $\tau_{\eta(x)-1}, \ldots, \tau_{\eta(x)-N}$ coincides with the distribution of the independent random variables $\gamma, \tau_1, \ldots, \tau_N$, where $\mathsf{P}(\gamma\geqslant t)$ equals (9).

4. We consider finally a random walk with alternating jumps of two kinds—from the sequence $\{\tau_j^1\}$ and the sequence $\{\tau_j^2\}$, both formed from independent random variables.

We put

$$X_0=0\,,\qquad X_{2k}=\textstyle\sum_{j=1}^{k}(\tau_j^1+\tau_j^2)\,,$$

$$X_{2k+1}=X_{2k}+\tau_{k+1}^1,\qquad \eta(x)=\min\{k: X_k\geqslant x\}\,,$$

$$\chi(x)=X_{\eta(x)}-x \quad\text{and}\quad \gamma(x)=x-X_{\eta(x)-1}\,.$$

Let A be the event that $\eta(x)$ is odd (the level x is attained by a jump from the first sequence).

Under these conditions we can also find the limiting distributions of $\chi(x)$ and $\gamma(x)$ without difficulty. Indeed, if $\tau^1+\tau^2$ is nonlattice, then by the fundamental renewal theorem

$$\mathsf{P}(\chi(x)>y;\,A=\textstyle\sum_{k=0}^{\infty}\int_0^x \mathsf{P}(X_{2k}\in dt)\mathsf{P}(\tau^1>x+y-t)$$

$$=\int_0^x \mathsf{P}(\tau^1>x+y-t)\,dH_{\tau^1+\tau^2}(t)\to\frac{1}{\mathsf{M}\tau^1+\mathsf{M}\tau^2}\int_y^\infty \mathsf{P}(\tau>v)\,dv\,.$$

Furthermore, the function

$$H_1(t)=\textstyle\sum_{k=0}^{\infty}\mathsf{P}(X_{2k+1}<t)$$

satisfies along with $H_{\tau^1+\tau^2}(t)$ the local renewal theorem:

$$\lim_{t\to\infty}\frac{H_1(t+h)-H_1(t)}{h}=\frac{1}{\mathsf{M}\tau^1+\mathsf{M}\tau^2} \tag{10}$$

and both satisfy the fundamental renewal theorem. To show (10) is suffices to note that

$$H_1(t)=\int_0^\infty \mathsf{P}(\tau^1\in du)\textstyle\sum_{k=0}^{\infty}\mathsf{P}(X_{2k}<t-u)=\int_0^\infty \mathsf{P}(\tau^1\in du)H_{\tau^1+\tau^2}(t-u)\,,$$

so that

$$\lim_{t\to\infty}\frac{H_1(t+h)-H_1(t)}{h}=\lim_{t\to\infty}\int_0^\infty \mathsf{P}(\tau^1\in du)\frac{H_{\tau^1+\tau^2}(t+h-u)-H_{\tau^1+\tau^2}(t-u)}{h}$$

$$=\int_0^\infty \mathsf{P}(\tau^1\in du)\frac{1}{\mathsf{M}\tau^1+\mathsf{M}\tau^2}=\frac{1}{\mathsf{M}\tau^1+\mathsf{M}\tau^2}\,.$$

On the basis of (10)

$$\mathsf{P}(\chi(x)>y;\,\bar{A})=\textstyle\sum_{k=0}^{\infty}\int_0^x \mathsf{P}(\chi_{2k+1}\in dt)\mathsf{P}(\tau^2>x+y-t)$$

$$\to\frac{1}{\mathsf{M}\tau^1+\mathsf{M}\tau^2}\int_0^\infty \mathsf{P}(\tau^2>v)\,dv\,,$$

where $\bar{A}$ is the complement of A. In exactly the same way one treats the case of alternating jumps of more than two types, where the jumps are taken by turns from sequences of independent random variables $\{\tau_j^1\}, \ldots, \{\tau_j^m\}$. If A_k is the event that the level x is attained by a jump from the sequence $\{\tau_j^k\}$, and if the distribution of $\tau^1 + \cdots + \tau^m$ is nonlattice, then we get as before

$$\lim_{x\to\infty} \mathsf{P}(\chi(x)>y;\, A_k)=\frac{1}{\mathsf{M}(\tau^1+\cdots+\tau^m)}\int_y^\infty \mathsf{P}(\tau^k>v)\,dv\,;$$

$$\mathsf{P}(A_k)=\frac{\mathsf{M}\tau^k}{\mathsf{M}(\tau^1+\cdots+\tau^m)}\,.$$

Analogous results will hold for the distribution of γ and also in the case in which $\tau^1 + \cdots + \tau^m$ is lattice.

Appendix 2. Factorization in the Ring $\mathfrak{B}$ and Some Theorems Associated with It

1. Let V be the collection of complex-valued functions $v(t)$ $(-\infty < t < \infty)$ having bounded variation. If we set

$$\|v\| = \text{Variation } v(t) = \int_{-\infty}^{\infty} |dv(t)|$$

and define multiplication $v_1 * v_2$ as the convolution

$$v_1 * v_2(x) = \int v_1(x-t)\,dv_2(t) \in V\,,$$

then V becomes a normed commutative ring. Along with V we consider the collection $\mathfrak{B}$ of Fourier–Stieltjes transformations

$$\mathfrak{v}(\lambda) = \int e^{i\lambda t}\,dv(t)\,,\quad \operatorname{Im}\lambda = 0$$

of functions from V. It is natural to put $\|\mathfrak{v}(\lambda)\| = \|v\|$. Then, as is well known, $\mathfrak{B}$ will be a ring w.r.t. ordinary multiplication and will be isometrically isomorphic to V.

We will also need the "shifted" ring $\mathfrak{B}(\mu)$ for various real μ. This is the collection of functions $\mathfrak{v}_\mu(\lambda) = \mathfrak{v}(\lambda - i\mu)$, where $\mathfrak{v} \in \mathfrak{B}$, which are now defined on the line $\operatorname{Im}\lambda = \mu$. In other words, $\mathfrak{v} \in \mathfrak{B}(\mu)$ if $\mathfrak{v}(\lambda + i\mu) \in \mathfrak{B}$. We can set $\|\mathfrak{v}_\mu\|_{\mathfrak{B}(\mu)} = \|\mathfrak{v}\|_{\mathfrak{B}}$, so that $\mathfrak{B} = \mathfrak{B}(0)$. A function $\mathfrak{v} \in \mathfrak{B}(\mu)$ admits the representation

$$\mathfrak{v}(\lambda) = \int_{-\infty}^{\infty} \exp\left[(i\lambda + \mu)t\right] dv(t)\,,\quad v \in V,\ \operatorname{Im}\lambda = \mu\,. \tag{1}$$

We now define the subring $\mathfrak{B}(\mu_-, \mu_+) \subset \mathfrak{B}(\mu_-)$. We will say that $\mathfrak{v} \in \mathfrak{B}(\mu_-, \mu_+)$ if $\mathfrak{v} \in \mathfrak{B}(\mu_-)$ and the function v in the representation (1) possesses the property

$$\int_{-\infty}^{\infty} \exp\left[(\mu_- - \mu_+)t\right] |dv(t)| < \infty\,.$$

In other words, $\mathfrak{B}(\mu_-, \mu_+) = \bigcap_{\mu \in [\mu_-, \mu_+]} \mathfrak{B}(\mu)$. It is clear that $\mathfrak{B}(\mu_-, \mu_+)$ will also be a ring and that functions from this ring admit a representation of the form $\mathfrak{v}(\lambda) = \int e^{i\lambda t + at}\,dv(t)$ for each λ in the strip $\mu_- \leqslant \operatorname{Im}\lambda \leqslant \mu_+$, where $a \in [\mu_-, \mu_+]$ and $v = v_a \in V$. In the interior of this strip $\mathfrak{v}(\lambda)$ is analytic, and is also continuous and bounded on the boundaries $\operatorname{Im}\lambda = \mu_-$ and $\operatorname{Im}\lambda = \mu_+$. The norm in $\mathfrak{B}(\mu_-, \mu_+)$ can be defined, for example, as the norm in $\mathfrak{B}(\mu_-)$.

Furthermore, we will say that $\mathfrak{v}\in\mathfrak{B}_+(\mu)$ ($\mathfrak{v}\in\mathfrak{B}_-(\mu)$), if in the representation (1) $\underset{(-\infty,0)}{\text{Variation}}\, v(t)=0$ ($\underset{(0,\infty)}{\text{Variation}}\, v(t)=0$). In the sequel we will denote such relations by using a briefer notation: for example, $\mathfrak{v}\in\mathfrak{B}_\pm(\mu)$ if

$$\mathfrak{v}(\lambda+i\mu)=\int_0^\infty e^{\pm i\lambda t}\, dv(t),\ v\in V.$$

Correspondingly, $V_\pm$ will denote the collection of functions from V which are constant for $t\lessgtr 0$. Clearly $\mathfrak{B}_\pm(\mu)$ and $V_\pm$ will also be rings.

We denote by $\Pi(\mu_-,\mu_+)$ the region $\mu_-\leqslant\operatorname{Im}\lambda\leqslant\mu_+$ and by $\tilde{\Pi}(\mu_-,\mu_+)$ the interior of this region. The line $\Pi(\mu_-,\mu_-)$ will be denoted by $\Pi(\mu_-)$, $\Pi(0)=\Pi$; the half-planes $\Pi(\mu,\infty)$ and $\Pi(-\infty,\mu)$ are written as $\Pi_+(\mu)$ and $\Pi_-(\mu)$ resp. Here we set $\Pi_\pm(0)=\Pi_\pm$. Elements of the ring $\mathfrak{B}_{\bullet}(\bullet)$ will be denoted throughout by Gothic letters and those of V by Latin ones.

Theorem 1. *If* $\mathfrak{v}\in\mathfrak{B}(\mu)$, $\mu\leqslant 0$, *then in the representation* $\mathfrak{v}(\lambda)=\int_{-\infty}^{\infty} e^{i\lambda t}\, dw(t)$, $w(0)=0$, *the function* $w(t)$ *has the form*

$$w(t)=w_1+w_2(t)\,,$$

where $w_1=\int_0^\infty e^{\mu t}\, dv(t)$, $|w_1|\leqslant\|v\|$ *and* $|w_2(t)|\leqslant e^{\mu t}$. $\underset{(t,\infty)}{\text{Variation}}\, v(u)\leqslant e^{\mu t}\|v\|$, $t>0$. *Here* $v\in V$ *is the function in the representation* (1).

Proof. The claim follows from the equality

$$w(t)=\int_0^t e^{\mu t}\, dv(t)=\int_0^\infty dv(t)\, e^{\mu t}-\int_t^\infty dv(t)\, e^{\mu t}\,. \quad \square$$

Denote by $\mathfrak{R}_{\bullet}(\bullet)$ the subring of $\mathfrak{B}_{\bullet}(\bullet)$ formed by those elements for which v in (1) is absolutely continuous.

Theorem 2. *If* $\mathfrak{v}\in\mathfrak{B}(\mu_-,\mu_+)$ *and* $\mathfrak{v}(i\mu_0)=0$, $i\mu_0\in\tilde{\Pi}(\mu_-,\mu_+)$, *then*

$$\mathfrak{r}(\lambda)=\frac{\mathfrak{v}(\lambda)}{\lambda-i\mu_0}\in\mathfrak{R}(\mu_-,\mu_+)\,,$$

and in the representation $\mathfrak{r}(\lambda)=\int_{-\infty}^{\infty} e^{i\lambda t} r(t)\, dt$ *the function* $r(t)$ *admits the estimate*

$$|r(t)|<e^{\mu_- t}\cdot\underset{(t,\infty)}{\text{Variation}}\, v\leqslant\|v\|\, e^{\mu_- t},\quad t>0\,.$$

Similar statements hold when $\mathfrak{v}\in\mathfrak{B}_+(\mu_-)$ *or* $\mathfrak{v}\in\mathfrak{B}_-(\mu_+)$.

Proof. For $\mu_+\geqslant\operatorname{Im}\lambda>\operatorname{Im} i\mu_0=\mu$ we have by (1)

$$\begin{aligned}\mathfrak{r}(\lambda)&=\frac{\mathfrak{v}(\lambda)}{\lambda-i\mu_0}=(-i)\frac{-1}{i\lambda+\mu_0}\mathfrak{v}(\lambda)=i\mathfrak{v}(\lambda)\int_0^\infty \exp\,[i\lambda t+\mu_0 t]\, dt\\ &=-i\int_{-\infty}^{\infty} e^{i\lambda t}\int_{-\infty}^{t} e^{\mu_- x}\, dv(x)\exp[\mu_0(t-x)]\, dx\\ &=-i\int_{-\infty}^{\infty}\exp[i\lambda t+\mu_- t]g(t)\, dt\,,\end{aligned}$$

where $g(t)=\exp[t(\mu_0-\mu_-)]\int_{-\infty}^{t}\exp[-x(\mu_0-\mu_-)]\, dv(x)$, $v\in V$.

The absolute integrability of the functions $g(t)$ and $\exp[(\mu_- - \mu_+)t]g(t)$ can be verified directly with the aid of the equality

$$\int_{-\infty}^{\infty} \exp[-t(\mu_0-\mu_-)]\, dv(t)=0 .$$

For example (for simplicity assume μ_0 is real and that v is monotonic),

$$\begin{aligned}\int_{-\infty}^{\infty} |g(t)|\, dt &= \int_{-\infty}^{\infty} e^{t(\mu_0-\mu_-)} \int_t^{\infty} \exp[-x(\mu_0-\mu_-)]\, dv(x)\\ &= \int_{-\infty}^{\infty} \exp[-x(\mu_0-\mu_-)]\, dv(x) \int_{-\infty}^{x} \exp[t(\mu_0-\mu_-)]\, dt\\ &= \frac{1}{\mu_0-\mu_-}\int_{-\infty}^{\infty} dv(t) .\end{aligned}$$

Finally, the estimate for $|r(t)|$ is justified by the inequality

$$\begin{aligned}|g(t)| &= |\exp[t(\mu_0-\mu_-)] \int_t^{\infty} \exp[-x(\mu_0-\mu_-)]\, dv(x)|\\ &\leqslant \int_t^{\infty} |dv(t)| . \quad \square\end{aligned}$$

2. *Factorizations.* **Definitions.**

1. We will say that a function $\mathfrak{v}$ analytic in $\tilde{\Pi}(\mu_-, \mu_+)$ and continuous on the boundaries $\operatorname{Im}\lambda=\mu_\pm$, $(\mu_- \leqslant \mu_+)$ admits a *factorization in the strip* $\Pi(\mu_-, \mu_+)$ if in this strip

$$\mathfrak{v}_+(\lambda)\mathfrak{v}(\lambda)=\mathfrak{v}_-(\lambda), \quad \lambda\in\Pi(\mu_-, \mu_+), \tag{2}$$

holds, where the $\mathfrak{v}_\pm(\lambda)$ are analytic in $\tilde{\Pi}_\pm(\mu_\mp)$, resp., and are also continuous and bounded on the boundaries $\operatorname{Im}\lambda=\mu_\mp$.

2. We will say that the function $\mathfrak{v}$ in Definition 1 admits a *canonical factorization* (c.f.) if in (2) we have in addition

$$\inf_{\lambda\in\Pi_\pm(\mu_\mp)} |\mathfrak{v}_\pm(\lambda)|>0 . \tag{3}$$

3. A *V-factorization* (*V*-f.) of the function $\mathfrak{v}\in\mathfrak{V}(\mu_-, \mu_+)$ in the strip $\Pi(\mu_-, \mu_+)$ will be the designation for the representation (2), where $\mathfrak{v}_\pm \in\mathfrak{V}_\pm(\mu_\mp)$.

4. A *canonical V-factorization* (c.*V*-f.) of $\mathfrak{v}\in\mathfrak{V}(\mu_-, \mu_+)$ in $\Pi(\mu_-, \mu_+)$ is a *V*-factorization satisfying the conditions of Definition 2.

By definition, two factorizations coincide if their components differ by constant factors.

If $\mathfrak{v}(\lambda)$ admits a c.f. (in particular, a c.V-f.), then the factorization (2) *is unique.*

Indeed, assume that $\mathfrak{u}_+(\lambda)\mathfrak{v}(\lambda)=\mathfrak{u}_-(\lambda)$ is a factorization different from (2). Then for $\lambda\in\Pi(\mu_-, \mu_+)$

$$\frac{\mathfrak{v}_+(\lambda)}{\mathfrak{v}_-(\lambda)}=\frac{\mathfrak{u}_+(\lambda)}{\mathfrak{u}_-(\lambda)},$$

or, what is the same,

$$\frac{\mathfrak{u}_+(\lambda)}{\mathfrak{v}_-(\lambda)}=\frac{\mathfrak{u}_+(\lambda)}{\mathfrak{v}_+(\lambda)}.$$

The right side here is analytic in $\tilde{\Pi}_+(\mu_-)$ and is continuous and bounded on $\operatorname{Im}\lambda=\mu_-$. The left side has analogous properties in $\tilde{\Pi}_-(\mu_+)$. Both functions coincide in $\Pi(\mu_-, \mu_+)$. This means that $\mathfrak{u}_+(\lambda)/\mathfrak{v}_+(\lambda)$ is an entire bounded function, which necessarily equals a constant. □

Let us find conditions sufficient for a c. V-f.

Theorem 3. *In order that* $\mathfrak{v} \in \mathfrak{B}(\mu)$ *admit a* c. V-f. *on* $\Pi(\mu)$ *it is sufficient that*

$$\mathfrak{u}(\lambda) = \ln \mathfrak{v}(\lambda) \in \mathfrak{B}(\mu)$$

(*we take* ln *here as a fixed branch of this function*). *One has* $\mathfrak{v}^{-1} \in \mathfrak{B}(\mu)$ *and it also admits a* c. V-f.

Proof. There exists a $u \in V$ such that

$$\mathfrak{u}(i\mu + \lambda) = \ln \mathfrak{v}(i\mu + \lambda) = \int e^{i\lambda t}\, du(t)\,, \quad \lambda \in \Pi\,. \tag{4}$$

Put

$$\mathfrak{u}^+(i\mu + \lambda) = \int_0^\infty e^{i\lambda t}\, du\,, \qquad \mathfrak{u}^-(i\mu + \lambda) = \int_{-\infty}^0 e^{i\lambda t}\, du$$

and

$$\mathfrak{v}_\pm(i\mu + \lambda) = \exp\{\mp \mathfrak{u}^\pm(i\mu + \lambda)\}\,. \tag{5}$$

Since

$$\exp(\mp \mathfrak{u}^\pm) = \sum \frac{(\mp \mathfrak{u}^\pm)^k}{k!} \quad \text{and} \quad \|\exp(\mp \mathfrak{u}^\pm)\| \leqslant \sum \frac{\|\mathfrak{u}^\pm\|^k}{k!} < \infty\,,$$

$$\mathfrak{v}_\pm(i\mu + \lambda) \in \mathfrak{B}_\pm\,, \qquad \mathfrak{v}_+ \mathfrak{v} = \mathfrak{v}_-\,.$$

Condition (3) is clearly also satisfied.

When the assumptions of Theorem 3 are fulfilled along with (5), then there exists another representation for the components $\mathfrak{v}_\pm(\lambda)$.

Theorem 4. *If* $\ln \mathfrak{v} \in \mathfrak{B}(\mu)$, *then for* $\operatorname{Im} \lambda > \mu$

$$\mathfrak{u}^+(\lambda) = \ln \mathfrak{v}_+(\lambda) = -\frac{1}{2\pi i} \int_{i\mu - \infty}^{\prime\, i\mu + \infty} \frac{\ln \mathfrak{v}(\gamma)\, d\gamma}{\gamma - \lambda}\,,$$

where $\int'$ *is taken in the principal-value sense.*

Proof. Let $l_\pm$ be the jumps of u at zero: $l_\pm = u(\pm 0) - u(0)$. Then

$$\int_0^{\pm\infty} e^{i\lambda t}\, du(t) \to l_\pm \quad \text{for } \operatorname{Im} \lambda \to \pm\infty\,. \tag{6}$$

We form a positively oriented contour $\mathcal{M}$ from the arc $\mathcal{M}'$ of the semicircle $|\lambda| = M$, $\operatorname{Im} \lambda > 0$, and the interval $[-M, M]$. According to the properties of the function $\mathfrak{v}_+$, for $\operatorname{Im} \lambda > 0$ we get

$$\ln \mathfrak{v}_+(i\mu + \lambda) = l_+ + \frac{1}{2\pi i} \int_{\mathcal{M}} \frac{\ln \mathfrak{v}_+(i\mu + \gamma) - l_+}{\gamma - \lambda}\, d\gamma\,. \tag{7}$$

From (6) it follows that

$$\int_{\mathcal{M}'} \frac{\ln \mathfrak{v}_+(i\mu+\gamma)-l_+}{\gamma-\lambda}\,d\gamma \to 0$$

as $M\to\infty$. Since M in (7) is arbitrary,

$$\ln \mathfrak{v}_+(i\mu+\lambda)=l_+ + \frac{1}{2\pi i}\int_{-\infty}^{\prime\,\infty} \frac{\ln \mathfrak{v}_+(i\mu+\gamma)-l_+}{\gamma-\lambda}\,d\gamma\,, \quad \operatorname{Im}\lambda>0\,,$$

where the prime indicates that the integral is taken in the principal-value sense. Similarly, for $\mathfrak{v}_-$ one has

$$0=-\frac{1}{2\pi i}\int_{-\infty}^{\prime\,\infty} \frac{\ln \mathfrak{v}_-(i\mu+\gamma)-l_-}{\gamma-\lambda}\,d\gamma\,, \quad \operatorname{Im}\lambda>0\,.$$

Adding the last two equalities and assuming the components $\mathfrak{v}_\pm$ are defined up to constant factors, we finish the proof. □

It is simplest to carry out factorization of rational functions. If

$$\mathfrak{v}(\lambda)=\frac{c\prod_{k=1}^{m}(\lambda-\lambda_k)}{\prod_{k=1}^{M}(\lambda-\Lambda_k)}\in\mathfrak{B},$$

then necessarily $m\leqslant M$ and $\operatorname{Im}\Lambda_j\neq 0$, $j=1,\ldots,M$ (the fraction is assumed irreducible). If $\mathfrak{v}$ admits a c.f. on Π, then, in addition, $\operatorname{Im}\lambda_j\neq 0$, $j=1,\ldots,m$.

We supply all zeros of the numerator and denominator lying in $\tilde{\Pi}_\pm$ with the indices $\pm$, resp. The number of these will be denoted by $m_\pm$ and $M_\pm$, resp. Then, if we set

$$\mathfrak{v}_-(\lambda)=\frac{\prod_{k=1}^{m_+}(\lambda-\lambda_k^+)}{\prod_{k=1}^{M_+}(\lambda-\Lambda_k^+)}, \tag{8}$$

where m_+ is such that $m_+\leqslant M_+$ and $m_-\geqslant M_-$, we obtain a factorization of the function $\mathfrak{v}$ on Π which will clearly not be unique.

If $\mathfrak{v}$ also admits a c.f. $\mathfrak{v}_+\mathfrak{v}=\mathfrak{v}_-$, then necessarily $m=M$ (since $\inf_{\lambda\in\Pi}|\mathfrak{v}|>0$) and $m_+=M_+$ (since $\inf_{\lambda\in\Pi}|\mathfrak{v}_-|>0$). Hence, for a c.f. (or equivalently, a c. V-f.) of a rational function, it is necessary and sufficient that $m=M$, the line Π contain no zeros or poles of the function, and that the numbers m_+ and M_+ of zeros and poles, resp. of $\mathfrak{v}$ in the upper half-plane coincide. Here

$$\mathfrak{v}_-(\lambda)=\prod_{k=1}^{m_+}\frac{(\lambda-\lambda_k^+)}{(\lambda-\Lambda_k^+)}.$$

3. *The discrete case.* An important rôle is played by the subring $\mathfrak{D}\subset\mathfrak{B}$ of functions which can be represented in the form $\mathfrak{v}(\lambda)=\mathfrak{d}(e^{i\lambda h})$. These are transformations of functions from V all of whose variation is concentrated at points of the form kh, $k=\cdots,-1,0,1,\ldots$. The number h can, without loss of generality, be assumed equal to 1. It is clear that the ring $\mathfrak{D}$ can be considered as the collection of absolutely convergent Fourier series

$$\mathfrak{d}(z)=\sum_{-\infty}^{\infty} d_k z^k, \qquad \|\mathfrak{d}(z)\|=\sum |d_k|,$$

defined on the unit circle $\Gamma=\{z:|z|=1\}$. In an entirely analogous way we can introduce the subrings $\mathfrak{D}(z_-, z_+)$ and $\mathfrak{D}_\pm(z_0)$ (these are the projections of the rings $\mathfrak{B}(-\ln z_+, -\ln z_-)$ and $\mathfrak{B}_\pm(-\ln z_0)$ on $\mathfrak{D}$). Functions from these subrings will be analytic in the interiors and continuous on the boundaries of the regions $\Gamma(z_-, z_+)=\{z: z_-\leqslant|z|\leqslant z_+\}$ and $\Gamma_\pm(z_0)=\{z:|z|\lesseqgtr z_0\}$, resp.

As before we will denote $\Gamma(z, z)$ by $\Gamma(z)$ $(\Gamma(0)=\Gamma)$; $\tilde{\Gamma}(\bullet)$ will stand for the interior of the region $\Gamma(\bullet)$; $\Gamma_\pm(0)=\Gamma_\pm$ (here, Γ_- includes the point at infinity).

Here it is natural to consider the *factorization* of a function $\mathfrak{d}\in\mathfrak{D}(z_-, z_+)$ in the annulus $\Gamma(z_-, z_+)$ and to call $\mathfrak{d}_+(z)\mathfrak{d}(z)=\mathfrak{d}_-(z)$; $z\in\Gamma(z_-, z_+)$, where $\mathfrak{d}_\pm(z)$ are analytic in the interior and continuous on the boundary of the regions $\Gamma_\pm(z_\mp)$, resp., its *representation*.

The *canonical factorization* requires, in addition, that $\mathfrak{d}_\pm(z)\neq 0$ for $z\in\Gamma_\pm(z_\mp)$.

The analogues of Theorems 1 and 2 are obvious. Those of Theorems 3 and 4 have the following form:

Theorem 3A. *The function* $\mathfrak{d}\in\mathfrak{D}(z_0)$ *admits a c.f. on* $\Gamma(z_0)$ *if*

$$\mathfrak{a}(z)=\ln \mathfrak{d}(z)\in\mathfrak{D}(z_0).$$

One has $\mathfrak{d}^{-1}\in\mathfrak{D}(z_0)$ *and it also admits a c.f.*

Theorem 4A. *If* $\ln \mathfrak{d}\in\mathfrak{D}(z_0)$, *then for* $|z|<z_0$

$$\mathfrak{d}_+(z)=\exp\frac{1}{2\pi i}\int_{\Gamma(z_0)}\frac{\ln \mathfrak{d}(t)\,dt}{t-z}.$$

The proofs of these theorems differ only through simplifications from those of Theorems 3 and 4.

4. Let $\mathfrak{R}$ be a commutative normed functional ring with identity $\mathbf{e}$. Under what conditions on an element $\mathfrak{k}$ of $\mathfrak{R}$ and a function Λ will $\Lambda(\mathfrak{k})$ be in $\mathfrak{R}$?

Denote by $S_\mathfrak{k}$ the spectrum of the element $\mathfrak{k}$, i.e., the set of numbers ζ for which there exists no inverse $(\mathfrak{k}-\zeta\mathbf{e})^{-1}$ in $\mathfrak{R}$ of $(\mathfrak{k}-\zeta\mathbf{e})$. It is clear that for an arbitrary polynomial $Q(\zeta)$ the "abstract" function $Q(\mathfrak{k})$ is in $\mathfrak{R}$. Furthermore, $\mathfrak{k}^{-1}\in\mathfrak{R}$ if $0\notin S_\mathfrak{k}$ or, equivalently, if $S_\mathfrak{k}$ belongs to the domain of regularity of $1/\zeta$. It's not hard to see that for a rational function $R(\zeta)=Q_1(\zeta)/Q_2(\zeta)$, $R(\mathfrak{k})\in\mathfrak{R}$ if $S_\mathfrak{k}$ is completely contained in the domain of regularity of $R(\zeta)$. A natural generalization of these facts is the following assertion (Gel'fand, Raikov and Šilov [28], corollary of Theorem 1 § 6):

Theorem 5. *Let* $\Lambda(\zeta)$ *be analytic in some simply-connected region* $\mathscr{D}$. *If* $S_\mathfrak{k}\subset\mathscr{D}$, *then* $\Lambda(\mathfrak{k})\in\mathfrak{R}$. *Here*

$$\Lambda(\mathfrak{k})=\frac{1}{2\pi i}\int_{\mathscr{T}}(\zeta e-\mathfrak{k})^{-1}\Lambda(\zeta)\,d\zeta,$$

where $\mathscr{T}\subset\mathscr{D}$ *is a closed contour containing* $S_\mathfrak{k}$.

The region $\mathcal{D}$ can be considered as lying on the multi-sheeted Riemann surface corresponding to Λ.

Hence, to determine the properties of $\Lambda(\mathfrak{v})$, $\mathfrak{v}\in\mathfrak{B}$, with the help of this theorem it is necessary to find the spectrum $S_{\mathfrak{v}}$.

Theorem 6. *Assume that in the representation*

$$\mathfrak{v}(\lambda)=\int e^{i\lambda t}\,dv(t)\,, \quad v\in V\,, \tag{9}$$

$v=v_c+v_d+v_s$ *is the decomposition of* v *into absolutely continuous, discrete and singular components. Then, if*

1) $$\inf_{\lambda\in\Pi}|\mathfrak{v}(\lambda)|>0$$

and

2) $$\|v_s\|<\inf_{\lambda\in\Pi}|\int e^{i\lambda t}\,dv_d(r)|\,,$$

we have $\mathfrak{v}^{-1}\in\mathfrak{B}$ (see Gel'fand, *et al.* [28], Theorem 4 § 32).

5. These two theorems allow us to find conditions sufficient for $\ln\mathfrak{v}\in\mathfrak{B}$ and, consequently, sufficient conditions for a c. V-f.

Theorem 7. *Assume* $\mathfrak{v}\in\mathfrak{B}$ *and that* $\Lambda(\zeta)$ *is analytic and single-valued in the simply-connected closed region* $\mathcal{D}$. *Furthermore, let the conditions*

$$[\mathcal{D}\Pi]_1:\ \mathfrak{v}(\lambda)\in\mathcal{D}\quad \textit{for all } \lambda\in\Pi$$

and

$$[\mathcal{D}\Pi]_2:\ \|v_s\|<\inf_{\lambda\in\Pi}|\zeta-\int_{-\infty}^{\infty}e^{i\lambda t}\,dv_d(t)|\ \textit{for arbitrary } \zeta\notin\mathcal{D}$$

be fulfilled. Then $\Lambda(\mathfrak{v})\in\mathfrak{B}$.

Proof. By Theorem 5 we must establish that the complement $\bar{\mathcal{D}}$ of $\mathcal{D}$ contains no points of the spectrum $S_{\mathfrak{v}}$. Let $\zeta\in\bar{\mathcal{D}}$. Then by $[\mathcal{D}\Pi]_1$ and the fact that $\mathcal{D}$ is closed

$$\inf_{\lambda\in\Pi}|\zeta-\mathfrak{v}(\lambda)|>0\,.$$

This along with $[\mathcal{D}\Pi]_2$ means that the conditions of Theorem 6 hold for the function $\mathfrak{v}^*=\mathfrak{v}-\zeta\mathbf{e}$ since for it we obviously have

$$\|\mathfrak{v}_s^*\|=\|\mathfrak{v}_s\| \text{ and } \int e^{i\lambda t}\,d\mathfrak{v}_d^*(t)=\int e^{i\lambda t}\,d\mathfrak{v}_d(t)-\zeta\,.$$

Consequently, $(\mathfrak{v}-\zeta\mathbf{e})^{-1}\in\mathfrak{B}$ and $\zeta\notin S_{\mathfrak{v}}$. □

We denote by $\mathcal{D}_\varepsilon$ the complement of an ε-neighborhood of the ray $\arg\zeta=-\pi$: $\mathcal{D}_\varepsilon=\{z:\inf|z-\zeta|\geqslant\varepsilon\}$, where the inf is taken over all ζ with $\arg\zeta=-\pi$ (Fig. 14)

Corollary 1. *In order for* $\mathfrak{v}^{\pm 1}$, $\mathfrak{v}\in\mathfrak{B}(\mu)$, *to admit c.V-f's on* $\Pi(\mu)$ *it is sufficient that for some* $\varepsilon>0$ *the function*

$$\mathfrak{v}^*(\lambda)=\mathfrak{v}(\lambda+i\mu)$$

satisfy conditions[1] $[\mathscr{D}_\varepsilon\Pi]_i$, $i=1,2$.

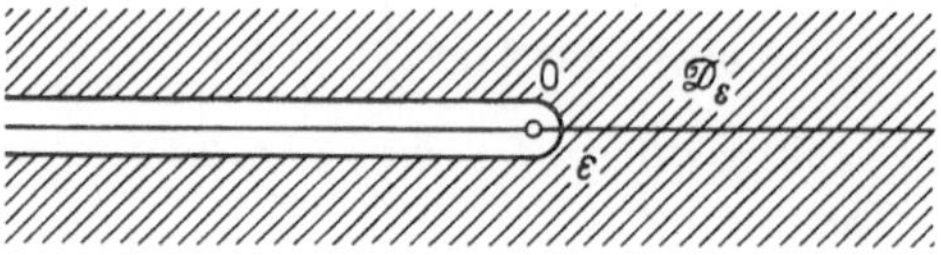

Fig. 14. The region $\mathscr{D}_\varepsilon$

Proof. The assertion follows from Theorems 3 and 7 since any branch of $\ln\zeta$ will be single-valued and analytic in $\mathscr{D}_\varepsilon$. We can take, for example, the principal branch, setting $\Lambda(\zeta)=\mathrm{Ln}\,(\zeta)$ (principal value). □

We dwell on two important special cases. In the first we have $\|v_s\|=0$ and $\int e^{i\lambda t}\,dv_d(t)=1$ in the representation (9). For such functions $\mathfrak{v}(\lambda)$, representable, obviously, in the form

$$\mathfrak{v}(\lambda)=1-\mathfrak{r}(\lambda)\,, \tag{10}$$

where $\mathfrak{r}(\lambda)\in\mathfrak{R}$ is the Fourier transformation of some absolutely integrable function, the condition $[\mathscr{D}_\varepsilon\Pi]_2$ is always satisfied. Moreover, by Lebesgue's theorem $\mathfrak{r}(\lambda)\to 0$ as $|\lambda|\to\infty$ and the set of values of $\mathfrak{v}(\lambda)$, $\lambda\in\Pi$, forms a closed contour. Hence, if $\mathfrak{v}(\lambda)\neq 0$, $\lambda\in\Pi$, then $\operatorname{ind}\mathfrak{v}=(1/2\pi)\int_{-\infty}^{\infty}d\arg\mathfrak{v}(\lambda)$ is defined. Since Theorem 7 also holds for functions with multiple branches, for the function (10) Corollary 1 is transformed into the following assertion:

For c.V-f.'s of $\mathfrak{v}^{\pm 1}$ *it is sufficient that* ind $\mathfrak{v}$ *be defined and that it be equal to zero.*

The second case is the discrete, in which $\|v_s\|=\|v_c\|=0$. By Theorem 6 it is sufficient for the inclusion $\mathfrak{d}^{-1}\in\mathfrak{D}$ $(\mathfrak{d}\in\mathfrak{D})$ that

$$\inf_{z\in\Gamma}|\mathfrak{d}(z)|>0\,. \tag{11}$$

Thus, as in the previous case, we can assert: *For c.f.'s of* $\mathfrak{d}^{\pm 1}$ *it is sufficient that*

$$\operatorname{ind}\mathfrak{d}=\frac{1}{2\pi}\int_{|z|=1}d\arg\mathfrak{d}(z)=0\,.$$

It is easy to verify that the formulated conditions are also necessary for c.f.'s in both cases. For example, in the case of (10) we immediately obtain

$$\inf_{\lambda\in\Pi}\mathfrak{v}=\inf_{\lambda\in\Pi}\frac{\mathfrak{v}_-}{\mathfrak{v}_+}>0\,,\quad \operatorname{ind}\mathfrak{v}=\operatorname{ind}\mathfrak{v}_- -\operatorname{ind}\mathfrak{v}_+=0\,,$$

---

[1] The condition $[\mathscr{D}_\varepsilon\Pi]_1$ can clearly be replaced by a less restrictive one allowing crossing "with returns" of the ray $-\infty\leqslant\zeta\leqslant 0$ by $\mathfrak{v}^*(\lambda)$.

since by the argument principle ind $\mathfrak{v}_\pm = 0$ because the functions $\mathfrak{v}_\pm$ are regular in $\tilde{\Pi}_\pm$ and have no zeros there.

6. *Factorizations of functions connected with* $1-f(\lambda)$, $f(\lambda)=\mathsf{M}\, e^{i\lambda\xi}\in\mathfrak{B}$. Of primary interest is the case in which $a=\mathsf{M}\xi<0$.

Theorem 8. *If $a<0$ and the function $F(x)=\mathsf{P}(\xi<x)$ has an absolutely continuous component, i.e., $\|F_c\|>0$ in the decomposition $F=F_c+F_s+F_d$, then the function*

$$\mathfrak{v}(\lambda)=\frac{1-f(\lambda)}{i\lambda}(i\lambda+1)\in\mathfrak{B}$$

and admits a c.V-f. The same assertion holds for $\mathfrak{v}^{-1}$.

Proof. Let

$$g(t)=\begin{cases}F(t), & t<0,\\ F(t)-1, & t\geqslant 0.\end{cases}$$

Since $\mathsf{M}\xi$ exists, g is absolutely integrable and

$$\int e^{i\lambda t}g(t)\,dt=\frac{1-f(\lambda)}{i\lambda}\in\mathfrak{R}\subset\mathfrak{B}.$$

Thus, $\mathfrak{v}(\lambda)\in\mathfrak{B}$ and for it

$$v(t)=-g(t)+\int_{-\infty}^{t} g(u)\,du.$$

Let us verify Condition $[\mathcal{D}_\varepsilon\Pi]_2$. Since $v_d=E-F_d$, for $\zeta\leqslant 0$

$$\inf_{\lambda\in\Pi}\left|\zeta-\int_{-\infty}^{\infty} e^{i\lambda t}\,dv_d(t)\right|\geqslant 1-\zeta-\sup_{\lambda\in\Pi}\left|\int e^{i\lambda t}\,dF_d\right|$$
$$=1-\zeta-\|F_d\|\geqslant 1-\|F_d\|=\|F_s\|+\|F_c\|>\|F_s\|.$$

Since this inequality is strict it remains valid for $\zeta\in\bar{\mathcal{D}}_\varepsilon$ for small enough ε.

To verify $[\mathcal{D}_\varepsilon\Pi]_1$ it is enough to note that $\mathfrak{v}(\lambda)$ is continuous on Π,

$$\mathfrak{v}(0)=-a>0 \quad\text{and}\quad \limsup_{|\lambda|\to\infty}|f(\lambda)<1. \quad\square$$

In a completely analogous way we can show that for $a>0$ the function

$$[(1-f(\lambda))/i\lambda]\,(i\lambda-1)$$

admits a c.V-f. and for $a=0$ the function

$$[(1-f(\lambda))/\lambda^2]\,(\lambda^2+1).$$

Theorems 5–7 in this exposition are contained in [28] (Gel'fand, *et al.*) and Theorems 3 & 4 on V-factorizations have been obtained by modifying arguments in [43].

Appendix 3. The Wiener-Lévy Theorems and the Asymptotic Behavior of the Coefficients of Absolutely Convergent Series

Assume, in keeping with the notation of Appendix 2, that $\mathfrak{D}$ is the ring of functions defined on the unit circle $\Gamma=\{z:|z|=1\}$ and representable as absolutely convergent Fourier series

$$\mathfrak{d}(z)=\sum_{-\infty}^{\infty} d_k z^k, \qquad \|\mathfrak{d}\|=\sum |d_k| .$$

By Theorem 6 of Appendix 2, $\mathfrak{d}^{-1}\in\mathfrak{D}$ iff $\inf_{z\in\Gamma}|\mathfrak{d}(z)|>0$. This says that the spectrum of $\mathfrak{d}$ coincides with the set of values of $\mathfrak{d}(z)$, $z\in\Gamma$, and by Theorem 5 Appendix 2 we get the following assertion (Wiener, Lévy):

(a) *Assume* $\mathfrak{d}(z)\in\mathfrak{D}$ *and that* $\Lambda(\zeta)$ *is a function analytic in a simply-connected domain* $\mathscr{D}$ *and located, generally speaking, on a multi-sheeted surface. Then, if the curve* $\zeta=\mathfrak{d}(z)$, $z\in\Gamma$, *can be considered as lying in* $\mathscr{D}$ *we have* $\Lambda(\mathfrak{d}(z))\in\mathfrak{D}$.

In a similar way we establish that

(b) *Assertion* (a) *remains valid if* $\mathfrak{D}$ *is replaced everywhere in it by* $\mathfrak{D}_+$ *(the subring for whose elements* $d_k=0$ *for* $k<0$*) and we require that the domain* $\mathscr{D}$ *contain the values of* $\zeta=\mathfrak{d}(z)$, $|z|\leqslant 1$.

Thus, when (b) is fulfilled there exists a sequence $\{\lambda_k\}$, $\sum|\lambda_k|<\infty$ such that for $z\in\Gamma$

$$\Lambda(\mathfrak{d}(z))=\sum_{n=0}^{\infty}\lambda_n z^n .$$

Set $\Delta(b_n)=b_n-b_{n+1}$ and let the symbol $\varepsilon(k)$ denote a slowly varying function (s.v.f.)

Theorem 1. *Assume in addition to* (b) *that*

1) $\sum_{n=1}^{\infty}\frac{1}{n}\sum_{k=n}^{\infty}|\Delta(kd_k)|<\infty$;

2) *There exists a s.v.f.* $\varepsilon(k)$ *and a constant* $\beta>0$ *such that*

$$|d_n|<\varepsilon(n)n^{-\beta}, \quad n=1,2,\ldots;$$

3) $d_{n+1}\sim d_n$ *when* $n\to\infty$ *and for some* $c>0$

$$|d_n|\geqslant c\varepsilon(n)n^{-\beta} .$$

Then under Conditions 1 *and* 2

$$\limsup_{n\to\infty}\frac{|\lambda_n|}{\varepsilon(n)n^{-\beta}}\leqslant c(\mathfrak{d},\Lambda)<\infty . \tag{1}$$

When 1) *and* 3) *hold*

$$\lambda_n\sim d_n\Lambda'(\mathfrak{d}(1)) .$$

Remark 1. Condition 1 is unnecessary[1] if $nd_n\downarrow$. In this case the series appearing there coincides with $\sum d_k<\infty$.

[1] *Note added in proof.* Recent work by Rogozin has shown that Condition 1 is always unnecessary (see Sibirsk. Mat. Ž. 14, 1304–1312 (1973) [In Russian]).

Remark 2. If we require fulfillment of (a) and add to 1) the requirement that

$$\sum_{n=1}^{\infty} \frac{1}{n} \sum_{k=n}^{\infty} |\Delta(kd_{-k})| < \infty ,$$

then the assertion of the theorem holds for arbitrary $\mathfrak{d}(z) \in \mathfrak{D}$ with negative powers of z (see [10]).

Proof. Put $\varphi(\theta) = \mathfrak{d}(e^{i\theta})$. We prove first that $\varphi(\theta)$ is differentiable in the interval $(0, 2\pi)$ and $\varphi'(\theta) \in L(0, 2\pi)$. To this end we use the following theorem (Bary, [1] pp. 95, 650 and 652 in the Russian edition):

If $b_n \downarrow 0$ and $\sum_1^{\infty} (b_n/n) < \infty$, then

$$b(\theta) = \sum_{n=0}^{\infty} b_n e^{in\theta} \in L(0, 2\pi) . \tag{2}$$

The series (2) *is the Fourier representation of the function* $b(\theta)$ *and it converges uniformly on any interval* $[\delta, 2\pi - \delta]$, $\delta > 0$.

It's easy to see that this assertion remains true if we require merely that

$$\sum_{n=1}^{\infty} \frac{1}{n} \sum_{k=n}^{\infty} |\Delta(b_k)| < \infty \text{ and } \liminf_{n \to \infty} |b_n| = 0 . \tag{3}$$

Indeed, put $B_n = \sum_{k=n}^{\infty} |\Delta(b_k)|$, $e_n = B_n - b_n$. From (3) it follows that $\lim_{n \to \infty} b_n$ exists and is zero since nonexistence of this limit would imply divergence of the series $\sum |\Delta(b_k)|$. Thus, $b_n = \sum_{k=n}^{\infty} \Delta(b_k)$, so that

$$\sum_1^{\infty} \frac{e_n}{n} \leqslant \sum_1^{\infty} \frac{B_n}{n} < \infty .$$

Consequently, the sequence b_n can be represented as the difference of two sequences $b_n = B_n - e_n$ satisfying the hypotheses of the theorem.

We turn now to the formal series for $\varphi'(\theta)$:

$$i \sum_{n=1}^{\infty} nd_n \cos n\theta - \sum_{n=1}^{\infty} nd_n \sin n\theta . \tag{4}$$

By Condition 1 and the absolute convergence of $\sum d_n$, the coefficients $b_n' = nd_n$ satisfy (3). Therefore, the sum of (4) is continuous inside $[0, 2\pi]$ and integrable. Carrying out the integration term-by-term we convince ourselves that for all $\theta \neq 0$

$$i \sum_{n=1}^{\infty} nd_n e^{in\theta} = \varphi'(\theta) \in L(0, 2\pi) .$$

From what has been said it follows that $\alpha(\theta) = \Lambda'(\varphi(\theta))$ is also differentiable and $\alpha'(\theta) \in L(0, 2\pi)$. Moreover, by (b) $\Lambda'(\mathfrak{d}(z)) \in \mathfrak{D}_+$. Hence, $\alpha(\theta) = \sum_{n=0}^{\infty} \alpha_n e^{in\theta}$, where $\alpha_n = o(n^{-1})$. Applying this estimate to the function $\Lambda(\varphi(\theta))$ we get $\lambda_n = o(n^{-1})$.

We now note that the formula

$$\frac{d}{dz} \Lambda(\mathfrak{d}(z)) = \Lambda'(\mathfrak{d}(z)) \mathfrak{d}'(z)$$

implies the equality

$$n\lambda_n = \sum_{k=0}^{n} kd_k \alpha_{n-k} . \tag{5}$$

We represent the right side of (5) as two sums

$$\sum_{k=1}^{n} kd_k\alpha_{n-k} = \sum_{1}^{[n/2]} kd_k\alpha_{n-k} + \sum_{[n/2]+1}^{n} kd_k\alpha_{n-k}\,. \tag{6}$$

One then has

$$\left|\sum_{1}^{[n/2]}\right| \leqslant \max_{[n/2]\leqslant k\leqslant n} |\alpha_k| \sum_{1}^{[n/2]} \varepsilon(k)k^{-\beta+1} = \begin{cases} o(\varepsilon(n)n^{-\beta+1}), & \text{if } \beta<2, \\ o(n^{-1+\delta}), & \text{if } \beta\geqslant 2 \end{cases}$$

for arbitrary $\delta>0$. Using the integral representation of s.v.f.'s (Feller [25], p. 272 ff.) we find that

$$\left|\frac{\varepsilon(n+k)}{\varepsilon(n)} - 1\right| < \frac{1}{2}$$

for arbitrary k, $|k|<n/2$ and all large enough n. Hence, for such n

$$\left|\sum_{[n/2]+1}^{n}\right| < 2^{\beta+1}\varepsilon(n)n^{-\beta+1} \sum_{k=0}^{[n/2]} |\alpha_k| \leqslant \varepsilon(n)n^{-\beta+1}2^{\beta+1} \sum_{k=0}^{\infty} |\alpha_k|\,.$$

As a result we find that for any $\delta>0$

$$\lambda_n = \gamma(n) + o(n^{-2+\delta})\,, \tag{7}$$

where for large enough n

$$|\gamma(n)| < \varepsilon(n)n^{-\beta}2^{\beta+1} \sum_{0}^{\infty} |\alpha_k|\,.$$

If $\beta<2-\delta$, this proves (1). If, however, $\beta\leqslant 2-\delta$, then applying (7) to the coefficients of the function $\alpha(\theta)=\Lambda'(\varphi(\theta))$, which clearly also fulfills Condition (b), we will have $\alpha_n=o(n^{-2+2\delta})$. Estimating the two sums in (6) anew, we can improve the last term in (7), for example, to $o(n^{-3+3\delta})$. If $\beta\geqslant 3-3\delta$ it is necessary to repeat the indicated operation once more, etc. After a finite number of steps (no greater than $[(\beta-1)/(1-\delta)]+2$) we arrive at

$$\lambda_n = \gamma(n) + o(n^{-\gamma})\,, \quad \gamma>\beta\,.$$

In order to verify the second assertion of the theorem it is now enough to show that in (6) as $n\to\infty$

$$\sum_{[n/2]+1}^{n} \sim nd_n\Lambda'(\mathfrak{d}(1))\,. \tag{8}$$

We choose N such that $\sum_{|k|>N}|\alpha_k|<\delta$. By Condition 3, for large enough n we have

$$\left|\frac{kd_k}{nd_n} - 1\right| < \delta$$

for all k, $|k-n|\leqslant N$. For such n

$$\sum_{k=n-N}^{n} kd_k\alpha_{n-k} = nd_n\left(\sum_{k=n-N}^{n} \alpha_{n-k} + \delta_{n,N}\right) = nd_n(\Lambda'(\mathfrak{d}(1)) + \delta'_{n,N})\,,$$

where

$$|\delta_{n,N}| < \delta \sum |\alpha_k| \quad \text{and} \quad |\delta'_{n,N}| < \delta(1+\sum|\alpha_k|)\,.$$

Furthermore, for sufficiently large n

$$\left|\sum_{[n/2]+1}^{n-N} kd_k\alpha_{n-k}\right| < 2^{\beta}\varepsilon(n)n^{-\beta+1} \sum_{N}^{[n/2]} |\alpha_k| < \frac{2^{\beta}}{c}\,\delta|nd_n|\,.$$

Hence, for arbitrary $\delta>0$ and all large enough n

$$\sum_{[n/2]+1}^{n} kd_k\alpha_{n-k} = nd_n(\Lambda'(\mathfrak{d}(1))+\delta''_{n,N}),$$

where

$$|\delta''_{n,N}|<\delta(1+\sum|\alpha_k|+(2^\beta/c)).$$

This yields (8). The theorem is proved. □

Appendix 4. Estimates for the Distributions of Sums of Independent Random Variables

Let $\xi_1, \xi_2, \ldots$ be a sequence of i.i.d. random variables with

$$\mathsf{P}(\xi_k<x)=F(x) \quad \text{and} \quad \mathsf{D}\xi_k=1.$$

As in Chapters 3 and 4 we put $\xi_k^+=\max(0,\xi_k)$, $Y_n=\sum_{k=1}^n \xi_k$ and $\overline{Y}_n=\max_{k\leqslant n} Y_k$.

We will assume everywhere in the sequel that $\mathsf{M}(\xi_k^+)^m=c_m<\infty$ for some $m>2$. Under these assumptions we will find estimates for $W_n(x)=\mathsf{P}(\overline{Y}_n>x)$. Write

$$L_m=\sup_{v\geqslant 1}\frac{v^m}{e^v}\int_1^v \frac{e^u}{u^m}\,du.$$

Theorem 1. *Assume* $a=\mathsf{M}\xi_k=0$. *For arbitrary* $x>0$ *and* $y>y_m\equiv e^{L_{m+1}}(2c_m n)^{1/m}$ *we have the inequality*

$$W_n(x)<n(1-F(y))+\left[\frac{2nc_m}{y^m}\right]^{x/y}\exp\left\{1+2n\left[\frac{m\ln y-\ln(2nc_m)}{y}\right]^2\right\}. \tag{1}$$

Putting $y=x(1+(1/m))^{-1}$ *we find that for* $x>y_m(1+(1/m))$

$$W_n(x)<\frac{nc_m e}{x^m}\{1+B_{x,n}\},$$

where

$$B_{x,n}=2\left[\frac{x^m}{2nc_m e}\right]^{-1/m}\exp\left\{1+\frac{5n}{x^2}\ln^2\frac{2nc_m e}{x^m}\right\}.$$

If, in addition, x *is so large that* $B_{x,n}\leqslant 1$, *then*

$$W_n(x)\leqslant\frac{2nc_m e}{x^m}.$$

Remark 1. We can take as c_m here a smaller quantity equal to $\int_{v_n}^{\infty} x^m dF(x)$ for which the sequence $v_n\to\infty$ as $n\to\infty$ (see the proof of the theorem).

Remark 2. It is not difficult to obtain a different kind of estimate for the number L_m. For example, because of the convexity of the function e^u/u^m for $u \geqslant 1$

$$\int_1^v \frac{e^u}{u^m}\,du = \frac{e^v}{v^m} - e + m\int_1^v \frac{e^u}{u^{m+1}}\,du \leqslant \frac{e^v}{v^m} - e + \frac{m(v-1)}{2}\left(\frac{e^v}{v^{m+1}} + e\right),$$

$$L_m \leqslant 1 + \frac{m}{2} + \frac{me}{2}\sup_{v \geqslant 1}\frac{v^{m+1}}{e^v} = 1 + \frac{m}{2} + \frac{m}{2}(m+1)^{m+1}e^{-m}.$$

Proof of Theorem 1. Put

$$A = \{\bar{Y}_n > x\} \quad \text{and} \quad B_j = \{\xi_j \leqslant y\}.$$

Then

$$W_n(x) = \mathsf{P}(A) \leqslant \mathsf{P}(\bigcup_{j=1}^n \bar{B}_j) + \mathsf{P}(A \cap_{j=1}^n B_j) \leqslant n(1-F(y)) + F^n(y)\mathsf{P}(A_y),$$

where $\bar{B}_j$ is the complement of B_j,

$$A_y = \{\bar{Y}_n^{(y)} > x\}, \qquad \bar{Y}_n^{(y)} = \max_{k \leqslant n} \sum_{j=1}^k \xi_j^{(y)}$$

and

$$\mathsf{P}(\xi_k^{(y)} < x) = \frac{F(x)}{F(y)}, \quad x \leqslant y.$$

But by Theorem 16, Chapter 4

$$\mathsf{P}(A_y) \leqslant e^{-\mu x}[\max(1, \mathsf{M}\exp(\mu\xi_1^{(y)}))]^n \tag{2}$$

for arbitrary μ. Since

$$\mathsf{M}\exp(\mu\xi_1^{(y)}) = \frac{\int_{-\infty}^y e^{\mu t}\,dF(t)}{F(y)} = \frac{R(\mu, y)}{F(y)},$$

we have

$$W_n(x) \leqslant n(1-F(y)) + e^{-\mu x}\max(1, R^n(\mu, y)).$$

Assuming $\mu > 1/y$, we now estimate

$$R(\mu, y) = \int_{-\infty}^y e^{\mu t}\,dF(t) = \int_{-\infty}^{1/\mu} + \int_{1/\mu}^y = R_1 + R_2.$$

One has

$$R_1 = \int_{-\infty}^{1/\mu}\left(1 + \mu t + \frac{\mu^2 t^2}{2}e^{\mu\theta(t)}\right)dF(t), \quad 0 \leqslant \theta(t)/t \leqslant 1. \tag{3}$$

Here

$$1 > \int_{-\infty}^{1/\mu} dF(t) = 1 - \int_{1/\mu}^{\infty} dF(t) > 1 - \mu^2\int_{1/\mu}^{\infty} t^2\,dF(t) > 1 - \mu^2,$$

$$0 > \int_{-\infty}^{1/\mu} t\,dF(t) > -\mu\int_{1/\mu}^{\infty} t^2\,dF(t) \geqslant -\mu,$$

and

$$\int_{-\infty}^{1/\mu} t^2 e^{\mu\theta(t)}\,dF(t) < e\int t^2\,dF(t) = e.$$

As a result we get $|R_1-1|<2\mu^2$. We now turn to the estimation of R_2 (in these calculations $v=\mu y>1$):

$$R_2=\int_{1/\mu}^{y} e^{\mu t}\, dF(t)=(1-F(t))\, e^{\mu t}|_{y}^{1/\mu}+\mu \int_{1/\mu}^{y} (1-F(t))\, e^{\mu t}\, dt$$
$$\leqslant e\left(1-F\left(\frac{1}{\mu}\right)\right)+\mu c_m \int_{1/\mu}^{y} \frac{e^{\mu t}}{t^m}\, dt \leqslant e c_m \mu^m + c_m \mu^m \int_{1}^{\mu y} \frac{e^u}{u^m}\, du$$
$$= c_m \mu^m\left[\frac{e^v}{v^m}+m\int_1^v \frac{e^u}{u^{m+1}}\, du\right] \leqslant c_m \mu^m \left[\frac{e^v}{v^m}+\frac{m\, e^v L_{m+1}}{v^{m+1}}\right]$$
$$\leqslant \frac{c_m \mu^m\, e^v}{v^m}\left[1+\frac{m L_{m+1}}{v}\right].$$

Hence, $R_2 \leqslant 2c_m y^{-m} e^v$ for $v \geqslant m L_{m+1}$. Now put $\mu=\mu_0=(1/y)\ln(y^m/2nc_m)$. Then $v=\ln(y^m/2nc_m) \geqslant m L_{m+1}$ for $y^m \geqslant 2nc_m e^{mL_{m+1}}$. Consequently, $R_2 \leqslant 2c_m y^{-m} e^v = 1/n$, so that

$$e^{-\mu_0 x} R^n(\mu_0, y) \leqslant \exp\left\{-\frac{x}{y}\ln\frac{y^m}{2nc_m}\right\}\left(1+2\mu_0^2+\frac{1}{n}\right)^n$$
$$\leqslant \left(\frac{2nc_m}{y^m}\right)^{x/y} \exp\left\{1+\frac{2n}{y^2}\ln^2\frac{y^m}{2nc_m}\right\}. \tag{4}$$

The first assertion of the theorem is proved. The rest follow immediately from (1). □

Now suppose $0>a=\mathsf{M}\xi>-\infty$ and that the remaining conditions of Theorem 1 hold[1]. Let z_m be the largest solution of the equation

$$\frac{(m-1)(a^2+1)\, e}{2}\frac{\ln y}{y}+\frac{2c_m}{(m-1)\ln y}=-a.$$

Theorem 2. *For arbitrary $x>0$ and $y>z_m$, $y^{m-1}>e^{mL_{m+1}}$,*

$$W_n(x) \leqslant n(1-F(y))+(y^{-m+1})^{x/y}.$$

For $x=y$ we find

$$W_n(x) \leqslant \frac{nc_m}{x^m}+\frac{1}{x^{m-1}}.$$

Proof. As in the proof of Theorem 1 we see that

$$W_n(x) \leqslant n(1-F(y))+F^n(y)\mathsf{P}(A_y).$$

By the corollary of Theorem 16, Chapter 4

$$F^n(y)\mathsf{P}(A_y) \leqslant e^{-\mu x}$$

for any μ such that

$$\int e^{\mu t}\, d\mathsf{P}(\xi_k^{(y)}<t) \leqslant \frac{1}{F(y)}$$

[1] Under the assumptions of Theorem 2 the requirement $\mathsf{D}\xi_k<\infty$ is too strong. It is sufficient to require the finiteness of $\mathsf{M}|\xi_k|^{1+\delta}$ for some $\delta>0$. The reader can easily show this after a glance at the proof.

or, equivalently,

$$\int_{-\infty}^{y} e^{\mu t}\, dF(t) \leqslant 1 .$$

We again split the integral up into two parts:

$$R(\mu, y) = \int_{-\infty}^{y} e^{\mu t}\, dF(t) = \int_{-\infty}^{1/\mu} + \int_{1/\mu}^{y} = R_1 + R_2 .$$

By (3)

$$R_1 \leqslant 1 + a\mu + e\frac{\mu^2}{2}(a^2+1) .$$

The estimate for R_2 does not change:

$$R_2 \leqslant 2c_m y^{-m} e^{v} \quad \text{for } v = \mu y \geqslant mL_{m+1} .$$

Put $\mu = \ln y^{m-1}/y$. Then the last inequality goes into

$$\ln y^{m-1} \geqslant mL_{m+1} .$$

For the chosen μ one has $R(\mu, y) \leqslant 1$ if

$$a + \frac{e(a^2+1)}{2}\frac{\ln y^{m-1}}{y} + \frac{2c_m}{\ln y^{m-1}} \leqslant 0 .$$

The theorem is proved. □

Estimates for $\mathsf{P}(\overline{Y} \geqslant x)$ when $a < 0$ can be obtained using Theorem 1.

Theorem 3. *Assume $\xi_k^0 = \xi_k - a$ satisfy the conditions of Theorem 1 and that y_m^0 and $B_{x,n}^0$ are the values of y_m and $B_{x,n}$ corresponding to ξ_k^0. Further let $x > y_m^0(1+1/m)$ be an integer and so large that $B_{x|a|/2,\,x}^0 \leqslant 1$. Then*

$$\mathsf{P}(\overline{Y} \geqslant x) \leqslant \frac{2c_m e}{x^{m-1}} B_m ,$$

where

$$B_m = 1 + \sum_{j=0}^{\infty} \frac{2^{j+1}}{(1-a2^j)^m} \leqslant 1 + \frac{2^m}{|a|^m(2^{m-1}-1)} .$$

Proof. Setting

$$W_n^0(x) = \mathsf{P}(\max_{k \leqslant n} (Y_k - ak) \geqslant x) ,$$

we have

$$\begin{aligned}
\mathsf{P}(\textstyle\bigcup_k \{Y_k \geqslant x\}) &= \mathsf{P}(\textstyle\bigcup_k \{Y_k - ak \geqslant x - ak\}) \\
&\leqslant W_x^0(x) + \mathsf{P}(\textstyle\bigcup_{j=1}^{\infty} \bigcup_{k>2^{j-1}x}^{2^j x} \{Y_k - ak > x - ak\}) \\
&\leqslant W_x^0(x) + \textstyle\sum_{j=1}^{\infty} W_{2^j x}^0(x - a2^{j-1}x) \\
&\leqslant \frac{2c_m e}{x^{m-1}} + 2c_m e \sum_{j=1}^{\infty} \frac{2^j x}{x^m(1-a2^{j-1})^m} .
\end{aligned}$$

The last inequality holds because $B_{x(1-2^{j-1}a),\,2^j x}^0 \leqslant 1$ when $B_{x|a|/2,\,x}^0 \leqslant 1$. □

To estimate the rates of convergence of the distributions of the waiting times w_n and w^0 under the assumptions of Chapters 1 and 4 (§ 23) we must be able to estimate

$$\mathsf{P}(Y_n \in [x - w_1, x)) .$$

Rough estimates of this probability are contained in Theorems 1–3. The following assertion is somewhat more exact.

Theorem 4. *Assume in addition to the assumptions of Theorem 1 that for all $y \geqslant y_m$ and some $l > 0$, $c < \infty$*

$$F(y+l) - F(y) \leqslant \frac{cl}{y^{m+1}} .$$

Then for $x > 0$ and $y \geqslant y_m$

$$\mathsf{P}(Y_n \in [x, x+l)) \leqslant \frac{cnl}{y^{m+1}} + \left(\frac{2nc_m}{y^m}\right)^{x/y} \exp\left\{1 + \frac{2n}{y^2} \ln^2 \frac{y^m}{2nc_m}\right\} .$$

For $y = x/2$ $(x > 2y_m)$

$$\mathsf{P}(Y_n \in [x, x+l)) \leqslant \frac{2^{m+1} cnl}{x^{m+1}} + \left(\frac{2^{m+1} nc_m}{x^m}\right)^2 \exp\{1 + C_{x,n}\} .$$

where

$$C_{x,n} = \frac{8n}{x^2} \ln^2 (x^m / 2^{m+1} nc_m) .$$

Corollary. *Assume that*

$$\alpha > 0 \quad \text{and} \quad x = \alpha n + z ,$$

whereby n is so large that

$$C_{x,n} \leqslant 1 \quad \text{and} \quad x > 2y_m .$$

Then

$$\mathsf{P}(Y_n - \alpha n \in [z, z+l)) \leqslant \frac{2^{m+1} n}{(\alpha n + z)^{m+1}} \left[cl + \frac{2^{m+1} nc_m^2 e^2}{(\alpha n + z)^{m-1}}\right] .$$

Proof. Define the functions $F^{(y)}(x)$ and $\tilde{F}^{(y)}(x)$ by the equalities

$$F^{(y)}(x) = \begin{cases} F(x) , & x \leqslant y , \\ F(y) , & x > y \end{cases}$$

and

$$\tilde{F}^{(y)}(x) = F(x) - F^{(y)}(x)$$

and denote by $F_k^{(y)}(x)$ and $\tilde{F}_k^{(y)}(x)$ the k-th convolutions of these functions. Then

$$\mathsf{P}(Y_n < x) = F_n^{(y)}(x) + \textstyle\sum_{k=1}^{n} \binom{n}{k} F_{n-k}^{(y)} * \tilde{F}_k^{(n)}(x)$$

and

$$\mathsf{P}(Y_n \in [x, x+l)) = F_n^{(y)}(x+l) - F_n^{(y)}(x)$$
$$+\sum_{k=1}^{n} \binom{n}{k} F_{n-k}^{(y)} * \tilde{F}_{k-1}^{(y)} * [\tilde{F}^{(y)}(x+l) - \tilde{F}^{(y)}(x)]$$
$$\leqslant F_n^{(y)}(x+l) - F_n^{(y)}(x) + \frac{cl}{y^{m+1}} \sum_{k=1}^{n} \binom{n}{k} \cdot \text{Variation } [F_{n-k}^{(y)} * \tilde{F}_{k-1}^{(y)}] .$$

The last sum does not exceed

$$\sum_{k=1}^{n} \binom{n}{k} F^{n-k}(y)(1-F(y))^{k-1} = \frac{1-F^n(y)}{1-F(y)} \leqslant n .$$

Hence,

$$\mathsf{P}(Y_n \in [x, x+l)) \leqslant F_n^{(y)}(\infty) - F_n^{(y)}(x) + \frac{cnl}{y^{m+1}} \cdot$$

Noting that (see (2))

$$F_n^{(y)}(\infty) - F_n^{(y)}(x) \leqslant F^n(y) \mathsf{P}(A_y) \leqslant e^{-\mu x} R^n(\mu, y) ,$$

and using the estimate (4) of Theorem 1, we obtain the claimed assertion. □

List of Basic Notation

a—mathematical expectation of some random variable
a.s.—almost surely
α—parameter of an exponental distribution
$\mathfrak{B}$—σ-algebra of a basic probability space
$C(0, t)$—space of functions continuous on $[0, t]$
$\gamma(t)$—"defect" variable w.r.t. the level t
$D(0, t)$—space of functions on $[0, t]$ without discontinuities of the second kind
$\mathfrak{D}$—ring of functions representable as power series converging absolutely on the unit circle
$\mathfrak{D}_+(\mathfrak{D}_-)$—subring of functions from $\mathfrak{D}$ representable as power series converging absolutely inside and on (outside and on) the unit circle
$\mathbf{e}$—the m-dimensional vector $(1, 0, \ldots, 0)$
$F(t)$—distribution function
$f(\lambda)$—characteristic function
$f_1(\mu) = f(-i\mu)$
$\left.\begin{array}{l}\langle E, \ldots\rangle_R \\ \langle G, \ldots\rangle \\ \langle G_I, \ldots\rangle\end{array}\right\}$—designations for service systems (§ 1)
$g(t) = \begin{cases} 1 - F(t), & t \geqslant 0 \\ -F, & t < 0 \end{cases}$
$I(A)$—indicator of the set A
$\eta(x) = \inf\{t\colon Y(t) \geqslant x\}$ (Chapter 2) or $\inf\{k\colon Y_k \geqslant x\}$
$\eta^*(x) = \min\{k \geqslant 1\colon Y_k \leqslant x\}$, $x \leqslant 0$
$H(t)$—renewal function
$\theta_n = \min\{k\colon Y_k = \overline{Y}_n\}$
$\theta^n = \max\{k \leqslant n\colon Y_k = \overline{Y}_n\}$
$\theta^n_* = \max\{k \leqslant n\colon Y_k = \overline{Y}^*_n\}$
θ—first time a process attains its maximum
$\mathbf{i}$—the n-dimensional vector $(1, 1, \ldots, 1)$
$\kappa(\xi) = \sup\{s\colon \mathsf{M}|\xi|^s < \infty\}$
$L(F, F_n)$—Lévy's distance
m—number of service channels in a multi-channel system
m_p—$[p + (\frac{1}{2})]$-th quantile of a distribution
$\mathfrak{M}$—σ-sub-algebra of $\mathfrak{B}$ generated by the sequence $\{\xi_n\}$
v_i^e, v_i^s—elements of governing sequences (§ 1)
$\xi_n = S_n - \tau_n^e$ or an element of an arbitrary sequence of random variables
$\mathsf{P}^*(A)$—the probability of the event A (where A refers, say, to a sequence $\xi_1, \ldots, \xi_n$), but for trajectories generated by the random variables $-\xi_n, -\xi_{n-1}, \ldots, -\xi_1$
q_n—queue length (number of occupied channels) at the moment of arrival of the n-th customer or request
$q_{n,k} = q_{n+k}$
$q_n(x)$—number of customers out of q_n whose service takes more time than x
$q^k(x)$—stationary (w.r.t. k) process, limit of $q_{n+k}(x)$ as $n \to \infty$
$q(t)$—queue length at time t
$q(t, x)$—number of customers out of $q(t)$ whose service takes more time than x
$q^c(t)$—stationary queue length process, limit of $q(u+t)$ as $u \to \infty$
$\mathbf{R}(\mathbf{x})$—vector obtained from $\mathbf{x}$ by putting its coordinates into increasing order
S_n—time spent on service of the n-th group of requests
$S(t)$—time required to serve requests arriving in the system up to time t
T—shift-transformation of sets from $\mathfrak{M}$
τ_i^e, τ_i^s—elements of governing sequences (§ 1)
U—transformation on random variables corresponding to the shift T
$\Phi(x)$—normal distribution function
$v_n = \min\{k \geqslant 0; w_{n-k} = 0\}$
V—ring of functions of bounded variation
$V_+(V_-)$—subring of V generated by functions whose variation on $(-\infty, 0)$ $((0, \infty))$ equals 0
$\mathbf{v}_n^k$—vectors defined in § 27
$\mathfrak{V}$—ring of Fourier–Stieltjes transformations of functions from V which is isometrically isomorphic to V
$\mathfrak{V}_+, \mathfrak{V}_-$—subrings of $\mathfrak{V}$ generated by functions from V_+, V_-, resp.
$\mathfrak{v}(\lambda) = \dfrac{1 - f(\lambda)}{i\lambda}(i\lambda + 1)$ or an arbitrary element of the ring $\mathfrak{V}$
$\chi(t)$—overshoot of the level t (with or without index)

w_n—waiting time until start of service of n-th request, or of the first request in the n-th group

$w_{n,k}=w_{n+k}$

w^k—stationary waiting time $(=\sup(0, \xi_k, \xi_k+\xi_{k-1}, \xi_k+\xi_{k-1}+\xi_{k-2}, \ldots))$

$\mathbf{w}_n$—vector of waiting times of n-th request treated in Chapters 5 and 6 (also realization of the sequence $\{w_{n,k}\}$ (§ 3))

$w_{n,i}$—time the n-th request has to wait until i channels become free of requests arriving before it

$w(t)$—virtual waiting time (of a request which arrives at time t)

$w_t(u)=w(t+u)$

$w^c(t)$—stationary virtual waiting time

$W_n(x)=\mathsf{P}(\overline{Y}^{(n)}>x)$, $\overline{Y}^{(n)}$ denotes $\overline{Y}$ in a double sequence.

$W(x)=\mathsf{P}(\overline{Y}>x)$

$\mathfrak{w}_z(\lambda)=1-zf(\lambda)$

$\mathfrak{w}_{z+}(\lambda)$, $\mathfrak{w}_{z-}(\lambda)$—components of the factorization of $\mathfrak{w}_z(\lambda)$

$\psi(\lambda)=\ln \mathsf{M}\, e^{i\lambda Y(1)}$

$\psi_1(\mu)=\psi(-i\mu)$

X_n—sum of n random variables

$X(t)=S(t)-t$ or an additive stochastic process

$\mathbf{x}$—m-dimensional vector

$x^+=\max(0, x)$

$\mathbf{x}^+=$vector whose coordinates x_i are replaced by x_i^+

Y_n—sum of n random variables

$\overline{Y}_n=\max(0, Y_1, \ldots, Y_n)$

$Y=\overline{Y}=\overline{Y}_\infty$

$y_n=\max(Y_1, \ldots, Y_n)$

$Y_n^*=\min(0, Y_1, \ldots, Y_n)$

$Y(t)$—additive stochastic process in Chapter 2 (with independent increments)

ω—element of Ω

$(\Omega, \mathfrak{B}, \mathsf{P})$—basic probability space

$\underset{d}{=}$—coincidence with respect to distributions

$\underset{d}{\geqslant}$: $\xi \underset{d}{\geqslant} \eta$ if $\mathsf{P}(\xi>x)\geqslant \mathsf{P}(\eta\geqslant x)$

$\prec$—convex inequality (§ 24)

$\square$—end of a proof

$\uparrow\downarrow$—monotone behavior of functions

$\Rightarrow$—weak convergence of distributions

Bibliographical Notes

Introduction

§ 1. The notation of Kendall [39] has gained wide-spread acceptance in the literature of queueing theory. However, it does not cover all types of systems and refers only to governing sequences in which $v^e \equiv v^s \equiv 1$ (for example, the systems $\langle G, 1, G/m, 1\rangle$ would be designated by $G/G/m$ in Kendall's notation). For the purposes of this book such a notation is insufficient.

Chapter 1

§ 3. The fact that the solution of the equation $w_{n+1} = \max(0, w_n + \xi_n)$ (see (8) § 2) converges in distribution to sup Y_k was noted for independent ξ_k by Lindley [44] and later under more general conditions by Loynes [45]. The last paper also contains Theorem 8.

The assertion of Theorem 7 in the case $\mathsf{M}\xi_n = 0$ follows from Borovkov [16] where an approach due to Loynes [45] was used.

§ 6. Eq. (27) and its solution were also investigated in Beneš' monograph [3].

§§ 7, 8. Formulae (47) § 7 and Theorem 15 are due to Beneš [3]. Problems of heavy and light traffic had been considered before but under much more special assumptions. The numerical example was prepared by A. I. Sahanenko.

§ 9. The results of this section are closely related to the usual approach of renewal theory (Smith [61]).

§ 11. The stability theorem follows as a special case from the results of Borovkov [17]. The examples relating to the necessity of (65) § 10 (see also § 21) were worked out by A. A. Mogul'skii. The numerical example was calculated by A. I. Sahanenko.

Chapter 2

§ 12. The complete or partial statement of Theorem 1 under special conditions on the process has also been obtained by J. Keilson [36] and A. V. Skorohod [60] and V. M. Zolotarev [74]. Theorem 1 in its complete form is given in the author's paper [4]. B. A. Rogozin [54] proved the necessity of the equality (3) § 12 for the continuity from below of a process with independent increments.

§ 13. The third assertion of Theorem 4 (on the form of the transformation of $\mathsf{P}(w(t)=0)$) is known, for example, from Beneš' book [3] (Chapter 4).

Concerning the busy period (Theorem 5), B. V. Gnedenko and I. N. Kovalenko obtain in [30] (Chapter 4) an equation satisfied by the Laplace transformation of the sought-for distribution. Theorem 5 gives the explicit solutions of this equation. A number of the results in Chapter 2 are contained in Takács' book [70] where the investigation of processes continuous from below with interchangeable (including those with independent) increments is carried out by elementary combinatorial methods.

Chapter 3

A large part of the results contained in this chapter has been previously obtained using different methods by Spitzer [62–65], Darling [22], Feller [25], Rosen [55] and others. Many of these results can be transferred to processes with independent increments (Rogozin [54] and Pečerskii and Rogozin [48]).

Chapter 4

§ 18. Theorems on the uniqueness of the V-factorization when $\mathsf{M}\xi$ exists are given in [5] and [9].

§ 20. Theorems 4 and 5 on the conditions for an explicit resolution of the V-factorization were obtained by the author in [6]. Similar results were also published later in [56].

§ 21. More restrictive conditions than those of Theorem 7 for the convergence of distributions of supremo are given by Rossberg [56]. The assertion of Theorem 8 was noted by N. P. Leont'eva.

§ 22. Some of the inequalities in Theorem 10 are given in [5], [9] and [52]. The first assertion of Theorem 11 is an estimate due to Cramér [20], see also Feller [25]. The second assertion of this theorem and Theorem 12 in somewhat weaker form were given by the author in [11].

§ 23. Theorem 16 overlaps with an inequality obtained by Täcklind [69], [20].

§ 24. The extremal character of deterministic service time and deterministic time between customer arrivals in the problem considered at the end of this section was discovered by B. A. Rogozin [53]. Similar comparison theorems are considered by H. and D. Stoyan [67].

§ 25. The systems in heavy traffic studied in this section have been considered by Kingman [41], Yu. V. Prohorov [51], B. V. Gnedenko and I. N. Kovalenko

[31] and others. Theorem 18 is due to Yu V. Prohorov [51]. Theorem 19 was obtained by the author in [11].

§ 26. The assertions of Theorems 21 and 22 were obtained by Leont'eva.

Chapter 5

§ 27. Theorem 1 is due to Loynes [45].

§ 28. The equation in Subsection 1 of this section for the stationary distribution of the waiting time for the systems $\langle G_I, 1, G_I/m, 1\rangle$ was obtained by Kiefer and Wolfowitz [40]. The simplified form of this equation given at the end of Subsection 1 was found by E. L. Presman [49]. Theorem 3 on continuous dependence on the distribution of the terms of the governing sequence was obtained by I. Ahmarov for this book.

§ 29. The presentation of Theorems 9, 10 and 11 on explicit formulae for the limiting distributions of the queue length and the waiting time for the systems $\langle G_I, E/m\rangle$ is based on Takács' book [71].

Chapter 6

§ 31. The stability theorems were obtained by the author in [18].

§§ 33–34. Theorems 9, 10 and 11 on explicit formulae for the systems $\langle E, 1, G_I/\infty, 1\rangle$ and $\langle G_I, 1, E/\infty, 1\rangle$ are borrowed from Takács [71]. Here also can be found the proof of a corollary of Theorem 6 related to the case $x=0$.

Chapter 7

§§ 35, 36. Results close to Theorems 1 and 2 but under more special conditions were obtained by P. Franken in [26]. The stability theorem 1.1 was given by the author in [18].

§§ 37, 38. The assertions of Theorems 6 and 7 contain a theorem of Sevast'yanov who used general theorems on Markov processes to study the asymptotic behavior of the probability that at time $t\to\infty$, k channels will be occupied and the customers located in them have already undergone service for times exceeding $x_1\leqslant x_2\leqslant\cdots\leqslant x_k$, resp.

Theorems 9 and 10 on explicit formulae for the systems $\langle G_I, 1, E/m, G_I\rangle_R$ are taken from Takács [71].

§ 39. The extremal character of two-point distributions in Subsection 3 (Theorem 14) was established by B. A. Rogozin [53]. Lemma 2 is a special case of a result by Hoeffding [33].

§ 40. In Subsection 2 we have reproduced the asymptotic analysis of the systems $\langle G_1, 1, E/m, 1\rangle_R$ given in the author's paper [7]. Results in this direction had also been obtained previously by O. V. Viskov and Yu. V. Prohorov [72], but are not as complete. Theorem 17 is a special case of one proved in [51]. There also is given the equality (2) of Chapter 8 for the queue length in the systems $\langle G\rangle_A$. As already mentioned in the text, Theorem 17 also holds for such systems.

§ 41. The relation between the systems $\langle G, G, E_k, 1\rangle_A$ and $\langle G, G, E_k, 1\rangle$ was noted by V. A. Kasparson [35].

Appendices

1. A detailed bibliography on renewal theory can be found in Cox' book [19]. It also contains a detailed discussion of the theorem cited in Appendix 1.

2. This appendix was compiled from results presented in the book by Gel'fand, Raikov and Šilov [28] and the author's papers [8 and 9]. The latter rely heavily on methods contained in Krein [43].

3. Here we present results contained in the author's paper [10].

4. In the proofs of the theorems here we used methods proposed by S. V. Nagaev in [47], as well as results of personal discussions on them.

Bibliography

1. Bary, N. K.: Trigonometric series. Fizmatgiz 1961 [in Russian]. Engl. transl.: Bary, N. K.: A treatise on trigonometric series. Oxford: Pergamon 1964.
2. Baxter, G.: An operator identity. Pacific J. Math. **8**, 649–663 (1958).
3. Beneš, V. E.: General stochastic processes in the theory of queues. Reading, Mass.: Addison–Wesley 1963.
4. Borovkov, A. A.: On the first passage time for a class of processes with independent increments. Teor. Verojatnost. i Primenen **10**, 360–364 (1965) [in Russian].
5. Borovkov, A. A.: On the distribution of the first overshoot. Reports of the 6th All-Union Congress on the theory of probability and mathematical statistics. Vil'nyus, 1962, pp. 7–23 [in Russian].
6. Borovkov, A. A.: Asymptotic methods in queueing theory. Winter school on the theory of probability and mathematical statistics in Užgorod, 1964; Kiev, 1964 [in Russian].
7. Borovkov, A. A.: Asymptotic analysis of some queueing systems. Teor. Verojatnost. i Primenen **11**, 675–682 (1966) [in Russian].
8. Borovkov, A. A.: New limit theorems for boundary problems connected with sums of independent random variables. Sibirsk. Mat. Ž. **3**, 645–694 (1962) [in Russian].
9. Borovkov, A. A.: Some theorems on a nonlattice random walk. Teor. Verojatnost. i Primenen **7**, (1962) [in Russian].
10. Borovkov, A. A.: Remarks on theorems of Wiener and Blackwell. Teor. Verojatnost. i Primenen **9**, 331–343 (1964) [in Russian].
11. Borovkov, A. A.: Some limit theorems in queueing theory I. Teor. Verojatnost. i Primenen **9**, 608–625 (1964) [in Russian].
12. Borovkov, A. A.: Some limit theorems in queueing theory II, Multi-server systems. Teor. Verojatnost. i Primenen **10**, 409–437 (1965) [in Russian].
13. Borovkov, A. A.: On factorization identities and the properties of the distribution of the supremum of sequential sums. Teor. Verojatnost. i Primenen **15**, 377–418 (1970) [in Russian].
14. Borovkov, A. A.: On conditions for convergence to diffusion processes and asymptotic methods in queueing theory. Reports of the International Congress of Mathematicians in Moscow, 1966, pp. 533–538; "Mir" 1968 [in Russian].
15. Borovkov, A. A., and Korolyuk, V. S.: On results of asymptotic analysis in boundary problems. Teor. Verojatnost. i Primenen **10**, 255–266 (1965) [in Russian].
16. Borovkov, A. A.: Some properties of the supremum of sums of stationarily related random variables. Teor. Verojatnost. i Primenen **17**, 147–150 (1972) [in Russian].
17. Borovkov, A. A.: Convergence of distributions of functionals of random processes and sequences on the real line. Trudy Mat. Inst. Steklov. **117**, (1972) [in Russian].
18. Borovkov, A. A.: Continuity theorems for multi-channel systems with refusals. Teor. Verojatnost. i Primenen **17**, 458–468 (1972) [in Russian].
19. Cox, D. R.: Renewal theory. London: Methuen 1962.
20. Cramér, H.: Collective risk theory. Stockholm: Esselta, Centraltryckeriet 1955.
21. Cramér, H.: On some questions connected with the mathematical theory of risk. Univ. Calif. Pub. in Stat. **2**, 5, 99–124 (1954).
22. Darling, D. A.: A unified treatment of finite fluctuation problems. Colloq. on Combinatorial Methods in Probability Theory. Math. Inst. Aarhus Univ., 3–6 (1962).
23. Doob, J. L.: Stochastic processes. New York: Wiley 1953.

24. Feller, W.: Introduction to the theory of probability and its applications Vol. I (3. Ed.) New York: Wiley 1968.
25. Feller, W.: Introduction to the theory of probability its applications Vol. II (2. Ed.) New York: Wiley 1971.
26. Franken, P.: Ein Stetigkeitssatz für Verlustsysteme. Operationsforschung und Math. Statistik II, 9–23 (1970) [Berlin, GDR].
27. Fuks, B. A.: Introduction to the theory of several complex variables. Fizmatgiz 1962 [in Russian].
28. Gel'fand, I. M., Raikov, D. A., and Šilov, G. E.: Commutative normed rings. Fizmatgiz 1960 [in Russian].
29. Gihman, I. I., and Skorohod, A. V.: Introduction of the theory of random processes. Philadelphia: Saunders 1969.
30. Gnedenko, B. V., and Kovalenko, I. N.: Introduction to queueing theory. Jerusalem. Israel Prog. for Sci. Transl. 1968.
31. Gnedenko, B. V., and Kovalenko, I. N.: Lectures on queueing theory. Kiev 1963 [in Russian].
32. Hinčin, A. Ya.: Papers on the mathematical theory of queueing. Fizmatgiz 1963 [in Russian].
33. Hoeffding, W.: The extrema of the expected value of a function of independent random variables. Ann. Math. Statist. **26**, 269–275 (1955).
34. Ibragimov, I. A., and Linnik, Yu. V.: Independent and stationarily related variables. Nauka 1969 [in Russian].
35. Kasparson, V. A.: Some results on the problem of service of an arbitrary stream of demands. Mat. Zam. [in Russian] (to appear).
36. Keilson, J.: The first passage time density for homogeneous skip-free walks on the continuum. Ann. Math. Statist. **34**, 1003–1011 (1963).
37. Kemperman, J. H. B.: Changes of sign in cummulative sums, I, II. Proc. Kon. Niderl. Ak. Series A. **64**, 3.
38. Kemperman, J. H. B.: A Wiener-Hopf type method for a general random walk with a two-sided boundary. Ann. Math. Statist. **34**, 1168–1193 (1963).
39. Kendall, D. G.: Stochastic processes occurring in the theory of queues and their analysis by the method of imbedded Markov chains. Ann. Math. Statist. **24**, 338–354 (1953).
40. Kiefer, J., and Wolfowitz, J.: On the theory of queues with many servers. Trans. Amer. Math. Soc. **78**, 1–18 (1955).
41. Kingman, F. G.: On queues in heavy traffic. J. Roy. Statist. Soc. Ser. B, **24**, 383–392 (1962).
42. König, D., Mathes, K., and Nawrotzki, K.: Verallgemeinerungen der Erlangschen und Engsetschen Formeln. Berlin: Akad.-Verlag 1967.
43. Krein, M. G.: Integral equations on the half-line with kernel depending on the difference of the arguments. Uspehi Mat. Nauk. **13**, 5 (83), 3–120 (1958) [in Russian].
44. Lindley, D. V.: The theory of queues with a single server. Proc. Cambridge Philos. Soc. **48**, 277–289 (1952).
45. Loynes, R. M.: The stability of a queue with non-independent inter-arrival and service times. Proc. Cambridge Philos. Soc. **58**, 497–520 (1962).
46. Miller, H. D.: A matrix factorization problem in the theory of random variables defined on a finite Markov chain. Proc. Cambridge Philos. Soc. **58**, 268–285 (1962).
47. Nagaev, S. V.: Some limit theorems for large deviations. Teor. Verojatnost. i Primenen **10**, 231–254 (1965) [in Russian].
48. Pečerskii, E. A., and Rogozin, B. A.: On the joint distribution of random variables related to the fluctuations of a process with independent increments. Teor. Verojatnost. i Primenen **14**, 431–444 (1969) [in Russian].
49. Presman, E. L.: On the waiting time in multi-server queueing systems. Teor. Verojatnost. i Primenen **10**, 70–81 (1965) [in Russian].
50. Presman, E. L.: Methods of factorization and boundary problems for sums of variables defined on a Markov chain. Izv. Akad. Nauk SSSR Ser. Mat. **33**, 861–900 (1969) [in Russian].
51. Prohorov, Yu. V.: Transient phenomena in queueing processes. Litovsk. Mat. Sb. **3**, 199–206 (1963) [in Russian].
52. Rogozin, B. A.: On the distribution of the size of the first overshoot. Teor. Verojatnost. i Primenen **9**, 498–515 (1964) [in Russian].
53. Rogozin, B. A.: Some extremal problems in queueing theory. Teor. Verojatnost. i Primenen **11**, 161–169 (1966) [in Russian].
54. Rogozin, B. A.: The distributions of some functionals related to boundary problems for processes with independent increments. Teor. Verojatnost. i Primenen **11**, 656–670 (1965) [in Russian].
55. Rosen, B.: On the asymptotic distribution of sums of identically distributed random variables. Ark. Mat. **4**, 323–332 (1962).

56. Rossberg, H. J.: Eine neue Methode zur Behandlung der Integralgleichung von Lindley und ihrer Verallgemeinerung durch Finch. Elektron. Informationsverar. und Kybernetik **3**, 4, 215–238 (1967).
57. Rossberg, H. J.: Über die Verteilung von Wartezeiten. Math. Nachr. **30**, 1–16 (1965).
58. Saaty, T.: Elements of queueing theory with applications. New York: McGraw-Hill 1961.
59. Sevast'yanov, B. A.: An ergodic theorem for Markov processes and its application to telephone systems with refusals. Teor. Verojatnost. i Primenen **2**, (1957) [in Russian].
60. Skorohod, A. V.: Random processes with independent increments. Nauka 1964 [in Russian].
61. Smith, W. L.: Renewal theory and its ramifications. J. Roy. Statist. Soc. Ser. B **20**, 243–302 (1958).
62. Spitzer, F.: A combinatorial lemma and its applications to probability theory. Trans. Amer. Math. Soc. **82**, 323–329 (1956).
63. Spitzer, F.: The Wiener-Hopf equation whose kernel is a probability density. Duke Math. J. **24**, 327–344 (1957).
64. Spitzer, F.: A Tauberian theorem and its probability interpretation. Trans. Amer. Math. Soc. **94**, 150–169 (1960).
65. Spitzer, F.: Principles of random walk. Princeton: Van Nostrand 1964.
66. Stone, C.: On the characteristic function and renewal theory. Trans. Amer. Math. Soc. **120**, 327–342 (1965).
67. Stoyan, H., and Stoyan, D.: Monotonieeigenschaften der Kundenwartezeiten im Modell GI/G/1. Z. Angew. Math. Mech. **49**, 729–734 (1969).
68. Syski, R.: Introduction to congestion theory in telephone systems. Edinburgh and London: Oliver and Boyd 1960.
69. Täcklind, S.: Sur le risque de ruine dans des jeux inéquitables. Skand. Aktuarietidskr. **25**, 1–42 (1942).
70. Takács, L.: Combinatorial methods in the theory of stochastic processes. New York: Wiley 1967.
71. Takács, L.: Introduction to the theory of queues. Oxford: Oxford Univ. Press. 1962.
72. Viskov, O. V., and Prohorov, Yu. V.: The probability of customer loss in heavy traffic. Teor. Verojatnost. i Primenen **9**, 99–104 (1964) [in Russian].
73. Wiener, N., and Hopf, E.: Über eine Klasse singulärer Integralgleichungen. Sitz. Akad. Wiss. Berlin 696–706 (1931).
74. Zolotarev, V. M.: The first passage time of a threshold and the behavior at infinity of a class of processes with independent increments. Teor. Verojatnost. i Premenen **9**, 724–733 (1964) [in Russian].

Author Index

Ahmarov, I. 271

Baxter, G. 273
Bary, N. K. 259, 273
Beneš, V. 29 ff., 58, 64, 76, 269, 270, 273
Blackwell, D. 244
Borel, E. 2, 23
Borovkov, A. A. 89, 95, 100, 102, 151, 155, 233, 259, 269, 270, 271, 272, 273

Čebyšev, P. L. 49, 134, 138, 140
Cox, D. R. 47, 272, 273
Cramér, H. 123, 186, 270, 273

Darling, D. A. 270, 273
Doob, J. L. 13, 14, 50, 54, 273

Erlang 112, 113, 214

Fatou 118
Feller, W. 78, 244, 260, 270, 274
Franken, P. 271, 274
Fuks, B. A. 153, 274

Gel'fand, I. M. 254, 255, 257, 272, 274
Gihman, I. I. 25, 148, 274
Gnedenko, B. V. 270, 274

Hinčin, A. Ya. 37, 58, 65, 159, 274
Hoeffding, W. 85, 271, 274

Ibragimov, I. A. 36, 39, 49, 50, 78, 139, 274

Kasparson, V. A. 272, 274
Keilson, J. 269, 274
Kemperman, J. H. B. 101, 274
Kendall, D. G. 269, 274
Kiefer, J. 271, 274
Kingman, F. G. 270
Kovalenko, I. N. 270, 274
Kolmogorov, A. N. 8, 22
König, D. 274
Krein, M. G. 272, 274

Laplace 65, 129, 231
Lebesgue 256
Leont'eva, N. P. 270, 271
Lévy, P. 65, 106, 123, 258, 267
Lindley, D. V. 269, 274
Linnik, Yu. V. 36, 39, 49, 50, 78, 139, 274
Loynes, R. M. 269, 271, 274
Laurent 109

Mathes, K. 274
Miller, H. D. 274
Minkowski 77

Nagaev, S. V. 272, 274
Nawrotzki, K. 274

Pečerskii, E. A. 270, 274
Pollaczek 159
Presman, E. L. 85, 271, 274
Prohorov, Yu. V. 25, 148, 270, 271, 272, 274

Raikov, D. A. 254, 255, 257, 272
Rogozin, B. A. 54, 102, 270, 271, 274
Rosen, B. 94, 270, 274
Rossberg, H. J. 270, 275

Saaty, T. 275
Sevast'yanov, B. A. 271, 275
Šilov, G. E. 254, 255, 257, 272
Skorohod, A. V. 25, 148, 269, 274, 275
Smith, W. L. 47, 269, 275
Spitzer, F. 65, 90, 91, 94, 270, 275
Stirling 94, 229
Stone, C. 246, 275
Stoyan, D. 270, 275
Stoyan, H. 270, 275
Syski, R. 275

Täcklind, S. 270, 275
Takács, L. 200, 218, 244, 270, 271, 275

Viskov, O. V. 272, 275
Volterra, V. 32, 194

Wald, A. 46, 98, 100, 182, 220
Wiener, N. 74, 89, 147, 258, 275
Wolfowitz, J. 271, 274

Zolotarev, V. M. 270, 275

Subject Index

absolutely continuous component of a distribution function 257
actual busy period 27
— waiting time 161
"angles" of discontinuity 171
approximation formulae for light and heavy traffic 33, 37
arbitrary service sequence 3
arcsin law 95
— —, local 95
assignment algorithms 165
asymptotic analysis of multi-channel systems 225
— behavior of coefficients of absolutely convergent series 258 ff.
— independence 15
— properties of distributions 92 ff.
autonomous service 2, 235

backward trajectory 83
Beneš' equation 29, 64, 76
— —, integral form of 29 ff.
— —, stationary solution of 33
boundary functionals 85
— problems for processes continuous from below 64
— — 64 ff., 85 ff.
busy period 11, 75, 78
— —, actual 27
— —, backward 11, 124
— (occupied) system 1

central limit theorem 78, 98
classification of systems 1
coefficients of drift and diffusion 65
communicating states 218
comparison theorems 141, 270
compound Poisson process 65, 67, 71, 76
conditional renewal function 35
connection between the waiting time and the queue length 161, 167, 172
— — —, stationary distributions of $w^c(t)$ and w^k 44

canonical factorization 251
— V-factorization 251
continuity from above or below 64, 80
convergence of processes 9, 25 ff.
convex minorant 142
Cramér's condition 186
— estimate 123

decay of a distribution 15
"defect" w.r.t. a barrier 177, 242
deterministic service, extremal property of 146
diffusion process 227
discrete arrival intervals 197
— time, 32, 37, 80, 184, 197, 200, 215
distance, Skorohod-Prohorov 25
double sequence 38, 118, 233
— tail 124, 245
— transformation 65, 67

ergodic Markov chain 202
— process 13, 236
— sequence 14, 62, 219
— theorems 7, 175
ergodicity 14, 62, 219
Erlang distribution 112 ff., 238
Erlang's formulae 214
estimates for distributions of sums of i.i.d. random variables 261 ff.
existence theorems for stationary solutions of recursion equations 161 ff.
exponential distribution 47, 112
— polynomials 105, 108
exponentially-distributed arrival times 47, 159
extended σ-algebra 208
extremal problems 141, 146, 222

factorization 86 ff., 103, 249, 251 ff.
— in the ring $\mathfrak{D}$ 253
— in the ring $\mathfrak{B}$ 249
— identity, first 88
— —, second 95
— of functions with rational quotients 105

factorization, resolvability of 106
—, V- 103, 251
first passage time 74, 243
Fourier–Stieltjes transformations, ring of 95
free (idle) system 1
fundamental renewal theorem 244 ff.
Gaussian process 39
general conditions for the convergence of the queue length 187
— — — the convergence of the waiting time 11, 27
governing algorithm 2
— by means of independent random variables 20
— process, prestationary 34
— — with independent increments 31
— —, homogeneous 24
— sequence 1, 23

heavy traffic 33, 37 ff., 147, 269
Hinčin's formula 37
Hinčin-Lévy representation 65
homogeneous governing process 24
— process 24, 68

identity, factorization 86, 88, 95
—, Spitzer's 91
—, Wald's 46, 98, 100, 182
independence of χ and η 100
independent increments, process with 31, 64, 68
— —, sequence with 33
index 109, 256
inequality for the distributions of $\bar{Y}_n$ and $\bar{Y}$ 139
— — — renewal function 244
infinity-divisible distributions 93
infinitely-many servers, systems with 185
inhomogeneous process 31
input process 207
— stream 1
integral renewal theorem 126, 243
intensity of a jump process 65
interrupted governing sequence 19
invariance of the $<$-relation under limit passages 145
invariant probability measure 104, 214
— set 14
inverse process 75

joint distribution of $\bar{Y}(t)$ and $Y(t)$ 73

Kolmogorov's inequalities 22
— theorem 8

Laplace, method of 231
large deviations 228
Laurent expansion 109
law of large numbers 18, 92, 160, 164
Lebesgue's theorem 256
Lévy-Hinčin representation 65
light traffic 33, 37, 269
limiting distribution for the queue length
— — — multi-channel systems 167, 172, 176, 182, 186, 193, 195, 203
— — — systems with autonomous service 236
— — — the waiting time in multi-channel systems 161
— — — the waiting time in systems with queues 8
local arcsin law 95
— power (l.p.) function 125
— renewal theorem 244
lower functionals 86

Markov chain 14, 104, 183, 199, 212, 217
measure-preserving shift transformation 208
methods of calculating stationary distributions 240
metric transitivity 13, 14, 28, 51, 162, 166, 222
Minkowski's inequality 77
Monte Carlo trials 42, 59, 131
multi-channel systems 2, 161, 225
— — with arbitrary initial conditions 206
— — — queues 161
— — — refusals 202
multiple-branched functions 256

negative tail (of a distribution) 137
nonlattice random variable 46
notation, list of 267
number of level crossings 88

overshoot of a barrier 151, 177, 214, 238

Poisson process 70
— —, compound 65, 76
— —, —, with drift 70 ff.
pole of a function 67, 111
Pollaczek-Hinčin formula 159
precision of renewal theorems 244 ff.
prelimiting waiting time 147
prestationary governing process 34
principal branch of a function 256
— part of a pole 67, 111
— pole of a rational function 111
probability of refusal 221 ff.
process, continuous from above 64 ff.
—, — — below 64 ff.
—, ergodic 13
—, governing 34

—, metrically transitive 13
—, service 1
—, stationary 9
—, waiting time 9, 23, 29
— with independent increments 31, 64
— — stationary increments 24, 45
Prohorov's theorem 148
properties of the distributions of $\chi(+0)$, $\chi^*(0)$ 124
— — — — — $w(t)$ 75
purely exponential distribution 112

queue length 29, 155, 169, 172, 202
— —, stationary 172

random walk in a strip 88, 101
— — with alternating jumps 248
rates of approach of distributions 139, 186
— — convergence 27, 48, 113, 120, 169
recursion equations describing systems 4, 161
refusal probability, interpreted in broader sense 221
refusals, systems with 2, 202
renewal density, unconditional 36
— function 98
— theory 243 ff.
representation of a factorization 254
resolvability of a factorization 106
Riemann surface, multi-sheeted 255
ring, normed commutative 249
—, "shifted" 249
—, V 249
—, $\mathfrak{B}$ 95, 249
—, $\mathfrak{D}$ 253
— of Fourier–Stieltjes transformations 95, 249

second factorization identity 95
sequence continuous from below 80
—, governing 23
—, proper 13
sequence, stationary 51, 162, 222
—, strictly stationary 7, 219
— with independent increments 33
— — stationary increments 24, 45
service process 1
shift operator 164
— transformation 203, 208
single-server systems with refusals 219
— — — — queues 2
Skorohod–Prohorov distance 25
slowly varying function 258
spectral function 115
spectrum of a function 254
Spitzer's identity 91
status of a system 1
stability theorems 52, 113, 169, 188
stationary distribution 44, 240
— queue length 172
Stirling's formula 94, 229
stream, arrival 1, 205
—, —, with independent increments 33
—, input 1
strictly stationary sequence 7, 219
— — process 24, 45
strong law of large numbers 18, 92, 164
symmetric random variables 93
system equations 2, 7, 22, 162, 169, 193
systems $\langle E, G, G, G\rangle$ 5
— $\langle E, G_I, E, G_I\rangle_A$ 241
— $\langle E, G_I, G_I, G_I\rangle_A$ 241
— $\langle E, G_I, G_I/m, 1\rangle$ 178
— $\langle E, 1, E/m, 1\rangle$ 182
— $\langle E, 1, G_I/m, 1\rangle$ 179
— $\langle E, 1, G_I/m, 1\rangle_R$ 211, 215
— $\langle E, 1, G_I/\infty, 1\rangle$ 195
— $\langle G\rangle_A$ 235
— $\langle G\rangle_R$ 219
— $\langle G_I/m\rangle$ 234
— $\langle G_I/m\rangle_A$ 234
— $\langle G, G, E, G\rangle$ 6
— $\langle G, G, E/m, G\rangle$ 168
— $\langle G, G, G, 1\rangle$ 6, 7
— $\langle G, G, G/m, 1\rangle$ 165
— $\langle G, G, G/m, 1\rangle_R$ 202
— $\langle G, G, G/\infty, 1\rangle$ 185
— $\langle G_I, G_I, E, G_I\rangle_A$ 240
— $\langle G_I, G_I, E/m, G_I\rangle_R$ 217
— $\langle G_I, G_I, G_I/m, 1\rangle$ 169, 175
— $\langle G_I, G_I, G_I/\infty, 1\rangle$ 193
— $\langle G, 1, G, G\rangle$ 5
— $\langle G, 1, G, 1\rangle_R$ 221
— $\langle G, 1, G/m, 1\rangle$ 161
— $\langle G_I, 1, E/m, G_I\rangle_R$ 217
— $\langle G_I, 1, E/m, 1\rangle$ 178
— $\langle G_I, 1, E/m, 1\rangle_R$ 225
— $\langle G_I, 1, E/\infty, 1\rangle$ 199
— $\langle G_I, 1, G_I/m, 1\rangle$ 169
— $\langle G_I, 1, G_I/m, 1\rangle_R$ 205, 211, 215
— described by recursion equations 4
— in heavy traffic 33, 37 ff., 147, 269
— — light traffic 33, 37 ff., 269
—, multi-channel 2, 161, 185, 202
—, single-channel 1, 4
— which can be described by recursion relations 4, 161
— with autonomous service 2, 235
— — infinitely-many channels (servers) 185
— — refusals 2, 202
supremum of sums of independent r.v.'s 103

tail behavior 124, 137
Tauberian theorems 98, 136
theorem, comparison 141
— on the queue length in multi-channel systems 167, 172
— — — queue length in single-channel systems 155

theorem on the stationary waiting time in multi-channel systems 161, 166
— — — stationary waiting time in single-channel systems 10
traffic, heavy 37 ff., 225, 269
—, light 37 ff., 269
transient regime 78
"transitional" asymptotic behavior 136
transitional phenomena 147, 160
triple transform 88
types of systems 1

unconditional renewal density 36
uniqueness of factorizations 103, 251
— theorems 103
upper functionals 86
— power function (u.p. function) 132

V-factorization 103, 251
variational problem 222
virtual waiting time 22, 155
Volterra equation of the first kind 32
— — — the second kind 194

waiting time 4, 51, 155, 169, 183
— —, prelimiting 147
— —, virtual 22
Wald's identity 46, 98, 100, 182
weak convergence 25
— — of distributions 21, 25, 192
— — — spectral functions 115
— mixing 36
weakly dependent increments 36
Wiener-Lévy theorems 258 ff.
Wiener process 147, 233